Chemical Synthetic Biology

Chemical Synthetic Biology

Editors

Pier Luigi Luisi
Università degli Studi di Roma Tre, Rome, Italy

and

Cristiano Chiarabelli
Università degli Studi di Roma Tre, Rome, Italy

A John Wiley & Sons, Ltd., Publication

This edition first published 2011
© 2011 John Wiley & Sons, Ltd

Registered office
John Wiley & Sons Ltd, The Atrium, Southern Gate, Chichester, West Sussex, PO19 8SQ, United Kingdom

For details of our global editorial offices, for customer services and for information about how to apply for permission to reuse the copyright material in this book please see our website at www.wiley.com.

The right of the author to be identified as the author of this work has been asserted in accordance with the Copyright, Designs and Patents Act 1988.

All rights reserved. No part of this publication may be reproduced, stored in a retrieval system, or transmitted, in any form or by any means, electronic, mechanical, photocopying, recording or otherwise, except as permitted by the UK Copyright, Designs and Patents Act 1988, without the prior permission of the publisher.

Wiley also publishes its books in a variety of electronic formats. Some content that appears in print may not be available in electronic books.

Designations used by companies to distinguish their products are often claimed as trademarks. All brand names and product names used in this book are trade names, service marks, trademarks or registered trademarks of their respective owners. The publisher is not associated with any product or vendor mentioned in this book. This publication is designed to provide accurate and authoritative information in regard to the subject matter covered. It is sold on the understanding that the publisher is not engaged in rendering professional services. If professional advice or other expert assistance is required, the services of a competent professional should be sought.

The publisher and the author make no representations or warranties with respect to the accuracy or completeness of the contents of this work and specifically disclaim all warranties, including without limitation any implied warranties of fitness for a particular purpose. This work is sold with the understanding that the publisher is not engaged in rendering professional services. The advice and strategies contained herein may not be suitable for every situation. In view of ongoing research, equipment modifications, changes in governmental regulations, and the constant flow of information relating to the use of experimental reagents, equipment, and devices, the reader is urged to review and evaluate the information provided in the package insert or instructions for each chemical, piece of equipment, reagent, or device for, among other things, any changes in the instructions or indication of usage and for added warnings and precautions. The fact that an organization or Website is referred to in this work as a citation and/or a potential source of further information does not mean that the author or the publisher endorses the information the organization or Website may provide or recommendations it may make. Further, readers should be aware that Internet Websites listed in this work may have changed or disappeared between when this work was written and when it is read. No warranty may be created or extended by any promotional statements for this work. Neither the publisher nor the author shall be liable for any damages arising herefrom.

Library of Congress Cataloging-in-Publication Data

Chemical synthetic biology / editors, Pier Luigi Luisi and Cristiano Chiarabelli.
   p. cm.
   Includes bibliographical references and index.
   ISBN 978-0-470-71397-6 (cloth)
   1. Biomolecules–Synthesis. I. Luisi, P. L. II. Chiarabelli, Cristiano.
   QD415.C47 2011
   572–dc22
                                                                                          2010045647

A catalogue record for this book is available from the British Library.

Print ISBN: 9780470713976
ePDF ISBN: 9780470977880
oBook ISBN: 9780470977873
ePub ISBN: 9781119990307

Set in 10.5 on 13 pt Sabon by Toppan Best-set Premedia Limited
Printed and bound in Singapore by Markono Print Media Pte Ltd

Contents

List of Contributors ix

Introduction 1
Pier Luigi Luisi

Part One: Nucleic Acids 5

1 Searching for Nucleic Acid Alternatives 7
   Albert Eschenmoser

2 Never-Born RNAs: Versatile Modules for Chemical
   Synthetic Biology 47
   *Davide De Lucrezia, Fabrizio Anella,
   Cristiano Chiarabelli, and Pier Luigi Luisi*

3 Synthetic Biology, Tinkering Biology, and
   Artificial Biology: A Perspective from Chemistry 69
   Steven A. Benner, Fei Chen, and Zunyi Yang

4 Peptide Nucleic Acids (PNAs) as a Tool in
   Chemical Biology 107
   Peter E. Nielsen

Part Two: Peptides and Proteins — 119

5 High Solubility of Random-Sequence Proteins Consisting of Five Kinds of Primitive Amino Acids — 121
Nobuhide Doi, Koichi Kakukawa, Yuko Oishi, and Hiroshi Yanagawa

6 Experimental Approach for Early Evolution of Protein Function — 139
Hitoshi Toyota, Yuuki Hayashi, Asao Yamauchi, Takuyo Aita, and Tetsuya Yomo

7 Searching for *de novo* Totally Random Amino Acid Sequences — 155
Cristiano Chiarabelli, Cecilia Portela Pallares, and Anna Quintarelli

Part Three: Complex Systems — 175

8 Synthetic Genetic Codes as the Basis of Synthetic Life — 177
J. Tze-Fei Wong and Hong Xue

9 Toward Safe Genetically Modified Organisms through the Chemical Diversification of Nucleic Acids — 201
Piet Herdewijn and Philippe Marliere

10 The Minimal Ribosome — 227
Hiroshi Yamamoto, Markus Pech, Daniela Wittek, Isabella Moll, and Knud H. Nierhaus

11 Semi-Synthetic Minimal Living Cells — 247
Pasquale Stano, Francesca Ferri, and Pier Luigi Luisi

Part Four: General Problems — 287

12 Replicators: Components for Systems Chemistry — 289
Olga Taran and Günter von Kiedrowski

13 Dealing with the Outer Reaches of Synthetic
Biology Biosafety, Biosecurity, IPR, and Ethical
Challenges of Chemical Synthetic Biology 321
Markus Schmidt, Malcolm Dando, and Anna Deplazes

14 The Synthetic Approach in Biology: Epistemological
Notes for Synthetic Biology 343
Pier Luigi Luisi

Index 363

List of Contributors

Takuyo Aita, Department of Functional Materials Science, Saitama University, Saitama, Japan

Fabrizio Anella, Department of Biology, University of Rome Tre – V.le G. Marconi 446 – 00146 Rome, Italy

Steven A. Benner, Department of Chemistry, University of Florida, Gainesville, FL 32611-7200, USA

Fei Chen, Foundation for Applied Molecular Evolution and The Westheimer Institute for Science and Technology, PO Box 13174, Gainesville FL 32604, USA

Cristiano Chiarabelli, Department of Biology, University of Rome Tre – V.le G. Marconi 446 – 00146 Rome, Italy

Malcolm Dando, Department of Peace Studies, Pemberton Building, University of Bradford, Bradford, West Yorkshire, BD7 1DP, UK

Anna Deplazes, UFSP Ethik, Universität Zürich, Klosbachstr. 107, 8032 Zürich, Switzerland; ETH Zürich, Institute of Biochemistry, Schafmattstrasse 18, 8093 Zurich, Switzerland

Nobuhide Doi, Department of Biosciences and Informatics, Keio University, 3–14–1 Hiyoshi, Kohoku-ku, Yokohama 223-8522, Japan

Albert Eschenmoser, Laboratory of Organic Chemistry ETH Hönggerberg, HCI-H309 CH-8093 Zürich, Switzerland

Francesca Ferri, Neuroscience Department, University of Parma, Via Volturno 39, 43100 Parma, Italy

LIST OF CONTRIBUTORS

Yuuki Hayashi, Department of Bioinformatic Engineering, Graduate School of Information Science and Technology, Osaka University, 2-1 Yamadaoka, Suita, Osaka 565-0871, Japan

Piet Herdewijn, Laboratory For Medicinal Chemistry, Rega Institute for Medical Research, Minderbroedersstraat 10, Leuven-3000, Belgium

Koichi Kakukawa, Department of Biosciences and Informatics, Keio University, 3-14-1 Hiyoshi, Kohoku-ku, Yokohama 223-8522, Japan

Günter von Kiedrowski, Department of Organic Chemistry I – Bioorganic Chemistry, Faculty of Chemistry, Ruhr University Bochum, Universitätstrasse 150, 44801 Bochum, Germany

Davide De Lucrezia, European Centre for Living Technology, University Ca' Foscari of Venice. Venice, Italy

Pier Luigi Luisi, Department of Biology, University of Rome Tre – V.le G. Marconi 446 – 00146 Rome, Italy

Philippe Marliere, Isthmus Sarl, 31 rue Saint Amand, F75015 Paris, France

Isabella Moll, Max F. Perutz Laboratories, Department of Microbiology, Immunobiology, and Genetics, Center for Molecular Biology, Dr. Bohrgasse 9/4, 1030 Vienna, Austria

Peter E. Nielsen, Department of Cellular and Molecular Medicine, Faculty of Health Sciences, The Panum Institute, University of Copenhagen, Blegdamsvej 3c, DK-2200, Copenhagen N, Denmark; Faculty of Pharmaceutical Sciences, Department of Medicinal Chemistry, Universitetsparken 2, DK-2100 Copenhagen Denmark

Knud H. Nierhaus, Max-Planck-Institut für Molekulare Genetik, AG Ribosomen, Ihnestr. 73, D-14195 Berlin, Germany

Yuko Oishi, Department of Biosciences and Informatics, Keio University, 3-14-1 Hiyoshi, Kohoku-ku, Yokohama 223-8522, Japan

Markus Pech, Max-Planck-Institut für Molekulare Genetik, AG Ribosomen, Ihnestr. 73, D-14195 Berlin, Germany

Cecilia Portela Pallares, Department of Biology, University of Rome Tre – V.le G. Marconi 446 – 00146 Rome, Italy

Anna Quintarelli, Department of Biology, University of Rome Tre – V.le G. Marconi 446 – 00146 Rome, Italy

LIST OF CONTRIBUTORS

Markus Schmidt, Organisation for International Dialogue and Conflict Management (IDC), Biosafety Working Group, Kaiserstr. 50/6, 1070 Vienna, Austria

Pasquale Stano, Biology Department, University of RomaTre, Viale G. Marconi 446, 00146 Roma, Italy

Olga Taran, Department of Organic Chemistry I – Bioorganic Chemistry, Faculty of Chemistry, Ruhr University Bochum, Universitätstrasse 150, 44801 Bochum, Germany

Hitoshi Toyota, Department of Biotechnology, Graduate School of Engineering, Osaka University, 2-1 Yamadaoka, Suita, Osaka 565-0871, Japan

Daniela Wittek, Max-Planck-Institut für Molekulare Genetik, AG Ribosomen, Ihnestr. 73, D-14195 Berlin, Germany

J. Tze-Fei Wong, Fok Ying Tung Graduate School and Department of Biochemistry, Hong Kong University of Science and Technology, Hong Kong, China

Hong Xue, Fok Ying Tung Graduate School and Department of Biochemistry, Hong Kong University of Science and Technology, Hong Kong, China

Hiroshi Yamamoto, Max-Planck-Institut für Molekulare Genetik, AG Ribosomen, Ihnestr. 73, D-14195 Berlin, Germany

Asao Yamauchi, Department of Biotechnology, Graduate School of Engineering, Osaka University, 2-1 Yamadaoka, Suita, Osaka 565-0871, Japan

Hiroshi Yanagawa, Department of Biosciences and Informatics, Keio University, 3-14-1 Hiyoshi, Kohoku-ku, Yokohama 223-8522, Japan

Zunyi Yang, Foundation for Applied Molecular Evolution and The Westheimer Institute for Science and Technology, PO Box 13174, Gainesville FL 32604, USA

Tetsuya Yomo, Department of Bioinformatic Engineering, Graduate School of Information Science and Technology, Osaka University, 2-1 Yamadaoka, Suita, Osaka 565-0871, Japan; Graduate School of Frontier Science, Osaka University, 2-1 Yamadaoka, Suita, Osaka 565-0871, Japan; ERATO, JST, 2-1 Yamadaoka, Suita, Osaka, 565-0871, Japan

Introduction

Pier Luigi Luisi
University of Roma 3, Biology Department, Viale G. Marconi 446, 00146 Roma, Italy

The novel and fashionable term synthetic biology (SB) is now used mostly to indicate a field aimed at synthesizing biological structures or life forms in the laboratory which do not exist in nature. Generally, existing microbial life forms are modified and the genomic content redirected towards novel modified organisms; for example, bacterial life that does not exist on Earth. First, important examples of these techniques can be found in recent issues in *Nature* [1] and *Science* [2]. This approach is the one which appears to have the greatest potentialities to do something socially useful; for example, novel bacteria for the production of hydrogen or methane to help our energy needs, or novel bacteria for the production of drugs and/or enzymes which are otherwise difficult to reach. All this is based on the hard hand of the bioengineering approach that thrives from classic DNA molecular biology. This is the major, most popular form of SB.

This book has a different slant, in the sense that it concerns work on SB which is not based on genetic manipulation, emphasizing instead a chemical approach, one which aims at the synthesis of molecular structures and/or multi-molecular organized biological systems that do not exist in nature. These man-made biological molecular or supramolecular structures that do not exist in nature can be obtained either by chemical or biochemical syntheses. For this, the term "chemical synthetic biology" has been coined [3]. This book deals with this aspect of SB and is based on original contributions, as well as on a couple of papers taken from the literature with suggestions and permissions from the authors.

One of the most beautiful examples of chemical SB, is the work by Albert Eschenmoser and coworkers at the ETH Zürich – presented in our book – on DNA having pyranose instead of ribose in the main chain. The basic question underlying this kind of work is: why this and not that? Why did nature choose that particular sugar and not another one? And this question impinges on a greater philosophical problem, that of the relation between determinism and contingency: is the way of nature the only possible one, as a kind of obligatory pathway? Or is it instead so that the choice of nature has been based on "chance" – and we have ribose instead of pyranose simply due to the vagaries of contingency? Or, looking at the work by Yanagawa in this book, why 20 amino acids to build proteins instead of, say, nine?

And the same question can be drawn for the work of Benner (why not different chemical modifications of nucleic acid?). Concerning nucleic acid, there is the approach by Henderwijn and Marliere, who argue that nucleic proliferation could be extended so as to enable the propagation *in vivo* of additional types of nucleic acid whose polymerization would not interfere with DNA and RNA biosynthesis.

In fact, chemical SB permits one to tackle the question "why this and not that?" by synthesizing in the laboratory the alternative forms – forms that do not exist in nature. And then by comparison with the natural forms, we can learn a lot on why nature had to proceed in one way instead of another one.

The question "why this and not that?" is particularly apparent in the project never-born proteins (NBPs), aimed at preparing families of totally random proteins which do not exist in nature and have never been subjected to evolution – with the important question: how and why have the "few" proteins which constitute our life been selected out? Such a project has been initiated at the Federal Institute of Technology in Zurich, Switzerland, to be pursued by my group transferred to the University of "Roma Tre", Italy, in particular, by Cristiano Chiarabelli and Davide de Lucrezia.

In this sense, it is clear that chemical SB is more naturally inclined towards basic science more than towards the engineering approach, which is rather devoted to making things to achieve a predetermined scope. Of course, the difference between the two approaches is often not so in a black-and-white form. One can also add that the bio-ethical problems, often connected to the genetic manipulation approach, are

INTRODUCTION

generally not so relevant in the chemical SB approach. These two important aspects of the field, namely the philosophical implications of SB and the bio-ethical aspects, are duly represented in this book (the last two chapters).

Also, the approach pioneered by Craig Venter and coworkers, aimed at synthesizing an entire genome by chemical methods [4], can be considered as one of the clearest examples of chemical SB. This contribution is missing in this book, not out of negligence of the editors, but because we were unable to get a contribution from this group. The very last, recent contribution of Venter [5] has caused much press clamor. It is, indeed, a Cyclopic work from the experimental point of view. From the conceptual point of view, it is no surprise of course that a synthetic DNA can replace the natural one. The clamor, I believe, is mostly due to the misconception, still so present in the simplistic press, that life is DNA. This equation, equating life and DNA, has been instrumental in a lot of misconception about the big question: "what is life?" Life is much more than DNA, as it is due to the dynamic interaction of thousands of molecular components – even in the bacterium of Venter – of which DNA is only one.

We also could not receive a contribution on another subject we really wanted, the chemical synthesis of a virus [6].

We were lucky enough to obtain a contribution by ProfessorNielsen on his famous PNA chemistry – another example of chemical structures not existing in nature, and one may wonder why.

Thus, there is a part of this book on proteic structures and one on nucleic acids.

After that, it is the time to address the interaction between the two classes of biopolymers. There is first of all the question of the genetic code, and Wong's school tackles the subject, whereas Nierhaus is asking whether ribosomes should really be so complex as they are in order to function, or whether one can construct some form of simpler ribosome; for example, one with a lower number of structural proteins.

The next part of this book is devoted to even more complex systems. And here the notion of a minimal cell is central. The procedure is based on the preparation and physico-chemical characterization of vesicles of given dimensions, and the entrapment of enzymes and DNA of known composition and concentration in their water pool, thus constituting a model for a biochemical cell. The main question here is what is the minimal and sufficient number of macromolecules which can endow the vesicle with cell-like functions? In particular, at which degree of complexity can cellular life arise? This kind of research is also important to show that life is an emergent property, which can arise "simply" from the interaction of nonliving chemical compounds. This project started in my laboratory in Zurich in the mid 1980s – the term "minimal cell" applied to liposomes appeared in a 1985 paper with Thomas Oberholzer and today around 10 groups around the world deal with this subject. Two chapters, those by Stano *et al.* and by Yomo and coworkers, are devoted to this subject.

The last part of the book is devoted to more general subjects. Thus, Schmidt and collaborators raise social questions in connection with SB in general, and in particular dwelling on bio-safety and bio-ethical issues. My own contribution is on epistemic aspects of SB, and tries to clarify the conceptual basis of the various approaches of SB, and also the difference between SB and other related fields, like artificial intelligence. Finally, the contribution of von Kiedrowski is an attempt to form a merger between chemical SB and system chemistry. The replicators as chemical systems which emulate the behavior of self-replication of nucleic acids offer a beautiful example of this endeavor. This merging is necessary and even inevitable when chemistry is moving topwards biology and where biology has to use the tools of chemistry to advance.

More generally, we believe that this book on chemical SB offers a space at the interface between chemistry, biology, and philosophy which is timely and useful.

REFERENCES

1. Church, G. (2005) Let us go forth and safely multiply. *Nature*, **438**, 423 and all articles in this special issue.
2. Ferber, D. (2004) Microbes made to order. *Science*, **303**, 158–161 and all articles in this special issue.
3. Luisi, P.L. (2007) Chemical aspects of synthetic biology. *Chemistry & Biodiversity*, **4**, 603–621.
4. Gibson, D.G., Benders, G.A., Andrews-Pfannkoch, C. *et al.* (2008) Complete chemical synthesis, assembly, and cloning of a *Mycoplasma genitalium* genome. *Science*, **319** (5867), 1215–1220. Epub 2008 Jan 24.
5. Gibson, D.G., Glass, J.I., Lartigue, C. *et al.* (2010) Creation of a bacterial cell controlled by a chemically synthesized genome. *Science*, **329**, 52–56.
6. Cello, J., Paul, A.V., and Wimmer, E. (2002) Chemical synthesis of poliovirus cDNA: generation of infectious virus in the absence of natural template. *Science*, **297** (5583), 1016–1018. Epub 2002 Jul 11.

Part One
Nucleic Acids

1

Searching for Nucleic Acid Alternatives

Albert Eschenmoser

Laboratory of Organic Chemistry ETH, Hönggerberg, HCI-H309, CH-8093 Zürich, Switzerland
The Scripps Research Institute (TSRI), 10550 North Torrey Pines Road, La Jolla, CA 92037, USA

References 39

"Back of the envelope" methods have their place in experimental chemical research; they are effective mediators in the generation of research ideas; for instance, for the design of molecular structures. Their qualitative character is part of their strength, rather than a drawback for the role they have to play. Qualitative conformational analysis of oligonucleotide and other oligomer systems on the level of idealized conformations is one such method; it has played a helpful role in our work on the chemical etiology of nucleic acid structure. This article, while giving a short overview of that work, shows how.

Chemists understand by *comparing*, not *ab initio*. To perceive and to create opportunities for drawing conclusions on the basis of comparisons is the organic chemist's way of interpreting and exploring the world at the molecular level. Comparing the properties of molecules of unknown constitution with those of already known constitution was the basic strategy in classical chemistry of structure determination by chemical methods, and the same still largely holds for more recent

---

Chemical Synthetic Biology, First Edition. Edited by Pier Luigi Luisi and Cristiano Chiarabelli.
© 2011 John Wiley & Sons, Ltd. Published 2011 by John Wiley & Sons, Ltd.
Reprinted in full with permission from the Swiss Chemical Society, Copyright 2005.

structure determinations in chemical laboratories by physical methods; it is only with regard to X-ray structure analysis that the statement ceases to hold in so far as chemistry is concerned. Model studies, an approach characteristically chosen by chemists when confronted with structural or transformational complexity, serve the purpose of creating opportunities to *compare* the behaviors of complex systems with those of simpler ones. Enzymic reactions and enzyme models are examples.

To reach an understanding of structures and structural transformations through *comparing* is not what chemists alone are aiming at; it is also true for biologists, dealing both with their own spatial and temporal resolution and with the conceptual resolution of their objects. Yet, biology is also a historic science: ever since the time of Charles Darwin, biologists – besides and beyond studying structure and function – have pondered the origin of their objects. Chemists do not have to; pursuing the question of origin with regard to the constituents of the periodic table is the job of cosmologists. However, with the progressive breakdown of borders between biology and (a very large part of) contemporary chemistry, the search for origins is bound to leap over from biology to chemistry; the quest for comprehending the evolution of living organisms will have its extension in the quest for an understanding of the origin of biomolecular structures. I do not mean the understanding of the origin of biomolecules in terms of their biosynthesis, but rather the origin of these biosyntheses themselves and, in the long term, of biogenesis, at the molecular level.

To systematically *compare* selected chemical properties of structural nucleic acid analogs with corresponding properties of the natural system is part of a project pursued in my research group(s) since 1986; the aim is to explore the potential of organic chemistry to arrive at an understanding of how and why Nature came to choose the specific structure type of the nucleic acids we know today as the molecular basis of genetic function. The specific property to be compared in this work is a given nucleic acid alternative's capacity for informational Watson–Crick nucleobase-pairing, the overall criterion for the selection for study of a given system being a structure's potential for constitutional self-assembly compared with that envisaged for the structure of the natural system itself [1].

There is another motive for pursuing in organic and medicinal chemistry laboratories the search for nucleic acid analogs: the worldwide hunt (also since the late 1980s) for nucleic acid substitutes that might be useful in medicinal antisense technology [2]. This kind of search is not subjected to the above-mentioned selection criterion that rather strictly narrows the choice of oligomer systems for study by demanding them to be potentially natural systems. Nevertheless, the antisense-technology-driven search for nucleic acid substitutes is not just very successful (chemically) in pursuing its own goals, some of its results importantly also complement etiology-oriented research on nucleic acid alternatives by corroborating and extending one of the major conclusions to be drawn from the latter, namely that, in contrast to what was believed before, the capability of informational Watson–Crick base-pairing is by no means limited to the structure type of the Watson–Crick double helix. Quite the contrary, it is widespread among oligomers containing backbones that may be quite different from one of the natural nucleic acids. Nielsen's peptide nucleic acid (PNA) exemplifies perhaps most instructively the degree of structural backbone variation Watson–Crick base-pairing is found to be compatible with. This wide spread of biology's genetic type of recognition process notwithstanding, an oligomer system's capability to show informational base-pairing demands rather stringent prerequisites to be fulfilled regarding the backbone structure and the three-dimensional relationship between backbone units and recognition elements (nucleobases). There is not just the question "base-pairing, yes or no?"; informational oligomer systems may "speak different base-pairing languages," in the sense that they can be capable of undergoing *intra*system base-pairing with themselves, while being unable to communicate with each other by *inter*system cross-pairing. To deduce from empirical and theoretical sources the structural constraints that control an oligomer system's capability of base-pairing in a given language would be important for eventually being able, on a qualitative level, to estimate in advance, if not to predict, whether or not a given type of oligomer structure could act as an informational oligomer. This challenge is clearly addressed to researchers in both the etiology-oriented and antisense-technology-driven branches of nucleic acid chemistry mentioned above.

From its very beginning, our own work was assisted by a set of such empirical qualitative criteria, a set that evolved with time and experience. In retrospect, it seems quite instructive to look at the balance between success and failure experienced so far in applying these criteria for "predicting" the base-pairing capability for new oligomer systems. I review them in this article by pointing to their origin, summarizing their essentials, commenting on their strengths and weaknesses as well as on their potential to be extended and applied for oligomer systems structurally different from those for which these rules were deduced and developed. I am, of course, not unaware of the fact that in contemporary chemistry the task of predicting for practical purposes the structure (and sometimes implicitly the properties) of molecules is to be pursued by theory-based and computer-assisted tools; yet this does not mean that efforts to reach such estimates by empirical arguments on a qualitative level have become obsolete. After all, the experimental chemist's pragmatic reasoning in terms of his classical formulae language coexists quite successfully with the kind of reasoning feasible on the level of quantum chemistry.

Our entry into the research field of nucleic acid chemistry was motivated by our interest in the broader context of an organic chemistry of biogenesis [3], and consisted in an experimental study on the formation and properties of oligo-dipeptamidinium salts (1982–1986, with Heinz Moser and Arthur Steiger [4]). At the outset of this project stood the recognition of the fact that the constitutional periodicity of a negatively charged oligonucleotide chain happens to be identical (six covalent bonds) with that of the chain of a positively charged oligo-dipeptamidinium salt; that is, of a polypeptide chain in which each *second* backbone amide group is replaced by a cationic amidinium group (Figure 1.1). The driving force behind the project was the question as to whether such an (etiologically not necessarily accidental) identity of constitutional periodicity in the two major types of biopolymer could give rise to a mutual catalysis in the formation of oligonucleotide and oligo-dipeptamidinium strands by oligomerization of their respective mono-

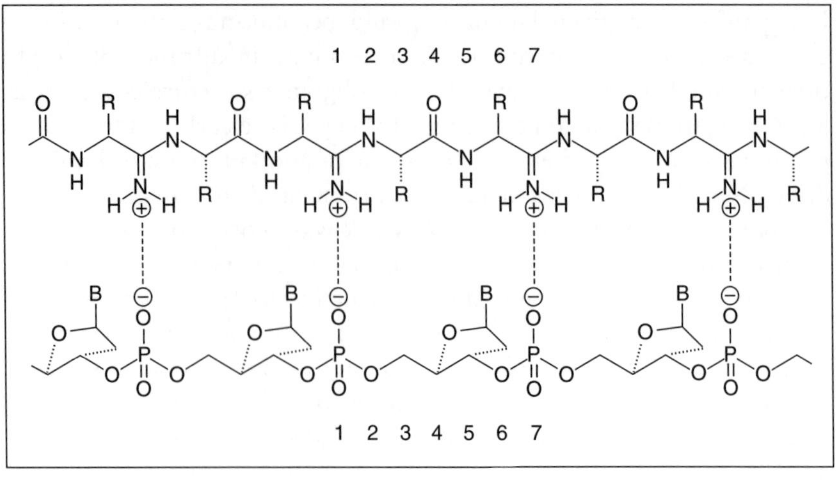

Figure 1.1 Six-bond periodicity in oligonucleotides and oligo-dipeptamidinium salts (facsimile from Ref. [4])

mers, given that oligo-dipeptamidinium salts can be considered to be oligomerization products of (the ammonium form of) dipeptide *nitriles* and could transform to corresponding oligopeptides under mild hydrolytic conditions. Experiments turned out to negate neither the feasibility of the oligomerization step (preferably catalyzed by thiols such as cysteine [5]), nor the hydrolytic conversion to oligopeptides, yet uncovered strand fission to be a competing process in that hydrolysis and, moreover, the overall stability of oligo-dipeptamidium salts toward such cleavage near neutral pH to be quite low [6].

One special aspect of that early project, namely the remarkable constitutional relationship between an oligo-*nucleotide* and an oligo-*dipeptide* chain, was to induce our next move. Only little was needed to realize that the complementarity of charges may be extended to the complementarity of other recognition elements; most appealingly, of course, nucleobases. Such a step called for nucleic acid alternatives composed of homochiral oligo-dipeptamidinium and also oligo-dipeptide chains in which *each second* α-amino acid would be tagged by a nucleobase (Figure 1.2a and b). We studied (with Peter Lohse [6]) the formation and the properties of such nucleobase-tagged oligo-dipeptamidinium salts with partial success, yet eventually succumbed to their relative instability. The literature existing at the time (1986) already contained reports on pioneering experimental efforts to prepare oligopeptides composed of nucleo-base-tagged α-amino acids; however, these reports were devoid of considerations according to which tagged oligomer units should be a *dipeptide* unit [7]. Encouraged by an ETH-built (only slightly "suffering") mechanical model of a double-helical hybrid duplex between an oligonucleotide chain and a nucleobase-tagged oligo-dipeptide chain, Gerhard Baschang of (former) CIBA-Geigy's newly formed "antisense group" synthesized in 1986 an octamer of an L-alanyl-glycyl-type oligomer in which each L-alanine was replaced by an L-serine-derived L-(1-thyminyl)-alanine (Figure 1.2b). Disappointingly, no base-pairing with poly-d-adenosine was observed of that (presumably earliest) sample of a constitutionally and stereochemically defined nucleobase-tagged oligo-dipeptide.[1] In the light of this experience, Peter Nielsen's pragmatically and ingeniously designed and, in turn, dramatically successful "PNA" a few years later amounted to an

---

[1] Unpublished; however, see Ref. [6]. In retrospect, we realize that the sequence length of an 8-mer (with thymin as nucleobase) is far too short to draw conclusions about an oligomer system's capability of base-pairing. Had we in the TNA series (see later) investigated the 8-mer only, we would have failed to observe the system's pairing capacity.

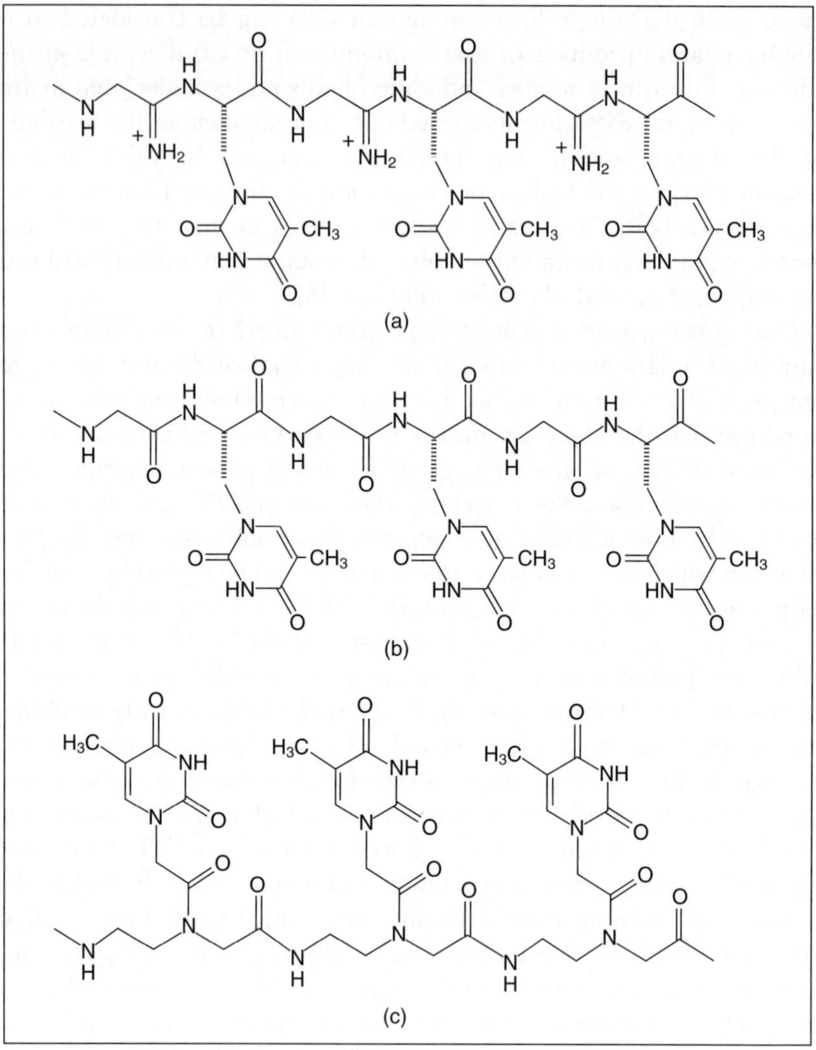

Figure 1.2 Six-bond periodicity in (a) oligo-nucleodipeptamidinium salts, (b) oligo-nucleodipeptides, and (c) Nielsen's PNA

overwhelming discovery [8]. Reassuringly, the periodicity of the nucleobase-tagging of PNA's backbone nicely corresponded to that of an oligo-*dipeptide*-type of oligomer structure (Figure 1.2c).

Gradually overlapping with our early steps toward oligo-dipeptide-based informational systems in the 1980s was our search for nucleic acid alternatives that contain phosphodiester backbones derived from hexopyranoses in place of ribofuranose units ("Why pentose and not hexose nucleic acids?" [9]; Figure 1.3). While our results of an extensive study of the (diastereoselective) aldolization of glycolaldehyde phosphate [10] had initiated these studies, it was mainly the following aspect that eventually led us to rigorously concentrate on this project: an assignment of etiological relevance to a given nucleic acid alternative will be more reliable the closer the structural and generational relationships between the alternative and the natural system are. The structure of an alternative should be derivable from an alternative aldose by the same type of chemistry that allows us to derive the structure of RNA from ribose. What, in addition, made such a study appealing from a purely chemical point of view was the following: oligonucleotide strands composed of hexopyranose sugar units would be amenable to qualitative conformational rationalization, in sharp contrast to the conformationally far more complex pentofuranose-based natural systems, the base-pairing capability of which had, at the time, never been rationalized at the level of the organic chemist's qualitative way of conformational reasoning.

Figure 1.4 reproduces the results of the kind of conformational analysis that allowed us to "predict" that pyranose analogs of DNA ("homo-DNA") might have the capability of informational base-pairing. The analysis was based on the following framework of assumptions and stipulations [9, 11]:

1. search for the ensemble of (formally) *least strained* conformers of an oligomer chain's constitutionally repeating monomer unit by restricting the rotational continuum around each covalent bond of the backbone to three ideally staggered (i.e. "idealized") conformations (bidentate oxygen centers containing two electron lone-pairs taken as tetrahedral centers);
2. *least strained* conformers of a monomer unit are those that have either no or the smallest possible number of, "1,5-repulsions" (i.e. repulsions resulting whenever in a five-center-chain *a–b–c–d–e* the two bonds *a–b* and *d–e* are positioned parallel to each other and, at the same time, neither *a* nor *e* is a hydrogen atom);
3. phosphodiester group conformations are allowed to be *gauche–gauche* or *gauche–trans*, but not *trans–trans*;[2]

---

[2] For a discussion of X-ray structures of phosphodiester groups, see Figure 9 in Ref. [10]. Whereas the *gauche–gauche* conformation appears intrinsically favored, the *trans–trans* conformation apparently has not been not observed. Stereoelectronic reasoning (generalized anomeric effect) led us originally to believe that the *gauche–trans* conformation should be disallowed; however, the structure of pyranosyl-RNA, for example, taught us to think differently (see below).

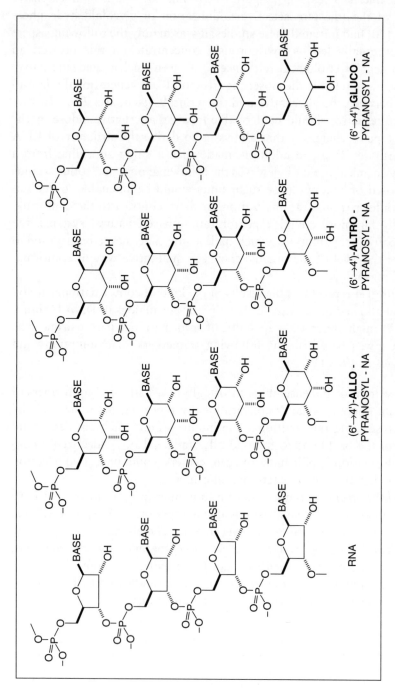

Figure 1.3 Three (potentially natural) hexopyranosyl alternatives of RNA studied in our laboratory [19–21]. The model system homo-DNA lacks the hydroxyl groups in positions 2' and 3'

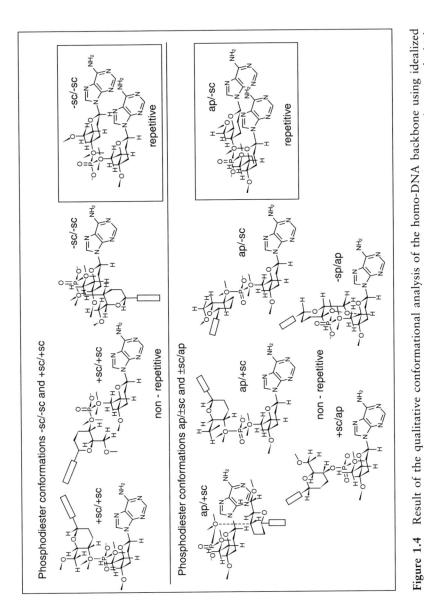

Figure 1.4 Result of the qualitative conformational analysis of the homo-DNA backbone using idealized conformations [9, 11]. The figure depicts the ensemble of least-strained monomer-unit conformers of which the two pairing conformations are framed (-sc/-sc = -g/-g and ap/-sc = -g/t; sc = synclinal, g = gauche, ap = antiplanar, t = trans)

4. an oligomer system can be expected to be a base-pairing system if the ensemble of (formally) *least strained* monomer-unit conformers contains at least one conformer that is in a "pairing conformation";
5. a "pairing conformation" of a monomer unit conformer is a conformation that repeats itself in both neighboring monomer units.

It is to be noted that, in such an analysis, based on idealized conformations, the criteria "conformational repetitivity of a monomer unit" and "pairing conformation" are assessable in such a straightforward way only if the constitutional periodicity of the oligomer backbone corresponds to an even number of bonds, as is the case, for example, in homo-DNA.

The analysis of the homo-DNA system (Figure 1.4) reveals an ensemble of *four* least-strained conformers, *one* of them conformationally repetitive, with the *gauche–gauche* (-*g*/-*g*) conformation of the phosphodiester group [12]. When the latter is allowed to assume the *trans–gauche* conformation, there are six additional least-strained conformers; again, with one of them (*t*/-*g*) conformationally repetitive. Therefore, a homo-DNA strand in a homo-DNA duplex can assume two types of conformation: the *gauche–gauche* phosphodiester type with (idealized) torsion angles $\alpha = -60°$, $\beta = 180°$, $\gamma = +60°$, $\delta = +60°$, $\varepsilon = 180°$ and $\zeta = -60°$; or the *trans–gauche* type with $\alpha = 180°$, $\beta = 180°$, $\gamma = 180°$, $\delta = +60°$, $\varepsilon = 180°$ and $\zeta = -60°$ (Figure 1.4). These are exactly the two structure types that an NMR-structure analysis of the homo-d(A5T5)$_2$ duplex shows to coexist in aqueous solution [13]. Duplex models, constructed out of homo-DNA monomer units in either the -*g*/-*g* or the *t*/-*g* type of pairing conformation with all torsion angles taken to be idealized and (for the sake of argument) with all bond lengths identical, would have a linear shape. In reality, the intrinsic nonideality of all structural parameters, assisted by the necessity of a duplex to reach optimal nucleobase stacking distances by either helicalization or by adapting axes inclination, is bound to result in a helical twist of the duplex structure, the degree and sense of which, however, the NMR-structure analysis was unable to determine.[3]

---

[3] A whole series of attempts in various laboratories to translate the X-ray refraction data of the beautifully crystalline and high-melting homo-DNA duplex of the self-complementary octamer sequence dd(CGAATTCG) into a consistent X-ray structure turned out to be unexpectedly difficult. According to a personal communication of Martin Egli (Vanderbilt University), it is only recently that the structure could be solved by making use of the new methodology of Se-labeling of oligonucleotide strands (unpublished; for the method, see Ref. [14]).

Perhaps the most interesting aspect of the outcome of this kind of qualitative conformational analysis of the homo-DNA structure is the following: out of the two possible (idealized) structures of a homo-DNA duplex, the (-g/-g)-structure, the one in which the (formal) nucleobase stacking distance is nearer to the optimal value, turns out to be identical in type with the structure observed by X-ray analyses as the A-type structure of DNA-duplexes [9, 13] and the structure of duplex RNA (Figure 1.5). This indicates that *content and outcome of the conformational analysis of the homo-DNA structure implies a rationalization of*

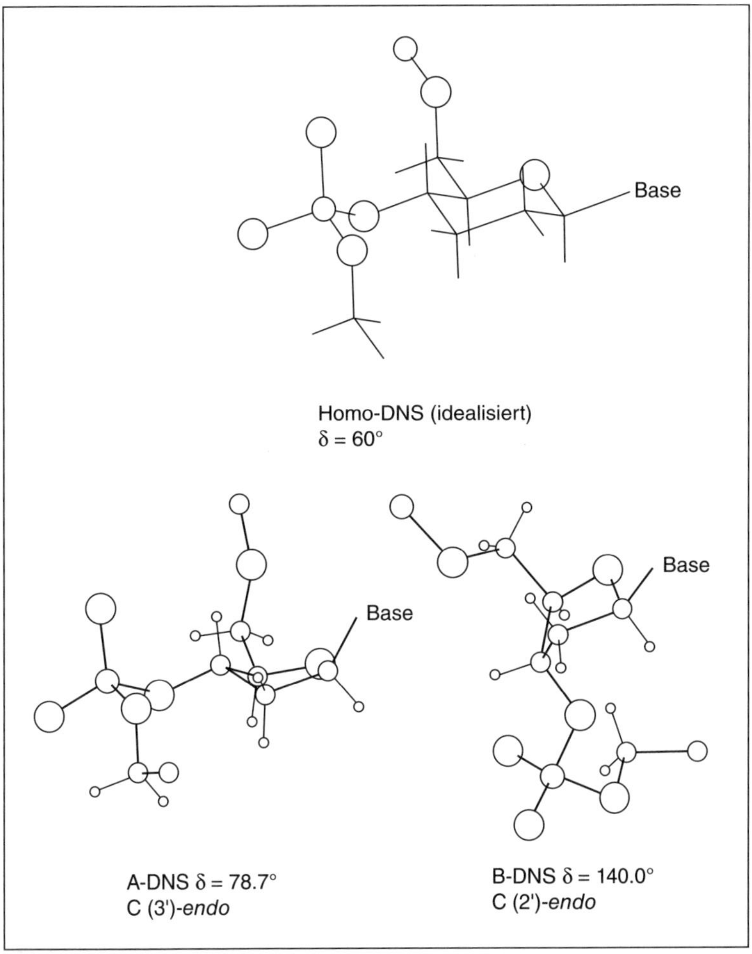

Homo-DNS (idealisiert)
$\delta = 60°$

A-DNS $\delta = 78.7°$
C (3')-*endo*

B-DNS $\delta = 140.0°$
C (2')-*endo*

Figure 1.5 Pictorial comparison of idealized (-g/-g)-pairing conformations of homo-DNA with experimental data of A- and B-type DNA duplexes [1]

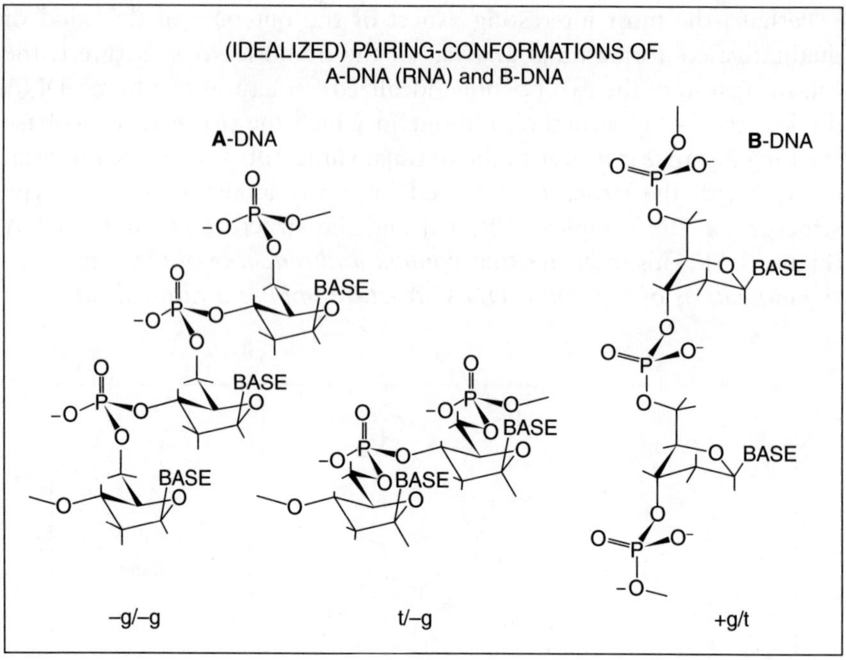

Figure 1.6 On the level of idealized conformations, the sugar conformations of A- and B-type DNA duplexes are related to each other by half-chair inversions of the furanose rings

the A-type structure of DNA and, at the same time, of RNA, at the level of qualitative organic stereochemistry (Figure 1.6). Furthermore, the finding that the A-type structure type of duplex DNA corresponds to a homo-DNA structure that is deduced by conformational analysis of a single strand must mean (assuming the latter finding would still hold for a conformational analysis of a DNA single strand) that DNA (and RNA) single strands are *preorganized* toward duplexation. Finally, the reasoning about homo-DNA and DNA in terms of idealized conformations provides a transparent rationalization of a well-known and biologically important fact, namely that duplex DNA has the option of assuming either the A-type or its characteristic B-type structure, whereas the A-type structure is mandatory for duplex RNA (1,5-repulsion between 2′-hydroxyl and one of the phosphodiester oxygen centers in B-type RNA; see Figure 1.7) and, therefore, that DNA is forced to structurally adapt to RNA (not vice versa) when the two are cross-pairing with each other. Analyzing and depicting the natural nucleic

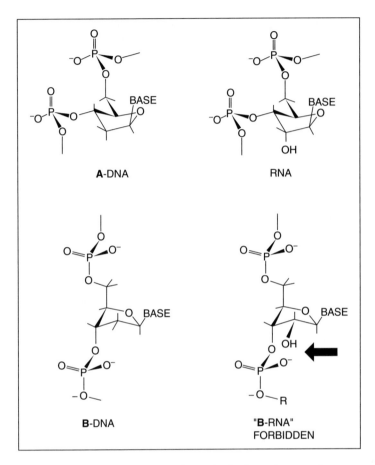

Figure 1.7 When RNA cross-pairs with DNA, the latter has to adapt structurally to the former, not vice versa. The steric hindrance in the (conformationally idealized) forbidden B-type RNA would consist in a repulsion between the 2′-hydoxyl and an oxygen atom of the phosphodiester group

acids' duplex structures in terms of idealized conformations within the framework of rules listed above allows the organic chemist to rationalize and appreciate these remarkable facts at a level of stereochemical reasoning that biologists tend to achieve (Figure 1.7).

The synthesis and first steps in our rather comprehensive exploration of the homo-DNA system (pioneered by Christian Leumann, Hans-Jörg Roth, Jürg Hunziker and Markus Boehringer [11, 15, 16]) had in our laboratory the status of a model study that served the purpose of developing experimental methodologies besides determining the influence of a simple insertion of a methylene group into DNA's furanose ring might have on the system's pairing capability. Homo-DNA was not considered to be a potentially natural nucleic acid alternative because the dideoxyhexose sugar building block is a generationally much more complex sugar in comparison with a nonreduced hexose; it is in this sense that the homo-DNA project represented only a model study.

The base-pairing properties of synthetic homo-DNA strands turned out to be remarkable, to say the least. Strands with complementary nucleobase sequences showed regular Watson–Crick pairing much stronger than in DNA, but did not display any cross-pairing with complementary DNA (or RNA) strands. In homo-DNA, there is strong purine–purine pairing of guanine with isoguanine and of 2,6-diaminopurine and xanthine in the Watson–Crick mode, as well as homo-purine–purine self-pairing of adenine and of guanine in the reverse-Hoogsteen mode. Comparison of thermodynamic parameters of homo-DNA and DNA duplexations clearly revealed homo-DNA's stronger Watson–Crick pairing to be of entropic rather than enthalpic origin and, therefore, presumably to reflect a higher degree of conformational strand preorganization toward duplex formation in homo-DNA than in DNA; and very plausibly so, considering the difference in flexibility between a pyranose chair and a furanose ring (Figure 1.8).

It was Christian Leumann who first drew the consequences of these observations by initiating and pioneering with his "bicyclo-DNA" project [17] an important strategy in antisense0oriented nucleic acid research (design of oligomer backbones that are conformationally preorganized toward duplexation with DNA and RNA), a strategy that turned out to be highly successful in his and a number of other laboratories (Figure 1.9) [2d, 18].

Figure 1.10 gives – in terms of the formulae of monomer units depicted in their (potential) pairing conformation – an overview of all the families of nucleic acid alternatives we investigated at ETH and TSRI. All (except homo-DNA) are taken from the close structural neighborhood of RNA and deemed to be potentially natural nucleic acid alternatives, since they relate generationally to their respective building blocks in the same way as RNA does to ribose, purines, pyrimidines, and phosphate. Grey scale formulae denote oligomer systems found experimentally to be Watson–Crick base-pairing systems, identical color indicates systems that are capable of communicating with each other through informational intersystem cross-pairing, and nongrey members denote oligomer systems that are devoid of any significant base-pairing capability in the Watson–Crick mode.

SEARCHING FOR NUCLEIC ACID ALTERNATIVES

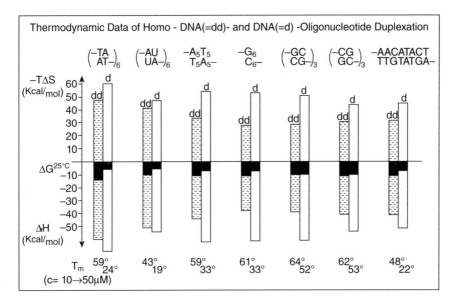

Figure 1.8 Comparison of thermodynamic data of duplexations in the homo-DNA versus the DNA series [11]. The larger negative ΔG values of the homo-DNA duplexations are entropic in origin. Homo-DNA single strands are more strongly preorganized toward duplexation than RNA single strands, and the latter (see Figure 1.7) more strongly than DNA single strands

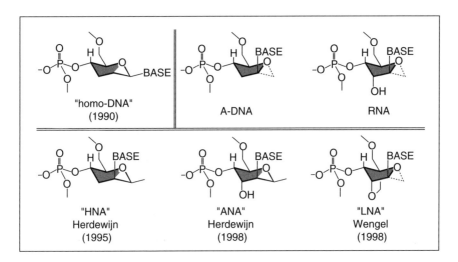

Figure 1.9 Pictorial juxtaposition (in the form of idealized conformations) of homo-DNA with the natural systems and important analogs of the latter (for these analogs, see Ref. [18])

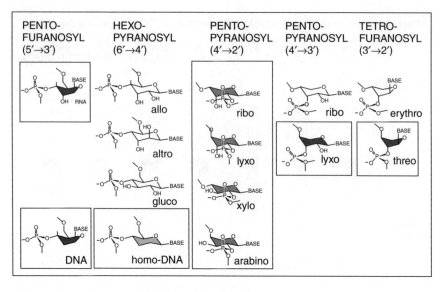

Figure 1.10 Overview of the families of nucleic acid alternatives investigated at ETH and TSRI (same shade denotes Watson–Crick base-pairing in the same base-pairing language; absence of color indicates lack of or only weak Watson–Crick base-pairing capability)

A remarkable and etiologically significant finding is the discrepancy between the base-pairing potential of the model system homo-DNA and the three members of the family of homo-DNA's "natural" analogs studied, namely the fully hydroxylated oligonucleotide systems derived from allose, altrose, and glucose [19–21]. The latter show neither efficient, nor consistent Watson–Crick base-pairing. Exploratory base-pairing tests with mono-hydroxylated models of the allose- and altrose-derived oligomers supported the hypothesis that the source of this crucial difference in pairing capability is steric hindrance involving the two additional hydroxyl groups of the fully hydroxylated hexopyranose unit, as well as neighboring nucleobases, precluding the population of the unit's pairing conformation in oligomers [22]. Clearly, RNA-alternatives derived from hexopyranoses in place of ribofuranose could not have acted as functional competitors of RNA in Nature's evolution of a genetic system [23].

Apart from homo-DNA, by far the most extensive of our experimental efforts invested in the nucleic acid etiology project were directed to the synthesis (pioneered by Stephan Pitsch and Sebastian Wendeborn [24]) and investigation of the structure and properties of the pyranosyl isomer of RNA ("p-RNA"), the nucleic acid alternative that is composed of the very same building blocks as natural RNA itself. In contrast to the backbone design for RNA alternatives in the hexopyranosyl series, where the phosphodiester group can link the sugar units in a way completely analogous to that in RNA (4′–6′-link in hexopyranose and 3′–5′-link in ribofuranose), the design of the backbone of a pyranosyl-RNA demanded a distinct deviation from that constitutional pattern in as far as a six-bond periodicity of the pento-pyranose backbone would only be compatible with a (2′–4′)-phosphodiester junction. Applying this criterion and defining the nucleoside to have the β-configuration (in order to have the nucleobase at the ribopyranose chair in an equatorial position) resulted in the p-RNA backbone depicted in Figure 1.11. p-RNA's pairing conformation, equally depicted in Figure 1.11, emerged in a remarkably unambiguous way from a conformational analysis carried out with the very same criteria used already for homo-DNA: the ensemble of nine *least-strained* monomer-unit conformers contains only a single member displaying a repetitive conformation and, therefore, embodying the pairing conformation of an p-RNA duplex (Figure 1.12). It was not only this singularity that allowed us to predict with high confidence that p-RNA will turn out to be a base-pairing system, it was above all our experience that the pairing model for homo-DNA, built

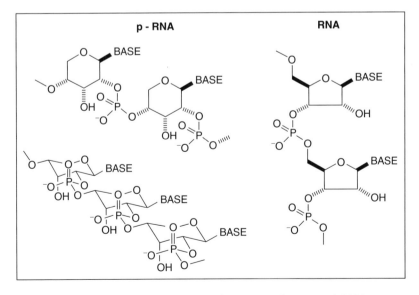

Figure 1.11 Constitution and pairing conformation of pyranosyl-RNA

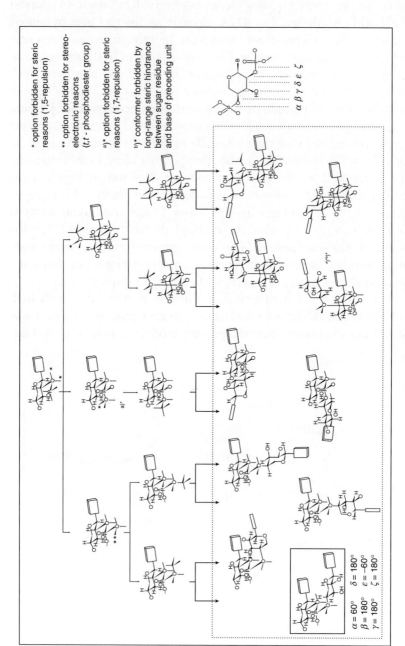

Figure 1.12 Qualitative conformational analysis of pyranosyl-RNA (facsimile from Ref. [24a])

SEARCHING FOR NUCLEIC ACID ALTERNATIVES 25

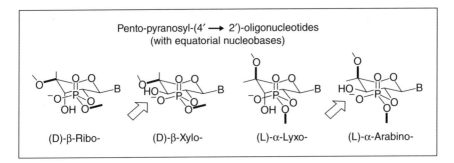

Figure 1.13 The four diastereomeric pentopyranosyl-RNA variants (with equatorial nucleobases) studied at ETH and TSRI. Arrows point to severe steric hindrance in (idealized) pairing conformation

with the same criteria as that of p-RNA had, in fact, found its counterpart in reality and, therefore, there was reasoned hope that the p-RNA model would do so, too. In fact, the agreement between the predicted model and the model that emerged from Bernhard Jaun's X-ray analysis and Romain Wolf's molecular mechanics-based modeling of the p-RNA–octamer duplex pr(CGAATTCG)$_2$ was exquisite [25].

While such unambiguous agreement between "theory" and experiment is of course welcomed, it is the encountering of clear-cut disagreement with specific details of a "theory's" predictions that reveals to us the limits of that "theory" and, therefore, should to be welcomed, too. We encountered such disagreement when we studied the entire family of the four possible diastereomeric pentopyranosyl-oligonucleotide systems (containing equatorial nucleobases) [26], namely (besides the beta-D-ribo-system) the β-D-xylo- [27], the α-L-lyxo- [28], and the α-L-arabino- [29] member (Figure 1.13). It was the arabinopyranosyl member that turned out to be the strongest base-pairing system by far of the entire family (Figure 1.14), while we had expected it to be a member that would definitely pair less strongly than p-RNA [24a].

The α-L-arabino-pyranosyl system is, in fact, one of the strongest oligonucleotide-type Watson–Crick pairing systems known today. The lesson which this discrepancy between expectation and fact taught us was (at least) twofold. First, in cases where the conformational analysis in terms of idealized conformations leads to an ensemble of monomer-unit conformers in which – according to the standard criteria – all members (the conformationally repetitive one(s) inclusive) are heavily strained, the task of judging on the level of idealized conformations as

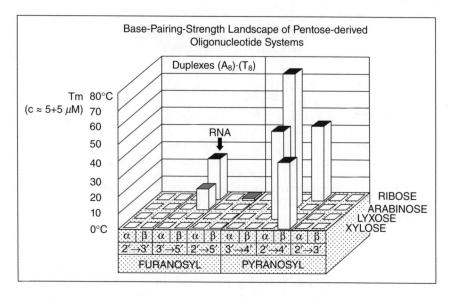

Figure 1.14 Comparison of base-pairing strength of pentopyranosyl-oligonucleotides and RNA (A8/T8-duplexes; in 0.15 M NaCl, 10 μM NaH$_2$PO$_4$, pH 7; see Refs [24, 26–29])

to whether a conformationally repetitive conformer is among the *least unstable* conformers is no longer straightforward; neither is, therefore, any prediction with regard to such a system's base-pairing capability. Second, the relative base-pairing strength of a system is intrinsically co-determined by *inter*-monomer-unit repulsions that are not taken into account by the routine analysis of the monomer unit conformations. However, perhaps the major lesson that our study of the pentopyranosyl-oligonucleotide family as a whole taught us was the one given by the α-L-*lyxo*pyranosyl system [26, 28]; it was the lesson that determined the next project in our work and came about as follows.

One of the most remarkable properties observed for pyranosyl-RNA was its capacity for template-controlled ligations, especially the capability of hemi-self-complementary p-RNA-tetramer-2′,3′-cyclo-phosphates to regio- and chiro-selectively self-template their gradual growth into duplexes of long complementary p-RNA sequences (pioneered by Stephan Pitsch and Martin Bolli [30, 31]). The study of this type of ligation chemistry demanded a careful proof of its regioselectivity, in connection of which iso-p-RNA-oligomers with (3′–4′)- instead of (2′–4′)-phosphodiester bridges were synthesized for comparison purposes (Figure 1.15). Not unexpectedly, such isomeric p-RNA strands

SEARCHING FOR NUCLEIC ACID ALTERNATIVES

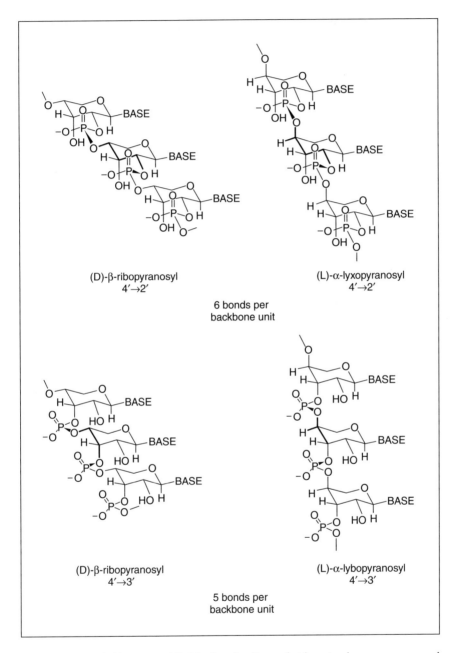

Figure 1.15 (4′–2′)- versus (4′–3′)-phosphodiester bridges in the pentopyranosyl RNA series

lacked the capability of either intrasystem or intersystem (with DNA) base-pairing (five-bond periodicity). However, the analogous modification in the α-L-lyxopyranosyl series, in which a (3′–4′)-phosphodiester bridge is *diaxially* attached to the pentopyranosyl chairs, produced oligomer strands that, though not completely unexpectedly, revealed themselves as being capable of intrasystem base-pairing and, surprisingly, of intersystem cross-pairing with DNA. It seemed clear that the reason for such a drastic six-bond-rule disdaining behavior must be lying in the unique *trans*-diaxial arrangement of the phosphodiester bridge [32].

Sugar units consisting of tetroses (which can form only furanoses) had originally been explicitly neglected as building blocks of nucleic acid alternatives because it is not possible to derive from them tetrofuranosyl-oligonucleotides with a backbone periodicity of six bonds. Our observations with the lyxopyranosyl system, however, made us rethink the matter and etiology-driven reasoning quickly associated the structure of an α-L-*threofuranosyl*-oligonucleotide strand in which the *trans*-phosphodiester bridge might be held in quasi-diaxial conformation by electrostatic reasons and, therefore, give rise to a behavior similar to that of an α-L-lyxopyranosyl oligonucleotide strand (Figure 1.16). Why should such a system be of etiological interest? It would be very much so, because the four-carbon sugar *threose* is a generationally simpler sugar than *ribose*.

α-L-Threofuranosyl-oligonucleotides ("TNA", Figure 1.17; pioneered by Uwe Schönig and Peter Scholz [33]) turned out to be a marvelous Watson–Crick base-pairing system of a pairing strength comparable to that of RNA and the capability to cross-pair with equal efficiency with the natural system (Figure 1.18). Bernhard Jaun's NMR-structure analysis of the duplex t(CGAATTCG)$_2$ showed an RNA-like right-handed helix [34] [and Jaun and Ebert (unpublished results)], and Martin Egli's two X-ray structure analyses of both an A- and B-type DNA duplex of a self-complementary dodecamer sequence containing a single TNA unit demonstrated how the quasi-diaxial phosphodiester bridge adjusts to the DNA double helix, revealing at the same time TNA's distinct preference for the A-type structure paralleling the observation that TNA cross-pairs more strongly with RNA than with DNA [35].

Among molecular biologists, TNA has become more popular than, for example, the pyranosyl isomer of RNA. This is as a simple consequence of the fact that TNA does not speak a "foreign" language as p-RNA does. In Jack Szostak's laboratory it was shown that TNA strands in the presence of a "tolerant" DNA-polymerase can template

SEARCHING FOR NUCLEIC ACID ALTERNATIVES

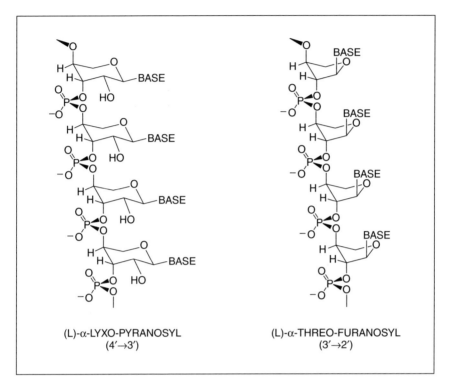

Figure 1.16 The conceptual leap from observing the properties of the (4′–3′)-lyxo-pyranosyl system to studying the (3′–2′)-threofuranosyl system

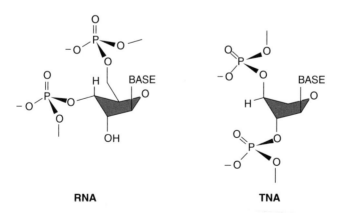

Figure 1.17 Juxtaposition of the (idealized) pairing conformations of RNA and TNA

	TNA	RNA	DNA	
TNA	42 56 53 31	28 57 57 39	32 47 43 25	A B C D
RNA	76 58 50 41	62 59 57 52	59 44 40 36	A B C D
DNA	68 41 36 26	47 43 41 35	55 48 43 36	A B C D

DUPLEX

A 3' - A_{16} - 2'
   2' - T_{16} - 3'

B 3' - $A_4T_3ATAT_2AT_2A$ - 2'
   2' - $T_4A_3TATA_2TA_2T$ - 3'

C 3' - $AT_2AT_2ATAT_3A_4$ - 2'
   2' - $TA_2TA_2TATA_3T_4$ - 3'

D 3' - ATTCAGCG - 2'
   2' - TAAGTCGC - 3'

T_m (10 μM)
1.0M NaCl, pH 7.0

	TNA	RNA	DNA
TNA	−14.5 −73.7 −59.2	−17.3 −93.9 −76.6	−13.3 −77.3 −64.0
RNA	−15.8 −95.2 −79.4	−20.0 −131.9 −111.9	−12.3 −73.9 −61.6
DNA	−11.0 −57.0 −46.0	−13.7 −98.0 −84.3	−16.8 −129.2 −112.4

DUPLEX

B 3' - $A_4T_3ATAT_2AT_2A$ - 2'
   2' - $T_4A_3TATA_2TA_2T$ - 3'

$\Delta G^{25°C}$
ΔH
$T\Delta S$

Figure 1.18 T_m-values and thermodynamic data of self- and cross-pairing in duplex formation within and in between the TNA-, RNA-, and DNA-series [33]

DNA-synthesis and, most remarkably, DNA strands can serve as template for the faithful synthesis of complementary TNA strands (so far up to 50-mer) from α-L-threofuranosyl-nucleotide-2'-triphosphates [36]. The authors of these findings consider such experiments as first steps in a project that aims at an *in vitro* evolution of TNA (with the help of contemporary enzymes) to the functional level of an artificial genotype possessing phenotypic properties. The model vision behind this work is the role assigned to RNA in a hypothetical world (the "RNA-world"), considered to have preceded our own, in which RNA is thought to have fulfilled both genotypic and phenotypic functions [37].

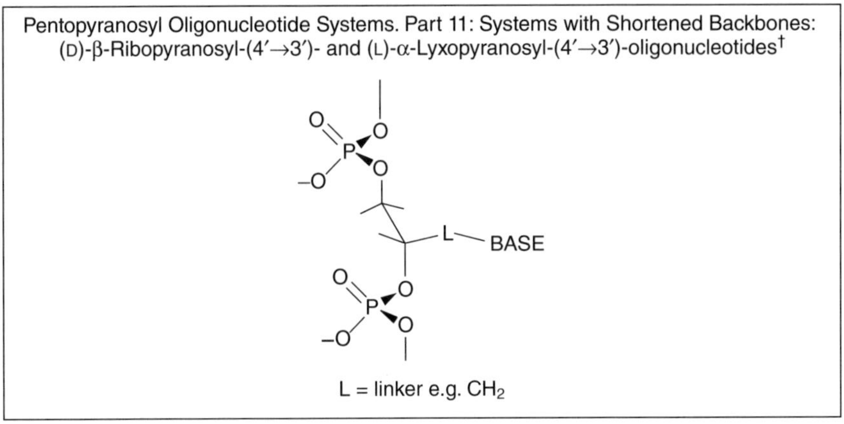

SEARCHING FOR NUCLEIC ACID ALTERNATIVES

> Pentopyranosyl Oligonucleotide Systems. Part 11: Systems with Shortened Backbones:
> (D)-β-Ribopyranosyl-(4'→3')- and (L)-α-Lyxopyranosyl-(4'→3')-oligonucleotides[†]
>
> [structure diagram showing phosphodiester groups connected through a carbon bearing an L—BASE substituent]
>
> L = linker e.g. CH$_2$

Figure 1.19 A hypothetical aliphatic oligonucleotide backbone system in which the vicinal phosphodiester groups might (as a result of a tendency to minimize electrostatic repulsion) assume an antiperiplanar conformation (excerpt from Ref. [32])

Can the structural and generational simplicity of an informational oligomer system such as TNA be pushed even further? In one of our papers [32], we had commented on such an extrapolation of the TNA structure to an even simpler system by hypothesizing that *acyclic* phosphodiester-based oligomers of the type depicted in Figure 1.19 might prove to behave like TNA, since the *trans*-antiperiplanar arrangement of the negatively charged phosphodiester groups – a violation of the *gauche* effect notwithstanding – might be favored by electrostatic repulsion. We refrained from testing the hypothesis experimentally because we supposed the simplicity of oligomers, such as the one derived from the C3-backbone ($n = 1$), is deemed to be structural, not generational (its backbone unit corresponds to a *reduced* sugar), and, therefore, to be a system that lacks the attribute of being potentially natural. Fortunately for nucleic acid chemistry, such a constraint is not what everybody adheres to, and so we know today, thanks to recent work by Eric Meggers [38], that the ($n = 1$)-oligomer of Figure 1.19 is an impressively efficient base-pairing system, able to cross-pair with the natural nucleic acids.

The available NMR-structure analyses for duplexes of the homo-DNA- [13], pyranosyl-RNA- [25], arabinopyranosyl-RNA- [39], and TNA- [34] [and Jaun and Ebert (unpublished results)] series revealed structural models that fall into three different groups where the degree

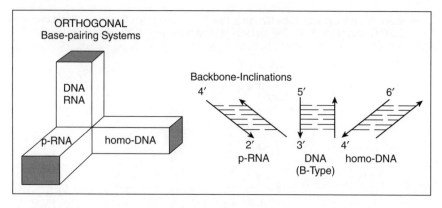

Figure 1.20 The three groups of base-pairing systems which exhibit strong intra-system base-pairing but do not cross-communicate with each other; they represent orthogonal base-pairing systems. This orthogonality is proposed to be related to the different backbone–base-pair axes' inclinations of these systems (excerpt from Ref. [32])

and orientation of the inclination between the (averaged) backbone axis and the (averaged) base-pair axis [40] is concerned. The ladder-like, feebly helical models of homo-DNA and the two pentopyranosyl-nucleic acids show a strong inclination of these axes, yet (significantly) in opposite orientation [40], whereas the compactly helical model of TNA resembles the right-handed helices of the natural nucleic acids of which, characteristically, the B-type helix of DNA has essentially no inclination. This difference in degree and orientation of backbone inclination is related to the base-pairing orthogonality of the three groups of nucleic acids (Figure 1.20), giving rise to barriers to conformational adjustments necessary for base-pairing, even though the inclination is (formally) efficiently adjustable through rotation around nucleosidic bonds (Figure 1.21) [32]. The degree of inclination of these axes (in any orientation) correlates with two further structural properties of informational oligomers: one is a system's propensity to pair in parallel, versus antiparallel, strand orientation [11]; the other is the nature of the base-stacking in duplexes, namely *intra*- versus *inter*-strand [25, 39, 40]. The more strongly inclined a system's backbone is, the more strongly forbidden is base-pairing in parallel strand orientation, and the more pronounced *inter*strand is the base-stacking.

Have we reached the boundaries of the structural landscape of the carbohydrate-derived phosphodiester-type base-pairing systems by what we know today? The answer is a clear *no* from a purely chemical point

SEARCHING FOR NUCLEIC ACID ALTERNATIVES

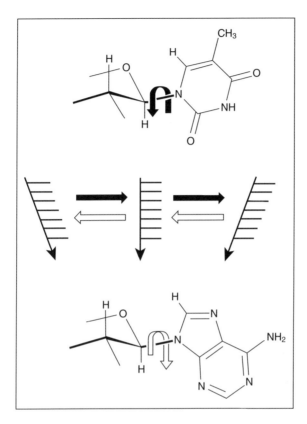

Figure 1.21 Rotation around the nucleosidic bonds in oligonucleotides changes the inclination between the (averaged) backbone–base-pair axes (rotation in the sense indicated by the picture on the top induces an inclination shift towards the right)

of view, and a cautious *very probably not* when referring to systems that fall into the category of potentially natural nucleic acid alternatives. What at this stage, however, seems more urgent than to go on and close this remaining gap of knowledge may be to abandon the selection criterion that determined the course of our work so far, namely that the structure of a nucleic acid alternative has to be taken from RNA's structural neighborhood in the sense that it must be derivable from a $(CH_2)_n$ aldose sugar by the same type of generational chemistry that allows the structure of RNA to be derived from ribose. There is that other class of fundamental biomolecules, the polypeptides and proteins, whose backbone can also serve as the skeleton of informational oligomers – as recent work on nucleobase-tagged peptide-like oligoamide backbones has amply shown [6, 41–47, 49].[4] The polypeptide instead of oligonucleotide type of structure could serve as reference in a search for nucleobase-tagged oligomers, whose repeating units would have to be tagged dipeptide units, as recognized at the very beginning of our work [41]. We would again demand candidate structures to be potentially natural, in the sense that they have to be derivable from their building blocks by the same type of basic chemistry that connects polypeptides with the proteinogenic α-amino acids and, moreover, that these nucleobase-tagged building blocks should be derivable from natural α-aminoacids by potentially primordial reactions.

The last part of this article deals with the question as to what extent the kind of conformational reasoning used in the search for oligonucleotide type nucleic acid alternatives could assist such a search in the field of nucleobase-tagged oligo-dipeptides and related systems and, moreover, could provide a qualitative rationalization of the facts already known in this field.

Figure 1.22 recapitulates the two types of (idealized) pairing conformations of the repeating monomer unit of homo-DNA, transcribed into the corresponding formulae of DNA (and RNA) backbones (see also Figure 1.6), the (-g/-g)-variant of the two representing the type of conformation characteristically observed for A-type duplexes of DNA and RNA. Important characteristics of this backbone conformation are the two 180° torsion angles β and ε (bonds in bold). In a further transcription of the conformation into a backbone variant that lacks the cyclic part of the natural systems, the bold bonds of the backbone formula define the positions at which double bonds could be accommodated without changing in any major way the overall shape of the backbone thread (Figure 1.23, formula a).[5] Interestingly, the *relative positions of these double bonds correspond to the relative positions of the (planar) amide bonds in oligopeptides*. This means that this specific (idealized) type of polypeptide backbone conformation corresponds to the -g/-g pairing conformation of the backbone in A-DNA (and RNA),

---

[4] For studies on δ-peptide analogs of pyranosyl-RNA with Gerhard Quinkert at the University of Frankfurt, see Ref. [49].

[5] A formal transformation of the idealized conformation of a saturated four-center chain containing two vicinal substituents in antiplanar arrangement into the idealized conformation of a 2,3-unsaturated chain with the two substitutents in *trans* arrangement would in fact require two 60° rotations around the resulting sp^3/sp^2 bonds in order to reach the (idealized) most stable conformations at the allylic positions (allylic hydrogen atoms synplanar with the double-bond axis). Allowing for this implicit conformational change would have to be part of a more detailed analysis of such an adaptation of the pairing conformation of an oligonucleotide to that of an oligopeptide. However, for the sake of pictorial simplicity and clarity, this aspect is not taken into account by Figures 1.22–1.24.

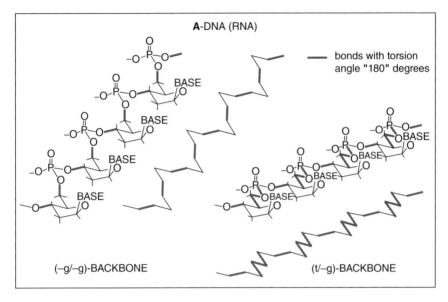

Figure 1.22 The two "allowed" pairing conformations of homo-DNA transcribed into the formula of A-type DNA (or RNA). Bond symbols in bold denote backbone torsion angles of 180°. The thread in the middle of the figure depicts the backbone-core of the (-g/-g)-conformer of A-type DNA. The thread conformation that corresponds to the t/-g conformation (lower right) happens to be identical to the conformation of the thread that would correspond to the (idealized) B-DNA conformation depicted in Figure 1.6

and that nucleobase-tagged oligo-dipeptides, depending on the structural details of the nucleobase-bearing chains alternatingly attached at tetrahedral positions of the backbone (see below), may represent systems that harbor the capacity of base-pairing with themselves and of cross-pairing with DNA and RNA.

As Figure 1.23 (formula b) indicates, there is another way of maintaining the conformational character of the A-DNA-type backbone within an aliphatic backbone thread, namely by putting electronegative atoms at the positions labeled in the figure as black spots. If such a backbone were to follow the organic chemist's "*gauche*-effect" in choosing its preferred conformations, the specific positioning of the electronegative centers would support the maintenance of the thread's specific folding and have an effect on the backbone's conformational preferences similar to that of the aforementioned double bonds. Taking the NH-group as the electronegative center, the oligomer backbone becomes that of an oligo-ethylenimine, pointing to the possibility that, again depending on the structural details of the attachment of the

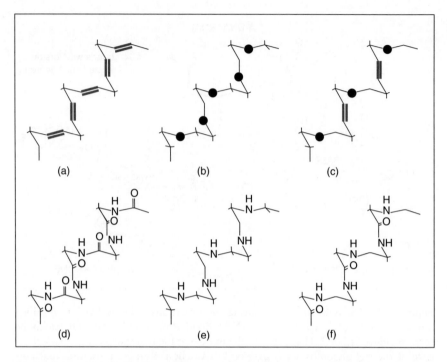

Figure 1.23 The idealized conformation of the backbone-core of A-DNA's (-*g*/-*g*)-conformer equated to the conformation of the backbones of aliphatic oligomer systems containing double bonds (or equivalents thereof) or electronegative centers (black dots)

nucleobase, such a type of structure might serve as the backbone of oligomer systems that cross-pair with DNA and RNA. A similar prediction could be made for a correspondingly nucleobase-tagged oligo-ethylenoxide backbone.

Finally, there is a third way of maintaining the special conformational relationship of an aliphatic backbone with the (-*g*/-*g*) conformation of A-type DNA, namely by any mixed positioning of double bonds (or amide bonds) and electronegative centers along the backbone thread as long the positional requirements stated above remain fulfilled. Figure 1.23 depicts with formula c a variant that shows regular alternation between an amide group and an imino group; this variant, interestingly enough, corresponds constitutionally to the (nontagged) backbone of Nielsen's PNA.

Figure 1.24 gives the generalized formula of nucleobase-tagged oligo-dipeptide oligomers drawn in the two conformations that correspond

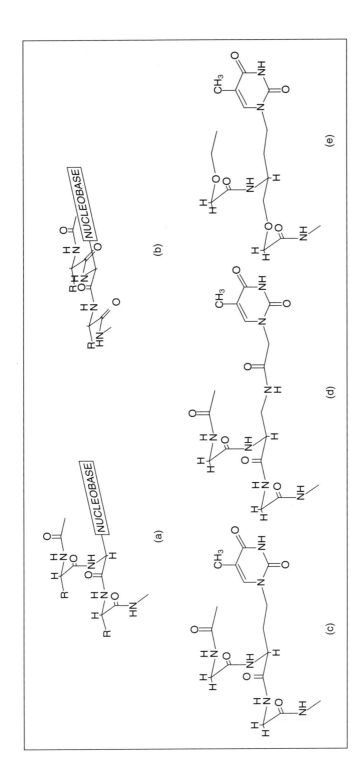

Figure 1.24 The two (idealized) backbone conformations of oligo-dipeptide base-pairing systems (a and b), and three experimentally studied systems taken from the literature, depicted in an (idealized) pairing conformation (c–e).

to the pairing conformations (-*g*/-*g* and *t*/-*g*) of homo-DNA and, consequently, A-type DNA and RNA (formulae a and b). A wide-ranging program of experimental screening of the base-pairing properties of such systems as a function of the structure of the nucleobase-bearing side-chain and, importantly, of the structure type of the nucleobases, is called for.

The residues labeled as R can be the residue of any α-amino acid, preferable one that facilitates water solubility. Only a few representatives of the potentially large diversity of such oligo-dipeptide systems are described in the literature [42]; however, data about the self-pairing capabilities of such systems (for example, of oligomers of type b in Figure 1.24) are absent and the reports about their capability to cross-pair with the natural systems are conflicting. DNA- and RNA- cross-pairing capabilities of such systems seem, indeed, to depend on the nature of the nucleobase-bearing side-chain, as Masayuki Fujii's well-studied and successful oligo-dipeptide system (Figure 1.24d) demonstrates [41]. Of special interest in this context is the oligomer of Figure 1.24e, which corresponds to the type of system pointed to in Figure 1.23c and which shows efficient cross-pairing with DNA and RNA [43], in remarkable contrast to the oligomer of Figure 1.24c, which is reported [43a] not to do so.

In a search for potentially primordial genetic systems within the landscape of nucleobase-tagged oligodipeptides, the main selection criterion for choosing the structure of systems to be studied will have to be a given oligomer system's generational simplicity relative to that of normal natural α-oligopeptides and their building blocks, the α-amino acids. Thereby, the exploration of the structural diversity of generationally simple heterocycles that could serve as alternative nucleobases and especially of the kind of "nucleosidation step" they would allow for will be challenges to be met.[6]

In recent years, a huge amount of empirical information about the structure and base-pairing properties of a large variety of structurally modified analogs of Nielsen's PNA has become available [46, 47].[7] To what extent these data fit into the kind of formal stereochemical exercise

---

[6] Important informational chemistry involving conventional nucleobases and peptidic backbones not based on the oligo-dipeptide (six-bond periodicity) concept exists: see the work of Ulf Diederichsen on nucleobase-tagged alanyl- and homoalanyl-PNAs [44] and that of Philip Garner on α-helical peptide nucleic acids (α-PNAs) [45].

[7] Most, if not all, of these variants do not belong to the landscape of potentially primordial oligomer systems. However, Nielsen's PNA itself has been considered in this context [46, 48].

discussed here, of connecting (idealized) pairing conformations of peptide-type oligomer systems with the "allowed" (idealized) pairing conformations of homo-DNA, is a question that might deserve a separate analysis.

REFERENCES

1. Eschenmoser, A. (1990) Kon-Tiki-Experimente zur Frage nach dem Ursprung von Biomolekülen, in *Materie und Prozesse. Vom Elementaren zum Komplexen* (ed. W. Gerok, W. Martienssen and H.W. Roesky), Verhandlungen Der Gesellschaft Deutscher Naturforscher und Ärzte **116**, Versammlung, Berlin, Wissenschaftliche Verlagsgesellschaft mbH, Stuttgart, p. 135; Eschenmoser, A. (1991) Warum Pentose- und nicht Hexose-Nucleinsäuren? *Nachrichten aus Chemie Technik und Laboratorium*, **39**, 795; Eschenmoser, A. and Loewenthal, E. (1992) Chemistry of potentially prebiological natural products. *Chemical Society Reviews*, **21**, 1.
2. (a) Uhlmann, E. and Peyman, A. (1990) Antisense oligonucleotides: a new therapeutic principle. *Chemical Reviews*, **90**, 543; (b) De Mesmaeker, A., Haener, R., Martin, P., and Moser, H.E. (1995) Antisense oligonucleotides. *Accounts of Chemical Research*, **28**, 366; (c) Hyrup, B. and Nielsen, P.E. (1996) Peptide nucleic acids (PNA): Synthesis, properties and potential applications. *Bioorganic and Medicinal Chemistry*, **4**, 5; (d) Herdewijn, P. (1996) Targeting RNA with conformationally restricted oligonucleotides. *Liebig's Annalen*, 1337; (e) Leumann, C.J. (2002) DNA analogues: from supramolecular principles to biological properties. *Bioorganic and Medicinal Chemistry*, **10**, 841.
3. Eschenmoser, A. (1982) Über organische Naturstoffsynthese: Von der Synthese des Vitamin B12 zur Frage nach dem Ursprung der Corrinstruktur. *Nova Acta Leopoldina, Neue Folge*, **55** (247), 5; Eschenmoser, A. (1985) Vitamin B12 und präbiotische Naturstoffchemie, in *Jahresbericht 1984*, Schweizerische Nationalfonds, Bern, 1985, p. 198; Ksander, G., Bold, G., Lattmann, R. *et al.* (1987) Chemie der α-Aminonitrile 1. Mitteilung Einleitung und Wege zu Uroporphyrinogen-octanitrilen. *Helvetica Chimica Acta*, **70**, 1115.
4. Moser, H., Fliri, A., Steiger, A. *et al.* (1986) Poly(dipeptamidinium)-Salze: Definition und Methoden zur präparativen Herstellung. *Helvetica Chimica Acta*, **69**, 1224.
5. Steiger, A. (1987) Oligomerisation von Dipeptidnitrilen. ETH-Dissertation Nr. 8367.
6. Lohse, P., Oberhauser, B., Oberhauser-Hofbauer, B. *et al.* (1996) Chemie von alpha-Aminonitrilen. XVII. Oligo(nukleodipeptamidinium)-Salze. *Croatica Chemica Acta*, **69**, 535.

7. Doel, M.T., Jones, A.S., and Taylor, N. (1969) An approach to the synthesis of peptide analogues of oligonucleotides (nucleopeptides). *Tetrahedron Letters*, **10**, 2285; De Koning, H. and Pandit, U.K. (1971) Unconventional nucleotide analogues. VI. Synthesis of purinyl- and pyrimidinyl-peptides. *Recueil des Travaux Chimiques des Pays-Bas*, **90**, 1069; Buttrey, J.D., Jones, A.S., and Walker, R.T. (1975) Synthetic analogues of polynucleotides – XIII: the resolution of DL-β-(thymin-1-yl)alanine and polymerisation of the β-(thymin-1-yl)alanines. *Tetrahedron*, **31**, 73; Jones, A.S. (1979) Synthetic analogues of nucleic acids – a review. *International Journal of Biological Macromolecules*, **1**, 194; Shvachkin, Y.P. (1979) Nucleoamino acids and nucleopeptides. 1. General concept and basic determinations. *Zhurnal Obshchei Khimii*, **49**, 5, 1157–1161; Cheikh, A.B. and Orgel, L.E. (1990) Polymerization of amino acids containing nucleotide bases. *Journal of Molecular Evolution*, **30**, 315.
8. Nielsen, P.E., Egholm, M., Berg, R.H., and Buchardt, O. (1991) Sequence-selective recognition of DNA by strand displacement with a thymine-substituted polyamide. *Science*, **254**, 1497.
9. Eschenmoser, A. and Dobler, M. (1992) Warum Pentose- und nicht Hexose-Nucleinsäuren? Teil I. Einleitung und Problemstellung, Konformationsanalyse für Oligonucleotid-Ketten aus 2',3'-Dideoxyglucopyranosyl-Bausteinen ('Homo-DNS') sowie Betrachtungen zur Konformation von A- und B-DNS (Why pentose and not hexose nucleic acids? Part I. Introduction to the problem, conformational analysis of oligonucleotide single strands containing 2',3'-dideoxyglucopyranosyl building blocks ('homo-DNA'), and reflections on the conformation of A- and B-DNA). *Helvetica Chimica Acta*, **75**, 218.
10. Müller, D., Pitsch, S., Kittaka, A. *et al.* (1990) Chemie von α-Aminonitrilen. Aldomerisierung von Glycolaldehyd-phosphat zu racemischen Hexose-2,4,6-triphosphaten und (in Gegenwart von Formaldehyd) racemischen Pentose-2,4-diphosphaten: *rac*-Allose-2,4,6-triphosphat und *rac*-Ribose-2,4-diphosphat sind die Reaktionshauptprodukte (Chemistry of α-aminonitriles. Aldomerisation of glycolaldehyde phosphate to *rac*-hexose 2,4,6-triphosphates and (in presence of formaldehyde) *rac*-pentose 2,4-diphosphates: *rac*-allose 2,4,6-triphosphate and *rac*-ribose 2,4-diphosphate are the main reaction products). *Helvetica Chimica Acta*, **73**, 1410.
11. Hunziker, J., Roth, H.J., Böhringer, M. *et al.* (1993) Warum pentose- und nicht hexose-nucleinsäuren? Teil III. Oligo(2',3'-dideoxy-β-D-glucopyranosyl) nucleotide ('homo-DNS'): Paarungseigenschaften (Why pentose- and not hexose-nucleic acids? Part III. Oligo(2',3'-dideoxy-β-D-glucopyranosyl)nucleotides ('homo-DNA'): base-pairing properties). *Helvetica Chimica Acta*, **76**, 259.
12. For a complete pictorial presentation of the sequence of stereochemical arguments leading to this result, see Quinkert, G., Egert, E., Griesinger, C.

(1996) *Aspects of Organic Chemistry: Structure*. Verlag Helvetica Chimica Acta, Basel, p. 144.
13. Otting, G., Billeter, M., Wüthrich, K. *et al.* (1993) Warum Pentose- und nicht Hexose-Nucleinsäuren? Teil VI. 'Homo-DNS': ^{1}H-, ^{13}C-, ^{31}P- und ^{15}N-NMR-spektroskopische Untersuchung von ddGlc(A-A-A-A-A-T-T-T-T-T) in wässriger Lösung (Why pentose- and not hexose-nucleic acids? Part IV. 'Homo-DNA': ^{1}H-, ^{13}C-, ^{31}P-, and ^{15}N-NMR-spectroscopic investigation of ddGlc(A-A-A-A-A-T-T-T-T-T) in aqueous solution). *Helvetica Chimica Acta*, 76, 2701.
14. Wilds, C.J., Pattanayek, R., Pan, C. *et al.* (2002) Selenium-assisted nucleic acid crystallography: use of phosphoroselenoates for MAD phasing of a DNA structure. *Journal of the American Chemical Society*, 124, 14910.
15. Böhringer, M., Roth, H.J., Hunziker, J. *et al.* (1992) Warum Pentose- und nicht Hexose-Nucleinsäuren? Teil II. Oligonucleotide aus 2′,3′-Dideoxy-β-D-glucopyranosyl-Bausteinen ('homo-DNS'): Herstellung (Why pentose and not hexose nucleic acids? Part II. Preparation of oligonucleotides containing 2′,3′-dideoxy-β-D-glucopyranosyl building blocks ('homo-DNS'): production) *Helvetica Chimica Acta*, 75, 1416.
16. Groebke, K., Hunziker, J., Fraser, W. *et al.* (1998) Warum Pentose- und nicht Hexose-Nucleinsäuren? Teil V. (Purin–Purin)-Basenpaarung in der homo-DNS-Reihe: Guanin, Isoguanin, 2,6-Diaminopurin und Xanthin (Why pentose- and not hexose-nucleic acids? Purine–purine pairing in homo-DNA: guanine, isoguanine, 2,6-diaminopurine, and xanthine). *Helvetica Chimica Acta*, 81, 375.
17. Tarköy, M., Bolli, M., Schweizer, B., and Leumann, C. (1993) Nucleic-acid analogues with constraint conformational flexibility in the sugar–phosphate backbone ('bicyclo-DNA'). Part 1. Preparation of (3S,5′R)-2′-deoxy-3′,5′-ethano-αβ-D-ribonucleosides ('bicyclonucleosides'). *Helvetica Chimica Acta*, 76, 481; Leumann, C.J. (2001) Design and evaluation of oligonucleotide analogues. *Chimia*, 55, 295; Leumann, C.J. (2002) DNA analogues: from supramolecular principles to biological properties. *Bioorganic and Medicinal Chemistry*, 10, 841.
18. Herdewijn, P. (1996) Targeting RNA with conformationally restricted oligonucleotides. *Liebig's Annalen*, 1337; Herdewijn, P. (1999) Conformationally restricted carbohydrate-modified nucleic acids and antisense technology. *Biochimica et Biophysica Acta*, 1489, 167; Wengel, J. (1999) *Accounts of Chemical Research*, 32, 301; Obika, S., Nanbu, D., Hari, Y. *et al.* (1998) Stability and structural features of the duplexes containing nucleoside analogues with a fixed N-type conformation, 2′-O,4′-C-methyleneribonucleosides. *Tetrahedron Letters*, 39, 5401; Lescrinier, E., Froeyen, M., and Herdewijn, P. (2003) Difference in conformational diversity between nucleic acids with a six-membered 'sugar' unit and natural 'furanose' nucleic acids. *Nucleic Acids Research*, 31, 2975.

19. Fischer, R.W. (1992) Allopyranosyl-Nukleinsäure: Synthese, Paarungseigenschaften und Struktur von Adenin-/Uracil-haltigen Oligonukleotiden. ETH-Dissertation Nr. 9971.
20. Helg, A.G. (1994) Allopyranosyl-Nukleinsäure Synthese: Paarungseigenschaften und Struktur von Guanin-/Cytosin-enthaltenden Oligonukleotiden. ETH-Dissertation Nr. 10464.
21. Diederichsen, U. (1993) A. Hypoxanthin-Basenpaarungen in Homo-DNA-Oligonucleotiden. – B. Zur Frage des Paarungsverhaltens von Glucopyranosyl-Oligonucleotiden. ETH-Dissertation Nr. 10122.
22. Hammer, R. (1992) Postdoctoral report ETH Zürich; Miculka, C. (1995) Postdoctoral report ETH Zürich and University of Frankfurt.
23. Eschenmoser, A. (1999) Chemical etiology of nucleic acid structure. *Science*, 284, 2118.
24. (a) Pitsch, S., Wendeborn, S., Jaun, B., and Eschenmoser, A. (1993) Why pentose- and not hexose-nucleic acids? Part VII. Pyranosyl-RNA ('p-RNA'). Preliminary communication. *Helvetica Chimica Acta*, 76, 2161; (b) Pitsch, S., Wendeborn, S., Krishnamurthy, R. *et al.* (2003) Pentopyranosyl oligonucleotide systems, 9th communication: the β-D-ribopyranosyl-(4'→2')-oligonucleotide system ('pyranosyl-RNA'): synthesis and resume of base-pairing properties. *Helvetica Chimica Acta*, 86, 4270.
25. Schlönvogt, I., Pitsch, S., Lesueur, C. *et al.* (1996) Pyranosyl-RNA ('p-RNA'): NMR and molecular-dynamics study of the duplex formed by self-pairing of ribopyranosyl-(C-G-A-A-T-T-C-G). *Helvetica Chimica Acta*, 79, 2316.
26. Beier, M., Reck, F., Wagner, T. *et al.* (1999) Chemical etiology of nucleic acid structure: comparing pentopyranosyl-(2'→4') oligonucleotides with RNA. *Science*, 283, 699.
27. Wagner, T., Huynh, H.K., Krishnamurthy, R., and Eschenmoser, A. (2002) Pentopyranosyl oligonucleotide systems, communication no. 12, the β-D-xylopyranosyl-(4'→2')-oligonucleotide system. *Helvetica Chimica Acta*, 85, 399.
28. Reck, F., Wippo, H., Kudick, R. *et al.* (2001) Pentopyranosyl oligonucleotide systems, communication no. 10, the α-L-lyxopyranosyl-(4'→2')-oligonucleotide system. *Helvetica Chimica Acta*, 84, 1778.
29. Jungmann, O., Beier, M., Luther, A. *et al.* (2003) Pentopyranosyl oligonucleotide systems, communication no. 13: the α-L-arabinopyranosyl-(4'→2')-oligonucleotide system: synthesis and pairing properties) *Helvetica Chimica Acta*, 86, 1259.
30. Pitsch, S., Krishnamurthy, R., Bolli, M. *et al.* (1995) Pyranosyl-RNA ('p-RNA'): base-pairing selectivity and potential to replicate. Preliminary communication. *Helvetica Chimica Acta*, 78, 1621; Bolli, M., Micura, R., Pitsch, S., and Eschenmoser, A. (1997) Pyranosyl-RNA: further observations on replication. *Helvetica Chimica Acta*, 80, 1901.

31. Bolli, M., Micura, R., and Eschenmoser, A. (1997) Pyranosyl-RNA: chiroselective self-assembly of base sequences by ligative oligomerization of tetranucleotide-2′,3′-cyclophosphates (with a commentary concerning the origin of biomolecular homochirality). *Chemistry and Biology*, **4**, 309.
32. Wippo, H., Reck, F., Kudick, R. *et al.* (2001) Pentopyranosyl oligonucleotide systems. Part 11: systems with shortened backbones: (D)-β-ribopyranosyl-(4′→3′)- and (L)-α-lyxopyranosyl-(4′→3′)-oligonucleotides. *Bioorganic and Medicinal Chemistry*, **9**, 2411.
33. Schöning, K.U., Scholz, P., Guntha, S. *et al.* (2000) Chemical etiology of nucleic acid structure: the α-threofuranosyl-(3′→ 2′) oligonucleotide system. *Science*, **290**, 1347; Schöning, K.U., Scholz, P., Wu, X. *et al.* (2002) The α-L-threofuranosyl-(3′→ 2′)-oligonucleotide system ('TNA'): synthesis and pairing properties. *Helvetica Chimica Acta*, **85**, 4111.
34. See Focus article by Hall, N. (2004) The quest for the chemical roots of life. *Chemical Communications*, 1247.
35. Wilds, C.J., Wawrzak, Z., Krishnamurthy, R. *et al.* (2002) Crystal structure of a B-form DNA duplex containing (L)-α-threofuranosyl (3′→2′) nucleosides: a four-carbon sugar is easily accommodated into the Bac. *Journal of the American Chemical Society*, **124**, 13716; Pallan, P.S., Wilds, C.J., Wawrzak, Z. *et al.* (2003) Why does TNA cross-pair more strongly with RNA than with DNA? An answer from X-ray analysis. *Angewandte Chemie, International Edition*, **42**, 5893.
36. Chaput, J.C., Ichida, J.K., and Szostak, J.W. (2003) TNA synthesis by DNA polymerases. *Journal of the American Chemical Society*, **125**, 856; Ichida, J.K., Zou, K., Horhota, A. *et al.* (2005) An in vitro selection system for TNA. *Journal of the American Chemical Society*, **127**, 2802; see also Kempeneers, V., Vastmans, K., Rozenski, J., Herdewijn, P. (2003) Recognition of threosyl nucleotides by DNA and RNA polymerases. *Nucleic Acids Research*, **31**, 6221.
37. Joyce, G.F. and Orgel, L.E. (1999) Prospects for understanding the origin of the RNA world, in *The RNA World*, 2nd edn (eds R.F. Gesteland, T.R. Cech and J.F. Atkins), Cold Spring Harbor Laboratory Press, Cold Spring Harbor, NY, p. 49.
38. Zhang, L., Peritz, A., and Meggers, E. (2005) A simple glycol nucleic acid. *Journal of the American Chemical Society*, **127**, 4174.
39. Ebert, M.O., Luther, A., Huynh, H.K. *et al.* (2002) NMR solution structure of the duplex formed by self-pairing of α-L-arabinopyranosyl-(4′2′)-(CGAATTCG). *Helvetica Chimica Acta*, **85**, 4055.
40. Micura, R., Kudick, R., Pitsch, S., and Eschenmoser, A. (1999) Chemistry of pyranosyl-RNA, Part 8. Opposite orientation of backbone inclination in pyranosyl-RNA and homo-DNA correlates with opposite directionality of duplex properties. *Angewandte Chemie, International Edition in English*, **38**, 680.

41. Fujii, M., Yoshida, K., Hidaka, J., and Ohtsu, T. (1998) Hybridization properties of nucleic acid analogs containing β-aminoalanine modified with nucleobases. *Chemical Communications*, 717.
42. Lenzi, A., Reginato, G., and Taddei, M. (1995) Synthesis of N-Boc-α-amino acids with nucleobase residues as building blocks for the preparation of chiral PNA (peptidic nucleic acids). *Tetrahedron Letters*, **36**, 1713; Lenzi, A., Reginato, G., Taddei, M., and Trifilieff, E. (1995) Solid phase synthesis of a self complementary (antiparallel) chiral peptidic nucleic acid strand. *Tetrahedron Letters*, **36**, 1717; Howarth, N.M. and Wakelin, L.P.G. (1997) α-PNA: a novel peptide nucleic acid analogue of DNA. *Journal of Organic Chemistry*, **62**, 5441; Garner, P. and Yoo, J.U. (1993) Peptide-based nucleic acid surrogates incorporating ser[CH$_2$B]-gly subunits. *Tetrahedron Letters*, **34**, 1275; Lewis, I. (1993) Peptide analogues of DNA incorporating nucleobase-Ala-Pro subunits. *Tetrahedron Letters*, **34**, 5697.
43. (a) Kuwahara, M., Arimitsu, M., and Sisido, M. (1999) Synthesis of δ-amino acids with an ether linkage in the main chain and nucleobases on the side chain as monomer units for oxy-peptide nucleic acids. *Tetrahedron*, **55**, 10067; Kuwahara, M., Arimitsu, M., and Sisido, M. (1999) Novel peptide nucleic acid that shows high sequence specificity and all-or-none-type hybridization with the complementary DNA. *Journal of the American Chemical Society*, **121**, 256; (b) Altmann, K.H., Schmit Chiesi, C. and Garcia-Echeverria C. (1997) Polyamide based nucleic acid analogs – synthesis of δ-amino acids with nucleic acid bases bearing side chains. *Bioorganic and Medicinal Chemistry Letters*, **7**, 1119; Garcia-Echeverria, C., Hüsken, D., Schmit Chiesi, C. and Altmann, K.H. (1997) Novel polyamide based nucleic acid analogs-synthesis of oligomers and RNA-binding properties. *Bioorganic and Medicinal Chemistry Letters*, **7**, 1123.
44. Diederichsen, U. (1996) Paarungseigenschaften von Alanyl-Peptidnucleinsäuren mit alternierend konfigurierten Aminosäureeinheiten als Rückgrat. *Angewandte Chemie*, **108**, 458; Diederichsen, U. and Schmitt, H.W. (1996) Self-pairing PNA with alternating alanyl/homoalanyl backbone. *Tetrahedron Letters*, **37**, 475; Diederichsen, U. and Schmitt, H.W. (1998) β-Homoalanyl-PNA: a special case of β-peptides with β-sheet-like backbone conformation; organization in higher ordered structures. *European Journal of Organic Chemistry*, **1998** (5), 827; Hoffmann, M.F.H., Brückner, A.M., Hupp, T. et al. (2000) Specific purine–purine base pairing in linear alanyl-peptide nucleic acids. *Helvetica Chimica Acta*, **83**, 2580; Brückner, A.M., Chakraborty, P., Gellmann, S.H., and Diederichsen, U. (2003) Molekulare Architektur mit funktionalisierten β-Peptid-Helices. *Angewandte Chemie*, **115**, 4532.
45. Garner, P., Dey, S., and Huang, Y. (2000) α-Helical peptide nucleic acids (αPNAs): a new paradigm for DNA-binding molecules. *Journal of the American Chemical Society*, **122**, 2405; see also α-Helical peptide nucleic

acids (αPNAs), in *Pseudo-Peptides in Drug Discovery* (ed. P.E. Nielsen), (2004) Wiley–VCH Verlag GmbH, Weinheim, Chapter 5, p. 193.
46. Nielsen, P.E. (1999) Peptide nucleic acid. A molecule with two identities. *Accounts of Chemical Research*, 32, 624; Nielsen, P.E., Koppelhus, U., and Beck, F. (2004) Peptide nucleic acid (PNA). A pseudo-peptide with DNA-like properties, in *Pseudo-Peptides in Drug Discovery* (ed. P.E. Nielsen), Wiley–VCH Verlag GmbH, Weinheim, p. 153.
47. Kumar, V.A. (2002) Structural preorganization of peptide nucleic acids: chiral cationic analogues with five- or six-membered ring structures. *European Journal of Organic Chemistry*, 2002 (13), 2021 (review); Kumar, V.A. and Ganesh, K.N. (2005) Conformationally constrained PNA analogues: structural evolution toward DNA/RNA binding selectivity. *Accounts of Chemical Research*, 38, 404.
48. Nelson, K.E., Levy, M., and Miller, S.L. (2000) Peptide nucleic acids rather than RNA may have been the first genetic molecule. *Proceedings of the National Academy of Sciences of the United States of America*, 97, 3868.
49. Karig, G., Fuchs, A., Büsing, A. *et al.* (2000) δ-Peptide analogues of pyranosyl-RNA, Part 1, nucleo-δ-peptides derived from conformationally constrained nucleo-δ-amino acids: preparation of monomers. *Helvetica Chimica Acta*, 83, 1049; Schwalbe, H., Wermuth, J., Richter, C. *et al.* (2000) δ-Peptide analogues of pyranosyl-RNA, Part 2, nucleo-δ-peptides derived from conformationally constrained nucleo-δ-amino acids: NMR study of the duplex formed by self-pairing of the (1′S,2′,4′S)-(phba)-nucleo-δ-peptide-(AATAT). *Helvetica Chimica Acta*, 83, 1079.

2

Never-Born RNAs: Versatile Modules for Chemical Synthetic Biology

Davide De Lucrezia[1], Fabrizio Anella[2], Cristiano Chiarabelli[2], and Pier Luigi Luisi[2]

[1]*University Ca' Foscari of Venice, European Centre for Living Technology, Venice, I-30124, Italy*
[2]*University of Roma 3, Biology Department, Viale G. Marconi 446, 00146 Roma, Italy*

2.1 Introduction 48
2.2 Random RNAs Readily Adopt a Compact and Thermostable Secondary Structure 50
2.3 Random RNAs can be Easily Designed and Engineered to Acquire Novel Functions 53
2.4 Random RNAs as a Modular Scaffold for Synthetic Biology 57
2.5 Conclusions 61
Acknowledgments 63
References 63

Synthetic biology aims to design and construct new biological parts, devices, and systems to be implemented in non-natural settings. Synthetic biology will fill the gap between description and understanding of biological systems, clarifying the fundamental principles of biological organization. To date, research in the field of synthetic biology has mainly focused on engineering extant life forms in order to introduce novel, desirable tracts. To unlock the full potential of system biology, parts (devices and systems) must be designed according to novel principles, such as modularity, insulation (orthogonality), standardization, and scalability. Within this framework, a new branch of synthetic biology termed "chemical synthetic biology" has emerged which aims at designing completely *de novo* parts (i.e. RNA, proteins, regulatory networks) to be exploited as novel functional scaffolds for synthetic biology by exploring the sequence space for novel biological entities (NBEs) that do not exist in nature. The rationality behind this approach relies on the observation that functional macromolecules selected from random libraries devoid of any evolutionary constraint can be considered as completely orthogonal (insulated) to extant organisms and, therefore, may be implemented into synthetic biology chassis, possibly reducing cross-talk and parasitic effects. NBEs (unlike redesigned natural macromolecules) can therefore be regarded as "virgin" whose functionality can be engineered without constrictions to meet the user's requirements. We will review the major features of never-born RNAs relevant for chemical synthetic biology such as structural diversity, evolvability, and modularity.

2.1 INTRODUCTION

Synthetic biology has emerged as a new discipline at the interface between biology and engineering. It aims to "(i) design and construct new biological parts, devices, and systems and (ii) re-design existing, natural biological systems for useful purposes" (http://www.syntheticbiology.org). Synthetic biology will certainly have an impact on basic knowledge, in that the implementation of existing genes, proteins, and metabolic pathways in non-native settings will help to shed light on their function and dynamic behavior. As stated by Richard Feymann, "what I cannot create, I do not understand" [1]. Within this framework synthetic biology will fill the gap between description and understanding of biological systems, clarifying fundamental principles of biological organization. On the other hand, synthetic biology will revolutionize technology and production paradigms in the twenty-first century and foster the development of new technological tools to produce innovative medicines (red biotechnology), generate new sources of energy and chemical processes (white biotechnology) and in prevention, and open new gateways in environmental risk research (green biotechnology).

To date, research in the field of synthetic biology has mainly focused on engineering extant life forms in order to introduce novel, desirable tracts. However, over the past few years, synthetic biologists have generated remarkable systems, including: an expanded genetic code in *Escherichia coli* [2]; various logic gates [3, 4]; rewired yeast mating and osmolarity response circuitry [5]; generating bistable switches in bacteria [6], yeast [7], and mammalian cells [8]; spatiotemporal control of cell signaling using light-switchable proteins [9]; genetic and metabolic oscillators [10–12]; producing artificial communication in bacteria [13] and yeast [14].

These remarkable achievements have largely been *one-offs*, since each one is a special case, and though they must be regarded as milestones in the respective field, they do not provide a comprehensive and coherent engineering roadmap for the next step [15].

Indeed, the rationality behind this approach neglects the inner complexity of biological systems and assumes that biological parts can be *cut and pasted* among different organisms, meanwhile retaining their functionality. To unlock the full potential of system biology, parts (devices and systems) must be designed according to novel principles such as modularity, insulation (orthogonality), standardization, and scalability [16, 17]. Within this framework, a new branch of synthetic biology termed "chemical synthetic biology" [18, 19] has emerged which aims at designing completely *de novo* parts (i.e. RNA, proteins, regulatory networks) to be exploited as novel functional scaffolds for synthetic biology by exploring the sequence space for novel biological entities (NBEs) that do not exist in nature. In this regard, "chemical synthetic biology" is concerned with the synthesis of chemical structures such as proteins, nucleic acids, vesicular forms, and others which do not exist in nature. The rationality behind this approach relies on the observation that functional macromolecules selected from random libraries devoid of any evolutionary constraint can be considered as completely orthogonal (insulated) to extant organisms and, therefore, may be implemented into synthetic biology chassis possibly reducing cross-talk and parasitic effects. NBEs (unlike redesigned natural macromolecules) can therefore be regarded as "virgin" whose functionality can be engineered without constrictions to meet the user's requirements. The possibility to freely customize parts' functionality (i.e. RNA) will deeply impact the design and implementation of devices and systems into standard chassis. In this regard, NBEs may pave the way for novel design paradigms in synthetic biology. In the following sections, we will review the major features of never-born RNAs relevant for chemical synthetic biology, such as structural diversity, evolvability, and modularity.

2.2 RANDOM RNAS READILY ADOPT A COMPACT AND THERMOSTABLE SECONDARY STRUCTURE

A stable and well-defined fold is a fundamental prerequisite for the biological activity of extant biopolymers, and this applies to RNAs as well. Accordingly, in order to exploit the potential of random RNA sequences – never-born RNAs – as a scaffold for chemical synthetic biology one must determine whether and to what extend random RNAs adopt a stable and well-defined fold. In addition, the question of whether a stable folding is a common feature of randomly synthesized biopolymers or the rare property of a few molecules is of great significance to the study of the origin of life, as well as to the contingency–determinism debate due to the potential role of RNA as ancestral molecule during evolution [20, 21]. It is worth noting that the notion of fold cannot be univocally defined, since RNA molecules longer than 30 nucleotides will always fold into some sort of secondary structure. Thus, fold becomes a matter of degree and can only be defined operationally; namely, it depends on the experimental criterion used. In our previous studies [22, 23] we used the RNA Foster (RNA FOlding Stability TEst) assay which relies on the specificity of the single-stranded nuclease S1. Basically, the RNA Foster assay allows one to determine the presence and thermal stability of secondary domains in RNA molecules by coupling enzymatic digestion with temperature denaturation (Figure 2.1). The assay is capable of quantitatively determining the fraction of folded RNAs f_{fold} as a function of the temperature. In previous studies we used the RNA Foster to test individually random RNAs 178 nucleotides long, 141 of which were completely random (randomized region equal to 79% of the sequence), in a range of temperature between 37 and 60 °C.

We selected a number of RNAs in a completely random fashion – that is, without any kind of selection bias – out of a library of 2.25×10^8 independent sequences in order to investigate their structural properties by means of the RNA Foster assay. In this regard, these RNAs are statistically significant, in the sense that their structural properties reasonably reflect the average structural features of random RNA molecules.

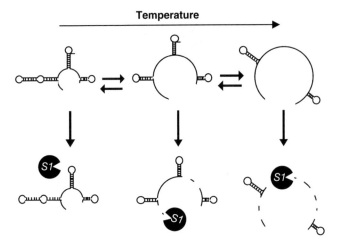

Figure 2.1 Schematic representation of the RNA Foster assay. RNA Foster assay employs S1 nuclease, a single-strand-specific nuclease, to cleave single-stranded regions, thus monitoring the presence of double-stranded domains and indirectly any possible tertiary structure. Folded RNAs are more resistant to S1 nuclease than unfolded ones, as the latter are degraded faster than the former. The capability of nuclease S1 to work over a broad range of temperatures can be used to probe RNA secondary domain stability at different temperatures. An increase in temperature destabilizes the RNA fold, inducing either global or local unfolding. The fraction of folded RNA f_{fold} at each experimental temperature is determined by measuring the amount of RNA remaining after S1 digestion, using electrophoresis and a suitable staining method. The f_{fold} at different temperatures is used to assess the Tm of the RNA molecule

The results obtained were used to determine the folding properties of RNA samples and to assess the melting temperature (Tm) for each RNA. Figure 2.2 plots the correlation between the fraction of S1-resistant RNAs and the temperature. All RNAs are insensitive to S1 digestion at 37 °C, suggesting that all RNA molecules possess a stable secondary structure at this temperature under the conditions observed during the assay. As the temperature increases, the number of RNA molecules sensitive to S1 increases as well. Half of the screened RNAs retain a stable secondary structure up to 50 °C, whereas above this temperature most of the RNAs lose their secondary structures and are readily degraded by S1. These results can be explained by assuming that most nucleotides of the RNA molecules are engaged in some kind of secondary structure at low temperature, so that a compact fold prevents RNAs from being cleaved by S1 nuclease. These secondary domains become

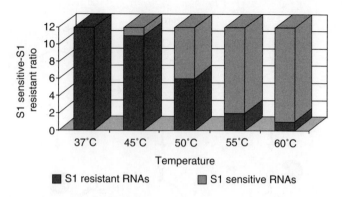

Figure 2.2 Fraction of folded RNA plotted against temperature. The plot shows the number of S1-resistant folded RNAs (blue series) out of 12 screened RNAs for five different temperatures. According to the experimental criterion, S1-resistant RNAs are considered folded (Adapted with permission from [23]. Copyright 2006 Wiley VCH Verlag GmbH & Co. KGaA)

looser as the temperature increases and, consequently, unstructured single-stranded regions become accessible to the nuclease S1 and are readily digested by it. It is remarkable that a "thermophilic" RNA could be found which retains its secondary structure up to 60 °C. It is noteworthy that only a limited number of RNAs have been screened and that the RNAs tested were randomly chosen without any selection bias; thus, the finding of an RNA which was stable up to 60 °C came unexpectedly. Subsequently, we investigated the correlation between thermal stability and GC content, since higher thermal stability is often associated with higher GC content. This study revealed no significant correlation, so that a well-defined and thermal stable fold is not univocally determined by sequence composition. These results show that the formation of secondary domains in random RNA molecules is a common and widely spread property, almost an intrinsic feature of random RNA molecules at 37 °C. Furthermore, half of the sampled RNAs maintain their folding up to 50 °C, suggesting that RNAs are prone to fold into compact secondary structures with a surprising thermal stability. In particular, a thermostable RNA has been found whose fold was stable up to 60 °C. Tested RNAs were randomly chosen and no kind of selection had been exerted on them prior to the analysis, so that this finding comes unexpectedly and suggests that thermostable RNAs might not be so rare in the RNA sequence space [24, 25]. In addition, no significant correlation has been found between the GC content and the thermal stability of RNAs. Although counterintuitive, these results suggest that the sequence composition does not directly determine the thermal stability of RNA folding.

Altogether, these results suggest that RNA molecules have an intrinsic tendency to fold into secondary structures with a surprising thermal stability even in the absence of any selective pressure (that is, when randomly synthesized) and regardless of their GC content. Accordingly, RNA sequence space can be regarded as a rich source of stable macromolecules that can be effectively exploited for synthetic biology.

2.3 RANDOM RNAS CAN BE EASILY DESIGNED AND ENGINEERED TO ACQUIRE NOVEL FUNCTIONS

Synthetic biology relies on the possibility to exquisitely design and engineer novel parts or devices to meet user-specific requirements. Molecular design, such as protein engineering, mainly relies upon three alternative strategies (Figure 2.3): *high-throughput screening* (HTS), *rational design* (RD), and *directed evolution* (DE) (also known as irrational design or applied molecular evolution).

The HTS approach [26, 27] relies on the generation of a vast library of candidate mutants that are individually tested for the desired function. HTS procedures can explore only an infinitesimal number of mutants with respect to all the theoretical ones. Indeed, the number of theoretically possible mutants is

$$M_H = a^H \left[\frac{L!}{(L-H)!H!} \right]$$

where M_H is the theoretical number of mutants for a given biopolymer of length L assuming H mutations and a possible monomers (four for nucleic acids) [28]. It is easy to calculate that for a ribozyme of only 200 nucleotides assuming three mutations per mutant there are more than $M_H = 4^3\{200!/[(200-3)!3!]\} \approx 84 \times 10^6$ theoretically possible mutants. The screening of such an astronomic number is far beyond technical reach, so that current screening procedures often fail to identify the best mutant (e.g. global optima) and are usually stacked to a limited region of the sequence space (e.g. local optima).

RD [29] relies on a detailed knowledge of the relationship between structure and function so that tailored, site-specific mutations can be performed to rationally alter the function of the target molecule

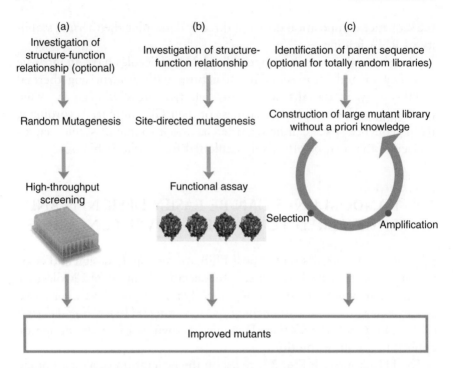

Figure 2.3 Schematic representation of the main approach to molecular design and engineering. (a) The HTS approach relies on the generation of a vast library of candidate mutants that are individually tested for the desired function. (b) RD relies on a detailed knowledge of the relationship between structure and function so that tailored, site-specific mutations can be performed to rationally alter the function of the target molecule. (c) DE mimics natural evolution by means of iterative cycles of mutation–selection–amplification which allows the simultaneous testing of a vast library of candidate mutants for a priori defined function without a priori knowledge

(e.g. enzymes or ribozymes). A comprehensive description of the structure–function relationship is rarely available, and this severely limits the deployment of rational design strategies.

Conversely, DE [30, 31] mimics natural evolution by means of iterative cycles of mutation–selection–amplification, which allows the simultaneous testing of a vast library of candidate mutants for an a priori defined function without a priori knowledge. The great advantage of DE techniques is that no prior structural knowledge of the target molecule is required, nor is it necessary to be able to predict what effect a given mutation will have on the target molecule's function.

Within this framework, DE represents by far the method of choice to employ to design and engineer synthetic biology parts at will. In order to effectively deploy DE there are two fundamental requirements. The first is the availability of a physical link between the genotype and the phenotype. The second relies on the availability of a suitable screening procedure to enrich the initial peptide population of those sequences satisfying the selection criteria.

Since the discovery in 1982 of the first ribozyme molecule [32, 33], RNA molecules have been extensively engineered owing to the fact that RNA is able both to store the genetic information and perform catalysis, fulfilling the fundamental criteria to deploy DE.

In particular, SELEX (Systematic Evolution of Ligands by EXponential enrichment) [34] has been used to engineer RNA molecules with various functions, such as binding and catalysis. The SELEX methodology foresees three major steps:

1. **Generation of molecular diversity** by introduction of genetic variations at DNA level through a variety of mutational techniques.
2. **Selection of best mutant molecules** capable of performing a specific function such as binding to a molecular target or catalyzing a specific reaction.
3. **Amplification of the selected molecules** by selectively recovering the best mutant by retro-transcription and subsequent amplification by polymerase chain reaction (PCR).

This three-step procedure is iteratively repeated to enrich the initial population of those mutants satisfying the selection criteria.

DE has been effectively employed to design novel ribozymes from random sequences. In their seminal work, Ellington and Szostak [35] reported the isolation of functional aptamer (i.e. RNA molecules that bind specific ligands – from the Latin word *aptus*, to fit) capable of tightly binding a small organic molecule. The starting pool consisted of 10^{13} DNA molecules of 155 base pairs comprising 100 random nucleotides flanked by constant regions. The corresponding RNAs were isolated for binding to a variety of dyes (baits) by means of affinity chromatography, where the stationary phase consisted of agarose beads covalently conjugated to the bait. RNAs that did not bind the bait were washed out from the columns, whereas binding RNAs were selectively eluted. The eluted RNAs were amplified and used for the next cycle of selection and eventually mutated by error-prone PCR to explore a larger region of the sequence space. Following the fourth cycle, selected RNAs were isolated and sequenced showing no significant homology to extant sequences. In addition, selected RNAs belong to distinct families with minimal sequence similarity with each other, proving that there are many independent sequence solutions to the binding problems. In another seminal study, Bartel and Szostak [36] successfully isolated new ribozymes from a pool of 10^{15} different molecules of 294 nucleotides comprising a long random stretch of 220 nucleotides structured in three random segments divided by only 6 bp nonrandom sequences (i.e. restriction sites). The selected strategy aimed at isolating novel RNA molecules capable of catalyzing the ligation of substrate oligoribonucleotides to the random 5'-end. The reaction sought was similar to the chain elongation during RNA polymerization in which the 3'-OH of an elongating strand condensates with the α-phosphate of an adjacent 5'-triphosphate, displacing a pyrophosphate (PPi), and yielding a 3',5'-phosphodiester bond. The initial RNA pool was bound to an agarose matrix and allowed to anneal to the substrate oligoribonucleotides. RNAs capable of intramolecular ligation were then eluted and selectively recovered by oligonucleotide affinity chromatography. Eluted RNAs were amplified and used for the next round of selection and eventually submitted to error-prone PCR to introduce mutations in the central core of random sequences. The selected ribozymes exhibited a reaction rate that is 7 million times faster than the uncatalyzed reaction after only four rounds of DE. These two pioneering studies prove the possibility of isolating functional RNAs from random libraries for both binding and catalysis. Since then a number of papers have reported the isolation and design of novel RNAs from random libraries, with a particular emphasis on target validation [37, 38], binding and inhibition of a wide variety of proteins, including growth factors [39], enzymes [40], receptors [41], and viral proteins [42]. Aptamers have also been isolated against a multitude of other targets, such as small organic molecules [43], peptides [44], and even complex multimeric structures [45]. In addition, ribozymes have been successfully used to perform metabolic engineering [46], gene therapy [47], antiviral therapy [48], and to catalyze a number of chemical reactions [49].

Noteworthy is that the fraction of novel functional RNAs isolated from a random library of sequences was estimated to be 1×10^{-10}. This figure is substantially higher than that estimated for functional proteins form random libraries (1×10^{-12}) [50]. The above examples witness the huge versatility potential and the ease with which they can be engineered to perform specific tasks that go beyond the natural and physiological role of extant RNA molecules. These features make random RNAs particularly suited as self-contained functional modules for synthetic biology.

2.4 RANDOM RNAS AS A MODULAR SCAFFOLD FOR SYNTHETIC BIOLOGY

One of the fundamental principles of synthetic biology is the concept of modularity, namely the design of self-contained parts, devices, and systems that can be connected and combined in any combination to achieve a specific function in a forward engineering fashion [17]. This implies that each part (or device or system) has a highly standardized function that can be cut and pasted to work in different contests without a significant alteration of the overall functionality. In order to guarantee "plug-and-play" compatibility of different components, the connections between the different parts need to be standardized [16] and the output of the interaction among different components needs to be predictable. Within this framework, RNA seems perfectly fitted to develop self-contained modules that can be assembled to yield more complex components with predictable features. One of the first researchers to exploit the inherent modularity of RNA structure was Jaeger, who coined the term RNA tectonics [51]: "a new LEGO game for supra-molecular chemists and biochemists [...] which refers to the construction of artificial RNA architectures with novel properties and takes advantage of the knowledge of folding and assembly rules governing the three-dimensional shape of complex natural RNA molecules." Jaeger and co-workers emphasize the modular structure of RNA that can be exploited to construct combinatorial libraries, using RNA modular units for creating molecules with dedicated shapes and properties. This approach moves from the observation that functional RNA molecules can be dissected into self-contained structured domains that self-assemble through non-covalent bounds to reconstitute a fully functional oligomer. This is the case of *Tetrahymena* ribozyme (Figure 2.4) belonging to the group I introns that catalyze their own excision from mRNA, tRNA, and rRNA precursors in a wide range of organisms. The core secondary structure consists of nine paired regions (P1–P9) that fold into two independent domains: P4–P6 (formed from the stacking of P4, P5, and P6 helices) and P3–P9 (formed from the P3, P7, P8, and P9 helices). Owing to the inherent modularity of the *Tetrahymena* fold, inactive mutant ribozymes lacking helices P5a, P5b, and P5c can be rescued by the addition in trans of these helices restoring the original function [52, 53] by means of oligomer formation mediated only by non-covalent interaction among tertiary structures. This study prompted the idea that independent folded RNA domains can be assembled in combinatorial fashion to achieve higher order structure with a priori defined functions and shapes.

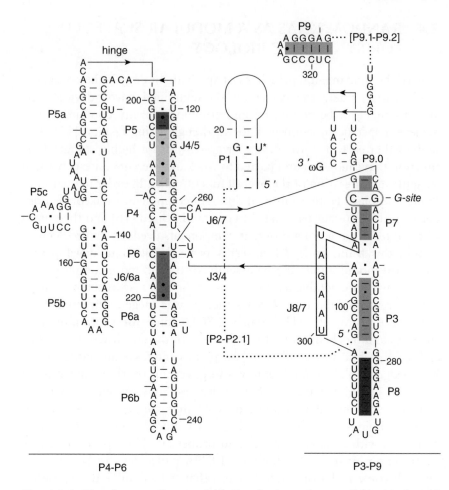

Figure 2.4 *Tetrahymena* ribozyme fold. *Tetrahymena* ribozyme fold with the self-contained independent P4–P6 (formed from the stacking of P4, P5, and P6 helices) and P3–P9 (formed from the P3, P7, P8, and P9 helices). Ribozymes lacking helices P5a, P5b, and P5c can be rescued by the addition in *trans* of these helices, restoring the original function by means of oligomer formation mediated only by non-covalent interaction among tertiary structures (adapted from http://cechlab.colorado.edu/ with author permission. Copyright (1998) American Association for the Advancement of Science, U.S.A.)

Within this framework, Jaeger and co-workers exploited self-contained RNA modules (named "TectoRNA") to form self-assembly nanoscale structures through tertiary interaction [54]. Researcher designed basic modular units that comprise a four-way junction with an interacting module on each helical arm that spontaneously assembles in one-dimensional arrays (Figure 3 in Ref. [55]). Modular assembly was obtained by either a GAAA loop or a specific GAAA loop receptor so that each tectoRNA was capable of interacting with two other tectoR-NAs via the formation of four-loop interaction. Using this approach, researchers demonstrated the possibility of assembling nano-objects containing more than 16 units. Researchers further expanded the repertoire of RNA modules by varying the length, helical twist, and flexibility of the linker region that separates the interacting motifs in each module, demonstrating that several tectoRNAs can associate at submicromolar concentrations [55, 56].

Modular design of novel ribozymes using physically separated RNA modules was also employed by Inoue and co-workers to develop an artificial RNA ligase by fusing RNA–RNA recognition motifs. The resulting ligase produces a 3′–5′ linkage in a template-directed manner for any combinations of two nucleotides at the reaction site [57]. Researchers used RNA modules with well-characterized three-dimensional (3D) structures to design self-folding RNA consisting of standard double-stranded helices connected by either a tetraloop–receptor interaction or consecutive base-triples [58]. Next they inserted a 30 nucleotide random region into the scaffold and performed DE to isolate novel ribozymes with ligase activity. Finally, researchers converted this *cis*-acting ribozyme to a *trans*-acting ligase by dissecting the 3D structure into two independent modules: the catalytic module and the substrate module (Figure 2.5). The substrate module could be further

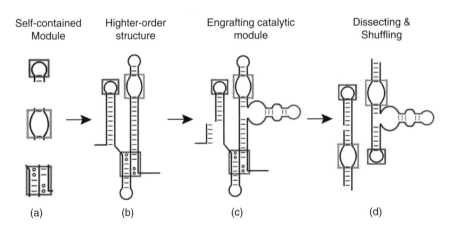

Figure 2.5 Design of novel ligase using physically independent RNA module. Scheme and structures of *de novo* ligase using self-folded RNA modules (a), combined in a higher order structure (b), further engineered to display ligase activity (c) and dissected into independent domains to allow substrate specificity redesign by module shuffling (d) (Adapted from [57]. Copyright 2004 National Academy of Sciences, U.S.A.)

engineered to alter specificity, proving the feasibility to construct versatile ribozymes exploiting combinatorial assembly of self-contained RNA modules. These examples demonstrate the possibility of combining different RNA parts to design higher order structures with specific functions empowering the central concept of synthetic biology: modularity.

Modularity is an intrinsic property of RNA molecules and it is intimately related to the folding process. Single-strand RNAs are flexible polymers with a virtual infinite number of spatial conformations. The folding process is mainly the result of base-pairing interactions occurring between far-off residues.[1] As a consequence, distant regions of the polymer are brought close with a net decrease of conformational freedom. Therefore, the entropy difference between the final state and the initial state is negative and can be quantitatively expressed as [59]

$$\Delta S_{closure} = R \ln(\Omega'/\Omega_t) < 0 \qquad (2.1)$$

where Ω' is the number of possible conformations of the folded state and Ω_t is the total number of possible conformations of the random coil polynucleotide. Consequently, the entropic contribution $-T\Delta S$ is a positive term in the free energy equation ruling the folding process [59]:

$$\Delta G_{folding} = \Delta H - T\Delta S \leq 0 \qquad (2.2)$$

Therefore, for a spontaneous folding, the entropy cost must be balanced and possibly overcome by the enthalpy change ΔH, which accounts for the heat released upon base-pairing (BP) and base-stacking (BS) interactions.

---

[1] RNA folding is mainly driven by interactions between nucleotides along the same chain. The four different kinds of bases interact with each other through hydrogen bonding allowed only between nucleotides with complementary hydrogen bonding profiles, which lead to extremely selective interactions according to Watson-Crick rules. This significantly reduces the number of theoretically possible interaction to two: AT and GC. As a consequence of the matching, all polar groups are tied up in interaction, with the result that the interaction surface becomes hydrophobic. When the matching involves several adjacent nucleotides the hydrophobic regions interact with each other, further stabilizing the structure (base stacking). In addition, phosphate groups will invariably point towards the polar solvent, i.e. water, reducing drastically the number of total possible conformation, namely Ω_t (see Equation (2.1)). The result is a reduction of the entropy cost of folding [59].

The free energy contributions of BP and BS range from −0.9 to −3.4 kcal mol^{-1} [59];; this value is even less than the free energy cost of four nucleotides loop closure ($\Delta G°_{37°C}$ = +4.5 kcal mol^{-1}) [59]. This implies that consecutive base pairs must form in order to stabilize even simple folds. The rate of consecutive base-pairs formation depends on the frequency of random collisions between complementary bases, which is a function of the number of nucleotides separating the two molecular partners [59]. From a kinetic standpoint the interaction of not too far-off residues is faster than and favored over the interaction of far-distant nucleotides. Besides, the entropy cost of bringing far-off regions close is certainly larger than the entropy cost to perform the same task for close regions [60]. The main conclusion is that kinetics goes along with thermodynamics during RNA folding. The consequence is that RNA folding is sequential as well as hierarchical, in the sense that the folding process moves from a fast[2] and stable structure to energy-demanding structures [60, 62]. Hence, the RNA 3D architecture results from the interaction of independent, separate, preformed, and stable domains. These exquisite features of the RNA folding process make RNA molecules extremely suited to implement a modular approach to the construction of a higher order structure starting from self-contained parts and devices.

2.5 CONCLUSIONS

In this chapter we reviewed the main features of never-born RNA (i.e. random sequences with no significant homology with extant sequences) that make this molecule particularly suited to customization and engineering, satisfying the basic principles of synthetic biology.

The first feature that makes never-born RNA an attractive option for chemical synthetic biology is that the sequence space related to random RNAs represents a huge resource of untapped diversity; indeed, if one considers an RNA of only 100 nucleotide length there are $4^{100} \approx 1.6 \times 10^{60}$

---

[2]The kinetics of RNA folding is limited by the rate of intramolecular stem formation, which is the slowest event. The unimolecular rate constant of an intramolecular stem has been determined as $k = f n \exp(-B/RT)$, where $f \approx 10^5 s^{-1}$ is the rate constant for base-pair formation of head-on residues, n is the number of base pairs in the helical stem, T is the absolute temperature, and B is the activation energy barrier for the loop closure [61]). For a hairpin with 5–12 nucleotides in the stem helix the formation time has been estimated in the millisecond region at 37°C [59].

theoretically different RNA sequences, which may be searched to isolate novel structures and functions. However, the actual diversity exploitable to isolate novel structures directly relates to the fraction of random RNA sequences that fold into a compact and stable structure, since a well-defined fold is a fundamental prerequisite for any biological activity. Accordingly, in order to exploit the potential of random RNA sequences as scaffold for chemical synthetic biology, one shall determine whether and to what extent random RNAs adopt a stable and well-defined fold. In our previous studies we demonstrate that RNA molecules have an intrinsic tendency to fold into secondary structure even in absence of any evolutive pressure; that is, when randomly synthesized. In addition, secondary domains are surprisingly stable, and thermostable RNA could be found even screening a small number of RNA molecules. This finding suggests that RNA sequence space is a rich source of molecules that readily adopt a stable and well-defined 3D structure that can be exploited effectively for synthetic biology. Another feature of RNAs is that they can be easily engineered to display a specific function. The most notable feature is that RNA molecules embody both the genotype and the phenotype, which allow the seamless exploitation of directed evolution. The possibility of accessing huge diverse libraries of random RNAs coupled with the power of selecting even rare sequences with the function of interest makes RNA the perfect substrate to develop custom synthetic parts for tailored applications. The examples reviewed in Section 2.3 showed that both binding and catalysis can be easily obtained by selecting novel functional RNAs from a large library of random sequences (up to 10^{15}); those functions can be further combined to control and program cellular function in natural systems [63]. Toward this direction, Collins and co-workers [64] designed ligand-responsive ribozymes that cleave specific mRNA in response to changing concentration of certain intracellular metabolites to fine-tune gene expression in biological systems. Although in this work researchers employed natural ribozymes, we anticipate that *de novo* design of ligand-responsive ribozymes may reduce part-chassis cross-talking and interference, leading to more robust synthetic control circuits.

The third characteristic that makes RNA an attractive scaffold for chemical synthetic biology relies on the inherent modularity of RNA fold that allows the combinatorial combination of a self-contained module to achieve higher order structure using a forward-engineering approach.

These features make never-born RNAs a rich source of molecular diversity and a malleable scaffold for chemical synthetic biology.

ACKNOWLEDGMENTS

We wish to acknowledge the fruitful discussions with Pasquale Stano, Giovanni Minervini, Alessandro Filisetti and Alex Graudenzi. We thank Elena and Matthew Lynch for manuscript revision.

REFERENCES

1. Hawking, S.W. (2001) *The Universe in a Nutshell*, Bantam Dell.
2. Wang, L., Brock, A., Herberich, B., and Schultz, P.G. (2001) Expanding the genetic code of *Escherichia coli*. *Science*, **292** (5516), 498–500.
3. Dueber, J.E., Yeh, B.J., Chak, K., and Lim, W.A. (2003) Reprogramming control of an allosteric signaling switch through modular recombination. *Science*, **301** (5641), 1904–1908.
4. Rackham, O. and Chin, J.W. (2005) A network of orthogonal ribosome center dot mRNA pairs. *Nat. Chem. Biol.*, **1** (3), 159–166.
5. Park, S.H., Zarrinpar, A., and Lim, W.A. (2003) Rewiring MAP kinase pathways using alternative scaffold assembly mechanisms. *Science*, **299** (5609), 1061–1064.
6. Gardner, T.S., Cantor, C.R., and Collins, J.J. (2000) Construction of a genetic toggle switch in *Escherichia coli*. *Nature*, **403** (6767), 339–342.
7. Becskei, A., Seraphin, B., and Serrano, L. (2001) Positive feedback in eukaryotic gene networks: cell differentiation by graded to binary response conversion. *EMBO Journal*, **20** (10), 2528–2535.
8. Kramer, B.P. and Fussenegger, M. (2005) Hysteresis in a synthetic mammalian gene network. *Proceedings of the National Academy of Sciences of the United States of America*, **102** (27), 9517–9522.
9. Levskaya, A., Weiner, O.D., Lim, W.A., and Voigt, C.A. (2009) Spatiotemporal control of cell signalling using a light-switchable protein interaction. *Nature*, **461** (7266), 997–1001.
10. Elowitz, M.B. and Leibler, S. (2000) A synthetic oscillatory network of transcriptional regulators. *Nature*, **403** (6767), 335–338.
11. Atkinson, M.R., Savageau, M.A., Myers, J.T., and Ninfa, A.J. (2003) Development of genetic circuitry exhibiting toggle switch or oscillatory behavior in *Escherichia coli*. *Cell*, **113** (5), 597–607.
12. Fung, E., Wong, W.W., Suen, J.K. *et al.* (2005) A synthetic gene-metabolic oscillator. *Nature*, **435** (7038), 118–122.
13. Bulter, T., Lee, S.G., Woirl, W.W.C. *et al.* (2004) Design of artificial cell–cell communication using gene and metabolic networks. *Proceedings of the National Academy of Sciences of the United States of America*, **101** (8), 2299–2304.

14. Chen, M.T. and Weiss, R. (2005) Artificial cell–cell communication in yeast *Saccharomyces cerevisiae* using signaling elements from *Arabidopsis thaliana*. *Nature Biotechnology*, **23** (12), 1551–1555.
15. Marguet, P., Balagadde, F., Tan, C.M., and You, L.C. (2007) Biology by design: reduction and synthesis of cellular components and behaviour. *Journal of the Royal Society Interface*, **4** (15), 607–623.
16. Heinemann, M. and Panke, S. (2006) Synthetic biology – putting engineering into biology. *Bioinformatics*, **22** (22), 2790–2799.
17. Endy, D. (2005) Foundations for engineering biology. *Nature*, **438** (7067), 449–453.
18. Luisi, P.L. (2007) Chemical aspects of synthetic biology. *Chemistry and Biodiversity*, **4** (4), 603–621.
19. Luisi, P.L., Stano, P., De Lucrezia, D. *et al.* (2009) Chemical synthetic biology. *Origins of Life and Evolution of Biospheres.*, **39** (3-4), 209–209.
20. Schwartz, A.W. (1995) The RNA world and its origins. *Planetary and Space Science*, **43** (1-2), 161–165.
21. Luisi, P.L. (2003) Contingency and determinism. *Philosophical Transactions of the Royal Society A: Mathematical, Physical and Engineering Sciences*, **361** (1807), 1141–1147.
22. De Lucrezia, D., Franchi, M., Chiarabelli, C. *et al.* (2006) Investigation of de novo totally random biosequences, Part III: RNA Foster: a novel assay to investigate RNA folding structural properties. *Chemistry and Biodiversity*, **3** (8), 860–868.
23. De Lucrezia, D., Franchi, M., Chiarabelli, C. *et al.* (2006) Investigation of de novo totally random biosequences, Part IV: folding properties of *de novo*, totally random RNAs. *Chemistry and Biodiversity*, **3** (8), 869–877.
24. Moulton, V., Gardner, P.P., Pointon, R.F. *et al.* (2000) RNA folding argues against a hot-start origin of life. *Journal of Molecular Evolution*, **51** (4), 416–421.
25. Miller, S.L. and Lazcano, A. (1995) The origin of life – did it occur at high temperatures? *Journal of Molecular Evolution*, **41** (6), 689–692.
26. Olsen, M.J., Stephens, D., Griffiths, D. *et al.* (2000) Function-based isolation of novel enzymes from a large library. *Nature Biotechnology*, **18** (10), 1071–1074.
27. Olsen, M., Iverson, B., and Georgiou, G. (2000) High-throughput screening of enzyme libraries. *Current Opinion in Biotechnology*, **11** (4), 331–337.
28. Bosley, A.D. and Ostermeier, M. (2005) Mathematical expressions useful in the construction, description and evaluation of protein libraries. *Biomolecular Engineering*, **22** (1-3), 57–61.
29. Curtis, E.A. and Bartel, D.P. (2005) New catalytic structures from an existing ribozyme. *Nature Structural & Molecular Biology*, **12** (11), 994–1000.

30. Johns, G.C. and Joyce, G.F. (2005) The promise and peril of continuous in vitro evolution. *Journal of Molecular Evolution*, **61** (2), 253–263.
31. Joyce, G.F. (1994) In vitro evolution of nucleic acids. *Current Opinion in Structural Biology*, **4** (3), 331–336.
32. Guerrier-Takada, C., Gardiner, K., Marsh, T. et al. (1983) The RNA moiety of ribonuclease P is the catalytic subunit of the enzyme. *Cell*, **35**, 13.
33. Kruger, K., Grabowski, P., Zaug, A. et al. (1982) Self-splicing RNA: autoexcision and autocyclization of the ribosomal RNA intervening sequence of *Tetrahymena*. *Cell*, **31** (1), 147–157.
34. Tuerk, C. and Gold, L. (1990) Systematic evolution of ligands by exponential enrichment: RNA ligands to bacteriophage T4 DNA polymerase. *Science*, **249**, 505–510.
35. Ellington, A.D. and Szostak, J.W. (1990) *In vitro* selection of RNA molecules that bind specific ligands. *Nature*, **346** (6287), 818–822.
36. Bartel, D.P. and Szostak, J.W. (1993) Isolation of new ribozymes from a large pool of random sequences. *Science*, **261** (5127), 1411–1418.
37. Shi, H., Hoffman, B.E., and Lis, J.T. (1999) RNA aptamers as effective protein antagonists in a multicellular organism. *Proceedings of the National Academy of Sciences of the United States of America*, **96** (18), 10033–10038.
38. Meyers, L.A., Lee, J.F., Cowperthwaite, M., and Ellington, A.D. (2004) The robustness of naturally and artificially selected nucleic acid secondary structures. *Journal of Molecular Evolution*, **58** (6), 681–691.
39. White, R.R., Shan, S., Rusconi, C.P. et al. (2003) Inhibition of rat corneal angiogenesis by a nuclease-resistant RNA aptamer specific for angiopoietin-2. *Proceedings of the National Academy of Sciences of the United States of America*, **100** (9), 5028–5033.
40. Bell, S.D., Denu, J.M., Dixon, J.E., and Ellington, A.D. (1998) RNA molecules that bind to and inhibit the active site of a tyrosine phosphatase. *Journal of Biological Chemistry*, **273** (23), 14309–14314.
41. Ulrich, H., Ippolito, J.E., Pagan, O.R. et al. (1998) *In vitro* selection of RNA molecules that displace cocaine from the membrane-bound nicotinic acetylcholine receptor. *Proceedings of the National Academy of Sciences of the United States of America*, **95** (24), 14051–14056.
42. Vo, N.V., Oh, J.W., and Lai, M.M.C. (2003) Identification of RNA ligands that bind hepatitis C virus polymerase selectively and inhibit its RNA synthesis from the natural viral RNA templates. *Virology*, **307** (2), 301–316.
43. Brockstedt, U., Uzarowska, A., Montpetit, A. et al. (2004) In vitro evolution of RNA aptamers recognizing carcinogenic aromatic amines. *Biochemical and Biophysical Research Communications*, **313** (4), 1004–1008.
44. Xu, W. and Ellington, A.D. (1996) Anti-peptide aptamers recognize amino acid sequence and bind a protein epitope. *Proceedings of the National*

Academy of Sciences of the United States of America, **93** (15), 7475-7480.
45. Chen, C.H.B., Chernis, G.A., Hoang, V.Q., and Landgraf, R. (2003) Inhibition of heregulin signaling by an aptamer that preferentially binds to the oligomeric form of human epidermal growth factor receptor-3. *Proceedings of the National Academy of Sciences of the United States of America*, **100** (16), 9226-9231.
46. Rangarajan, S., Raj, M.L.S., Hernandez, J.M. *et al.* (2004) RNase P as a tool for disruption of gene expression in maize cells. *Biochemical Journal*, **380**, 611-616.
47. Malerczyk, C., Schulte, A.M., Czubayko, F. *et al.* (2005) Ribozyme targeting of the growth factor pleiotrophin in established tumors: a gene therapy approach. *Gene Therapy*, **12** (4), 339-346.
48. Benitez-Hess, M.L. and Alvarez-Salas, L.M. (2006) Utilization of ribozymes as antiviral agents. *Letters in Drug Design & Discovery*, **3** (6), 390-404.
49. Tsukiji, S., Pattnaik, S.B., and Suga, H. (2003) An alcohol dehydrogenase ribozyme. *Nature Structural Biology*, **10** (9), 713-717.
50. Keefe, A.D. and Szostak, J.W. (2001) Functional proteins from a random-sequence library. *Nature*, **410** (6829), 715-718.
51. Westhof, E., Masquida, B., and Jaeger, L. (1996) RNA tectonics: towards RNA design. *Folding and Design*, **1** (4), R78-R88.
52. Doudna, J.A. and Cech, T.R. (1995) Self-assembly of a group-I intron active-site from its component tertiary structural domains. *RNA*, **1** (1), 36-45.
53. Vanderhorst, G., Christian, A., and Inoue, T. (1991) Reconstitution of a group-I intron self-splicing reaction with an activator RNA. *Proceedings of the National Academy of Sciences of the United States of America*, **88** (1), 184-188.
54. Jaeger, L. and Leontis, N.B. (2000) Tecto-RNA: one-dimensional self-assembly through tertiary interactions. *Angewandte Chemie, International Edition in English*, **39** (14), 2521.
55. Chworos, A., Severcan, I., Koyfman, A.Y. *et al.* (2004) Building programmable jigsaw puzzles with RNA. *Science*, **306** (5704), 2068-2072.
56. Jaeger, L., Westhof, E., and Leontis, N.B. (2001) TectoRNA: modular assembly units for the construction of RNA nano-objects. *Nucleic Acids Research*, **29** (2), 455-463.
57. Ikawa, Y., Tsuda, K., Matsumura, S., and Inoue, T. (2004) De novo synthesis and development of an RNA enzyme. *Proceedings of the National Academy of Sciences of the United States of America*, **101** (38), 13750-13755.
58. Ikawa, Y., Fukada, K., Watanabe, S. *et al.* (2002) Design, construction, and analysis of a novel class of self-folding RNA. *Structure*, **10** (4), 527-534.

59. Fernandez, A. and Cendra, H. (1996) *In vitro* RNA folding: the principle of sequential minimization of entropy loss at work. *Biophysical Chemistry*, **58** (3), 335–339.
60. Tinoco, I. and Bustamante, C. (1999) How RNA folds. *Journal of Molecular Biology*, **293** (2), 271–281.
61. Anshelevich, V., Vologodskii, A., Lukashin, A., and Frank-Kamenetskii, M. (1984) Slow relaxational processes in the melting of linear biopolymers: a theory and its application to nucleic acids. *Biopolymers*, **23**, 39–58.
62. Das, R., Kwok, L.W., Millett, I.S. *et al.* (2003) The fastest global events in RNA folding: electrostatic relaxation and tertiary collapse of the *Tetrahymena* ribozyme. *Journal of Molecular Biology*, **332** (2), 311–319.
63. Isaacs, F.J., Dwyer, D.J., and Collins, J.J. (2006) RNA synthetic biology. *Nature Biotechnology*, **24** (5), 545–554.
64. Isaacs, F.J., Dwyer, D.J., Ding, C.M. *et al.* (2004) Engineered riboregulators enable post-transcriptional control of gene expression. *Nature Biotechnology*, **22** (7), 841–847.

3

Synthetic Biology, Tinkering Biology, and Artificial Biology: A Perspective from Chemistry

Steven A. Benner, Fei Chen, and Zunyi Yang

Foundation for Applied Molecular Evolution and The Westheimer Institute for Science and Technology, PO Box 13174, Gainesville FL 32604, USA

3.1 Introduction 70
   3.1.1 Multiple Meanings for "Synthetic Biology" 70
   3.1.2 Challenges in Contemporary Synthetic Biology 71
   3.1.3 This Is Not the First Time That Biology Has Been Declared to Be Engineerable 73
   3.1.4 What Do Opposite Meanings of "Synthetic Biology" Have in Common? 74
   3.1.5 The Value of Failure, and the Analysis of Failure 75
3.2 Attempting to Synthesize An Artificial Genetic System 77
   3.2.1 Synthesizing Artificial Genetic Systems 79
   3.2.2 Failure Changing the Sugar 79
   3.2.3 Failure Changing the Phosphates 83
3.3 Building Genetics from the Atom up 88
   3.3.1 Could Base Pairing behind Darwinian Evolution Be So Simple? 89
   3.3.2 Synthetic Genetic Systems to Make Synthetic Protein Systems 91
   3.3.3 Synthetic Genetics Supports Human Health Care 91

Chemical Synthetic Biology, First Edition. Edited by Pier Luigi Luisi and Cristiano Chiarabelli.
© 2011 John Wiley & Sons, Ltd. Published 2011 by John Wiley & Sons, Ltd.

3.3.4 The Next Challenge. Can Artificial Synthetic Genetic Systems Support Darwinian Evolution? 95
3.3.5 Is This Synthetic Life? 97
3.4 Does Synthetic Biology Carry Hazards? 99
3.5 Conclusions 102
Acknowledgments 103
Statement of Conflicts of Interest 103
References 104

Chemical theory cannot yet support an engineering vision that allows DNA sequences, proteins and other biomolecules to be used as interchangeable parts in artificial constructs without "tinkering", a process that alters the structure of molecules without much predictability. Indeed, attempting to do biomolecular design as a part of a program in synthetic biology pursues challenges at the limits of existing theory. These force scientists across uncharted terrain where they must address unscripted problems. In areas where chemical theory is inadequate, the attempt fails, and fails in a way that cannot be ignored. Thus, synthesis drives discovery and paradigm change in ways that analysis cannot. Further, if the failures in a synthetic biological exercised are analyzed, new theories can emerge. Here, we illustrate this by synthesizing artificial genetic systems capable of Darwinian evolution, an ability theorized to be universal to life.

3.1 INTRODUCTION

3.1.1 Multiple Meanings for "Synthetic Biology"

Many languages have "contranyms," words and phrases that have two nearly opposite meanings. If you run "fast," you are moving at great speed; if you hold "fast," you are not moving at all. A "citation" from Harvard *University* is good, but a "citation" from the Harvard University *police* is bad.

"Synthetic biology" is a contranym. In one version popular today in many engineering communities, synthetic biology seeks to use *natural* parts of biological systems (such as DNA fragments and protein "biobricks") to create assemblies that do things that are *not* done by natural biology (such as the manufacture of a specialty chemical or digital computation). Here, engineers hope that the performance of molecular parts drawn from living systems can be standardized, "specs" can be written, and parts can then be reassembled in architectures with predictable outcomes, just as engineers can mix and match standardized transistors to give integrated circuits with predictable performance. For this, the whole must be the sum of its parts.

Among chemists, "synthetic biology" means the opposite. The "synthetic biology" of chemists seeks to use *unnatural* molecular parts to do things that *are* done by natural biology. Chemists hold that if they can reproduce biological behavior *without* making an exact molecular replica of a natural living system, then they have demonstrated their understanding of the intimate connection between molecular structure and biological behavior. If taken to its limit, this synthesis would provide a chemical understanding of life. Although central to chemistry, this research paradigm was perhaps best expressed by a physicist, Richard Feynman, in the phrase: "What I cannot create, I do not understand." (quoted in Hawking [1]).

Waclaw Szybalski was thinking of a still different meaning when he coined the term "synthetic biology" in 1974 [2]. Szybalski noted that recombinant DNA technology would soon allow the construction of new cells with rearranged genetic material. He realized that this deliberate synthesis of new forms of life provided a way to test hypotheses about how the rearranged material contributed to the function of natural cells.

Szybalski also undoubtedly had the experience of chemistry in mind. In 1974, structure theory in chemistry was the most powerful theory in science. It became so largely because chemistry possessed technology that allowed chemists to synthesize new chemical matter to study. This supports powerful processes for testing hypotheses and models, power that Szybalski saw that biotechnology was offering to biology. Such a power was (and remains) unavailable to (for example) astronomy, planetary science, and social science.

In 1974, "synthetic organic chemistry" had already become a tool to explore biology. For example, in the previous decade, "biomimetic chemists" had created small designed molecules that reproduced the elementary behaviors of biomolecules, such as their ability to bind to ligands or to catalyze reactions. Jean-Marie Lehn in Strasbourg shared a Nobel Prize for his work developing molecules able to do exactly this. One of his signature structures from the 1960s is shown in Figure 3.1.

3.1.2 Challenges in Contemporary Synthetic Biology

Today, the chemist's vision for synthetic biology goes far beyond Lehn's vision. Chemists hope that molecular design supported by structure

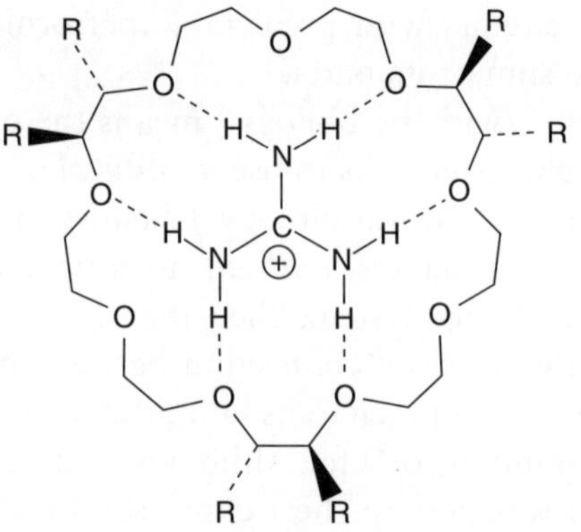

Figure 3.1 Synthesis in biology was first used to help understand the connection between biomolecular structure and behavior by making unnatural molecules that bind to small molecules. This synthetic receptor was created by Jean-Marie Lehn and his colleagues to mimic the ability of natural receptors to bind to small cationic ligands, and was cited in his Nobel Prize lecture

theory will yield unnatural molecular species able to mimic not just binding and catalysis of specific biomolecules, but also the highest forms of biological behavior, including macroscopic self-assembly, replication, adaptation, and evolution. Any theory that enables such design will have demonstrated an ability to account for these features of "life," especially if chemists can make a totally synthetic version of life without exactly reproducing the chemistry of a natural terran organism.

With the term "synthetic biology" now in jeopardy by "trademark creep," it might be appropriate to coin a new term to describe this process. "Artificial biology" has been proposed, but computer engineers have already used this term to mean something different.

Given these nearly opposite uses of the same term, spectators are naturally puzzled. "What's the fuss?" some have asked. "Isn't synthetic biology just more 'Flavr Savr'® tomatoes?"

The question is raised in analogous form by molecular biologists who see in synthetic biology "contests," which attract student participation worldwide, nothing more (and nothing less) than the very same cloning that has been done since the 1970s.

Nor do molecular biologists attempting to understand life entirely understand the hullabaloo over the (difficult to repeat) use of DNA hybridization and ligation to compute a solution to the "traveling salesman problem" [3]. There is no obvious reason to do digital computation with DNA. After all, the rate at which DNA molecules hybridize in solution is limited by the bimolecular rate constant for molecular diffusion, about $10^8\,\text{M}^{-1}\,\text{s}^{-1}$. In layman's language, this means that the half-life with which a DNA molecule finds its complement cannot be faster than ~0.01 s when the complement has micromolar concentration. In practice, the rate is much slower. Compare this with the limit on the rate at which semiconductors compute: the speed of light in the conducting material. At $3 \times 10^{10}\,\text{cm}\,\text{s}^{-1}$, communication across a meter of space is ~0.000 000 003 s. Even with the possibility of improved parallelism with DNA computation, there is no contest.

Francis Collins, the director of the National Institutes of Health in the USA, captured a similar sentiment. Collins is reported to have mused about the "new" field of synthetic biology as applied to virus synthesis: "This was completely a no-brainer. I think a lot of people thought, 'Well, what's the big deal? Why is that so exciting?'" (quoted in Regis [4]).

3.1.3 This Is Not the First Time That Biology Has Been Declared to Be Engineerable

Salesmanship accompanying some discussions today of "synthetic biology" has also engendered a certain level of cynicism [5]. Those whose professional lives started before the age of the internet remember more than one time where biology, it was claimed, had at last entered the realm of engineering. It was not so then and, in the broadest vision put forward by the engineering community, it is not so now.

For example, a quarter century ago, *Science* published an article entitled "Protein engineering" by Kevin Ulmer, Director of Exploratory Research at GeneX, a biotechnology company [6]. The 1983 paper is both in substance and form like the breathless reporting in today's popular science magazines covering efforts to make and use biobricks.

In 1983, Ulmer said that the goal of this new engineering biology was to "control in a predictable fashion" the properties of proteins to be building blocks in industrial processes. This new era of engineering would set aside "random mutagenesis techniques" in favor of a "direct approach to protein modification." Ulmer referred to protein domains encoded by exons and the use of repressors with altered enzymes to assemble new regulatory pathways.

Ulmer's 1983 vision failed. GeneX is no longer in business. Twenty-five years later, we are still struggling to engineer the behavior of individual proteins. Chemical theory has improved, but not enough to make proteins engineerable.

The 1980s' engineering vision failed for reasons discussed in a 1987 review by Jeremy Knowles, then the Dean of Harvard College. Knowles published another paper in *Science*, this one entitled "Tinkering with enzymes" [7]. Knowles was a chemist and, therefore, understood that "scale" matters. Molecules, 1 to 0.1 nm in size, behave differently from transistors, even small transistors having sizes between 1 and 0.1 μm. This creates difficulties in transferring microengineering concepts to molecular (nano) engineering. The same is true in the next jump downwards in scale; molecules at 1 to 0.1 nm scale behave differently from quantum species operating on the pico- or femto-meter distance scales.

Knowles' "Tinkering" comments are apt even today. Referencing Ulmer's paper, Knowles dryly wondered whether the engineering vision was not, perhaps, a bit "starry-eyed." He acknowledged that "gee-whiz" experiments that put things together to "see-what-happens" could aid in understanding.

But Knowles made the point that is still true: *Nothing of value comes unless the tinkering is followed by studies of what happened*. Especially if the synthetic effort fails. Absent that, modern synthetic biology, at the molecular, DNA, protein, or cell level, will be "tinkering" without value of any consequence.

Analysis of failure is generally less enthusing (and more laborious) than the initial design, as any observer of current synthetic biology "contests" can see. The analysis of failure requires discipline, a discipline that is difficult to teach. Thus, it helps to remember the dictum: *It is just as hard to solve an unimportant problem as an important one*. One is more likely to analyze a failure to the depth needed to learn from that failure if the goal is felt to be very important.

We understand much more now about the behavior of molecules, biological and otherwise, than we understood a quarter century ago *because* gene and protein tinkers studied their failures. Accordingly, the ball has been moved, from Lehn-like small molecules to proteins to synthetic genes, protein assemblies, cells, and assemblies of cells. We are still doing what might be called "tinkering biology," but we are doing it farther down the field.

3.1.4 What Do Opposite Meanings of "Synthetic Biology" Have in Common?

One lesson in particular might be learned from these "pre-internet" failures in synthetic biology. *Synthesis is a research strategy, not a field* [8, 9]. Synthesis sets forth a grand challenge: *Create an artificial chemical system capable of Darwinian evolution.* Or: *Create a set of DNA bricks that can be assembled to form an adding machine.* Or: *Rearrange a set of regulatory elements to make a cell that detects nerve gas.* Or: *Assemble enzyme catalysts extracted from a different organisms to assemble a pathway that makes an unnatural chemical that is part of an anti-malarial drug.* Attempting to meet these challenges, scientists and engineers must cross uncharted territory where they must encounter and solve unscripted problems guided by theory. If their guiding theory is adequate, the synthesis works. If it is not, the synthesis fails.

This exercise is different from observation, analysis, and probing, which are other strategies used in science. Here, as often as not, observations are often either discarded or rationalized away when they contradict a (treasured) theory. We see this also in computational modeling using numerical simulations. "Modeling," it is said "is doomed to succeed." If a model does not give a desired answer, it is tweaked until it does.

Selection of data to get the "right" answer has a long tradition in biology. A well-known example is Gregor Mendel, who evidently stopped counting round and wrinkled peas when the "correct" ratio (which is 3:1) was reached. Objective observations have an uncanny ability to confirm a desired theory, even if the theory is wrong.

Self-deception is far more difficult when doing synthesis. If, as happened with the Mars Climate Orbiter, the guidance software is metric and the guidance hardware is Imperial, one can ignore the incongruent observations arising from a false theory (as was done) as the craft was in transit to Mars. But when the spacecraft gets to Mars, if the theory is wrong, then the spacecraft crashes (and it did).

For this reason, synthesis as a research strategy can drive discovery and paradigm shift in ways that observation and analysis cannot. Indeed, this is one way to distinguish science from nonscience: Science is a human intellectual activity that incorporates a mechanism to avoid self-deception [9]. Synthesis provides such a mechanism.

3.1.5 The Value of Failure, and the Analysis of Failure

The failures encountered as modern synthetic biologists attempt to rearrange atoms to create artificial genetic species [10], regulatory elements to give synthetic circuits [11], or enzymes to give synthetic pathways [12] carry a clear message: we need to learn more about the behavior of physical matter on the "one to one-tenth" nanometer scale. This is not a declaration of defeat. Rather, it is a challenge, one that begins by recognizing that our guiding theory is still inadequate to hand biology over to engineers.

Nowhere is the value of failure set within a synthetic challenge more evident than within recent efforts of Craig Venter, Hamilton Smith, and others to construct a cell where *all* of its constituent genes come from elsewhere [13, 14]. In a real sense, this grand challenge is the apotheosis of the 1970s' version of bioengineering. More Flavr Savr® tomatoes, but now thousands of times repeated.

At first, the challenge appeared to be simple enough. Since the time of Szybalski, scientists had been able to move a single natural gene from one organism to another. Scientists had long been able to move two natural genes from two other organisms into a different organism. By a kind of argument that corrupts the idea behind mathematical induction (if one can do *n* genes, and if one can do $n + 1$ genes, one should be able to do any number of genes), it seemed that simple iteration would allow scientists to get *all* genes in an organism from somewhere else.

When these scientists set out to do this, they had high hopes. I had dinner with Hamilton Smith, in October 2006. They were just "6 months away" from getting a synthetic cell constructed in this way. We met again 18 months later at Janelia Farms; they were still 6 months away. Three years later, the final announcement of success [14] was made just as this article was going to press. The estimated price tag of $40 million (US) shows just how great this challenge was.

The difficulties arose because the operating theory used to guide this synthesis was missing something. In the case of the synthetic cell, Venter noted that a single nucleotide missing in a critical gene prevented success for many months. Again, this is not a reason for despair. The *purpose* of synthesis is to bring those missing "somethings" to light. As they pursued their grand challenge, Smith, Venter, and their colleagues had to solve unscripted problems, driving knowledge in ways impossible through analysis and observation alone.

Here, as always, problem selection is important. Selecting problems at the limits of what is possible is a poorly understood art. Further, those selections change over time. For example, in the 1960s it was sufficient as a challenge to try to develop organic molecules that would bind to small molecules, such as the synthetic receptors that earned Jean-Marie Lehn his Nobel Prize (Figure 3.1).

The most useful challenges in synthetic biology are those that are most likely to generate the most consequential pursuits. These challenges are just at the limits of the do-able, and perhaps just a bit farther. As Medewar said, science is the "art of the soluble" [15]. However, the selection of a synthetic challenge also reflects choices personal to scientists in a laboratory. This review describes choices made in my laboratory by my coworkers and me in a type of synthetic biology that was chemical in its viewpoint.

3.2 ATTEMPTING TO SYNTHESIZE AN ARTIFICIAL GENETIC SYSTEM

Our efforts in synthetic biology began in the mid 1980s, immediately after my group synthesized the first gene encoding an enzyme [16]. This synthesis followed the synthesis of a gene encoding interferon [17]. Today, total gene synthesis is routine, being done by commercial supply houses for just pennies a nucleotide.

These genes were (and are) little different from what Nature delivered. Accordingly, our next "grand challenge" was to synthesize an *unnatural* genetic system, something different from what billions of years of biological evolution has delivered to us. Only by doing this could we demonstrate that structure theory understood the connection between molecular structure and genetics, including reproduction, adaptation, and evolution. The first book in synthetic biology, originally entitled *Redesigning Life*, appeared in 1987 [18].

It took 20 years to create an artificial chemical system capable of Darwinian evolution, two decades that record more failures than successes. Again, no problem. This is what synthesis is for. Let me highlight some of our work to illustrate what synthetic efforts directed towards this challenge taught us through failure. More details are available in a new book *Life, the Universe, and the Scientific Method* (www.ffame.org) [19].

Our selection of this particular challenge focused on a broad question: How might we use synthesis to develop a better understanding of the concept of "life"? This question lies, of course, at the abstract core of biology.

But it also has practical implications. For example, NASA and ESA are sending missions to Mars, Titan, and elsewhere, looking for life in environments that are more or less like Earth. On Mars, the environment is more like Earth, as liquid water most likely lies beneath the Martian surface.

On Titan, however, the environment is less like Earth. The most abundant matter on Titan's surface that might serve as a biosolvent is liquid methane at 94 K (−179 °C); the liquid water beneath the surface may exist as water–ammonia eutectics. Nevertheless, organic species are abundant on Titan. Thus, as many have argued, if life is an intrinsic property of organic species in complex mixtures [20], then Titan should hold life.

Accordingly, my laboratory set up a set of programs in the 1980s to pursue four approaches towards understanding the concept of "life." These are illustrated in Figure 3.2.

Any definition of "life" must be embedded within a "theory of life" [9, 21]. One such theory is captured by a "NASA definition" of life as "a self-sustaining chemical system capable of Darwinian evolution" [22]. This definition–theory captures the opinion of its creators about what is possible within biomolecular reality. It excludes, for example, nonchemical and Lamarckian systems from our concept of "life" [23]. Should we encounter them (and many science fiction stories describe them in various forms), we would be forced to concede that our definition–theory of life is wrong.

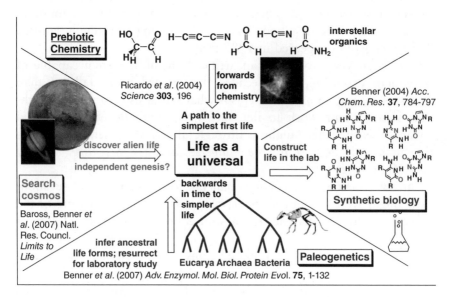

Figure 3.2 Four approaches to understand life as a universal

The NASA definition–theory of life offers a clear direction for exploratory synthetic biology. If life were so simple, then a target would be an artificial chemical system capable of Darwinian evolution. If the NASA definition–theory of life is on point, this artificial system should be able to recreate all of the properties that we value in life.

Like the Venter–Smith team two decades later, we also had high hopes when we set out. After all, the existing theory at that time, constructed at the molecular level, seemed to associate Darwinian evolution with some quite simple molecular structures, even simpler than those required to assemble an artificial cell from natural genes taken from elsewhere. Indeed, the first-generation theory of nucleic acids, adumbrated by Watson and Crick a half century ago, is so simple that it is taught in middle school.

3.2.1 Synthesizing Artificial Genetic Systems

Consider, for example, the double helix structure of DNA, the molecule at the center of natural Darwinian evolution, modeled by James Watson and Francis Crick in their epic 1953 paper. In the double helix, two DNA strands are aligned in an antiparallel fashion. The strands are held together by nucleobase pairing that follows simple rules: A pairs with T and G pairs with C. Behind the double helical structure lie two simple rules for molecular complementarity, based in molecules described at atomic resolution. The first rule, size complementarity, pairs large purines with small pyrimidines. The second, hydrogen bonding complementarity, pairs hydrogen bond donors from one nucleobase with hydrogen bond acceptors from the other (Figure 3.3).

In the first-generation model for the double helix, the nucleobase pairs were central. In contrast, the backbone "bricks," made of alternating sugar and negatively charged phosphate groups, were viewed as being largely incidental to the molecular recognition event at the center of natural genetics and Darwinian evolution.

3.2.2 Failure Changing the Sugar

If this simple "first-generation" model for the double helix were correct and complete, then we should be able to synthesize a different molecular system with different sugars and/or phosphates (but the same nucleobases) to get an unnatural synthetic system that could mimic the

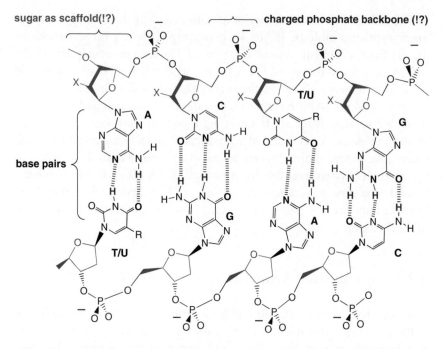

Figure 3.3 The rules governing the molecular recognition and self-assembly of DNA duplexes are so simple that many are tempted to believe that molecular recognition in chemistry is, in general, similarly simple. Hence, various individuals seek "codes" for protein folding or drug binding

molecular recognition displayed by natural DNA and RNA. We might even be able to get this artificial genetic system to have children and, possibly, evolve.

In our efforts, much failure ensued. For example, we decided to replace the ribose sugars by flexible glycerol units to give a "flexible" kind of synthetic DNA (Figure 3.4). This followed a suggestion of Joyce et al. [24], who had noted some of the difficulties in identifying processes that, on Earth before biology, might have generated ribose and 2′-deoxyribose, the "R" and the "D" in RNA and DNA respectively. Glycerol, from a prebiotic perspective, was certainly more accessible to a pre-life Earth; it is a major component of organic material delivered to Earth today by meteorites [25].

Unfortunately, the system failed to deliver rule-based recognition having any quality. As summarized in Table 3.1, synthetic molecules that replaced one ribose by a glycerol all bound less tightly to their

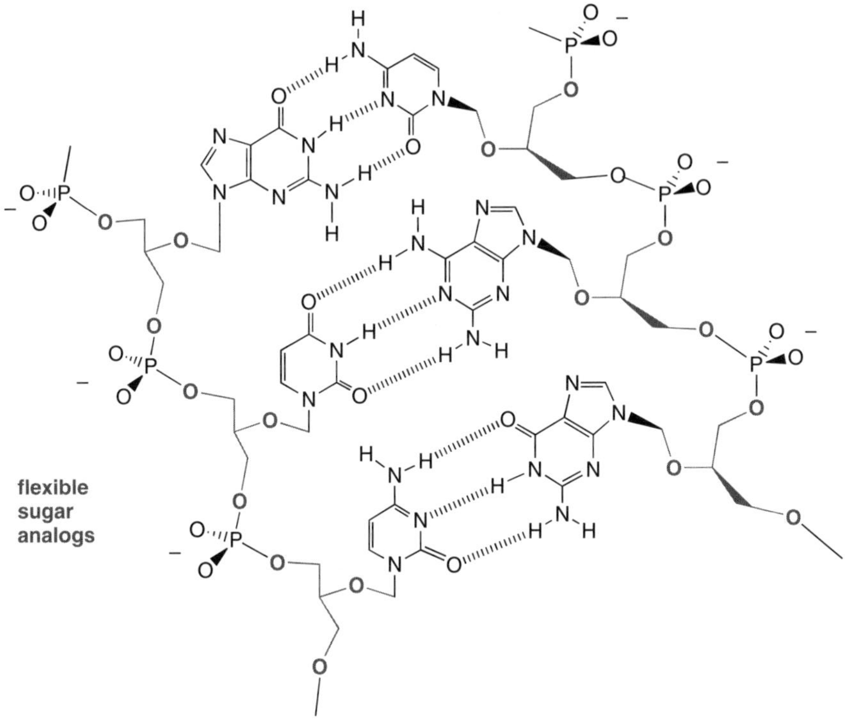

A PERSPECTIVE FROM CHEMISTRY

Figure 3.4 The failure of these flexible glycerol DNA molecules led us to re-evaluate our view of the role of sugars in double helix formation

Table 3.1 Melting temperatures for flexible glycerol synthetic DNA

CTTTTTTTG	40°	CAAATAAAG	37°
GAAAAAAAC		GTTTATTTC	
CTTTtTTTG	25°	CAAAtAAAG	25°
GAAAAAAAC		GTTTATTTC	
CTTtTtTTG	13°	CAAtAtAAG	12°
GAAAAAAAC		GTTATATTC	
CTTttTTTG	11°	CAAttAAAG	11°
GAAAAAAAC		GTTAATTTC	
CTTTTTTTG	21°	CTTTtTTTG	12°
GAAAGAAAC		GTTTGTTTC	
CTTTTTTTTTTG	55°	CttttttttttG	<0°
GAAAAAAAAAAAC		GAAAAAAAAAAAC	

complement. Putting in two flexible glycerols lowered the melting temperatures of the duplexes even more. Faced with this failure, we went further, synthesizing DNA analogs where all of the 2′-deoxyribose units were replaced by glycerol units. Making the molecule entirely from glycerols destroyed all of the molecular recognition needed for DNA-like genetics.

This failure taught us the inadequacy of our then-existing theory to account for genetics. Two hydrogen bonds joining the nucleobase pairs were simply not enough to hold together two strands where the ribose was replaced by synthetic glycerol. Conversely, these experiments showed that the sugars were not entirely incidental to the molecular recognition event.

When three hydrogen bonds held the nucleobase pairs together, things work better. Further, based on a new-found appreciation of the contribution of the sugar to the ability of DNA to support Darwinian evolution, synthesis went further. Smaller carbohydrates and carbohydrates whose conformation was locked were synthesized and found to work better than standard DNA, at least by some metrics [26, 27]. These themes were further developed by the synthesis of more backbone-modified DNA species by many luminaries in modern synthetic chemistry, including Albert Eschenmoser [28], Piet Herdewijn [29], and Christian Leumann [30].

Thus, a theory that taught that the backbone sugars of DNA were incidental to Darwinian evolution failed to support a synthetic endeavor. This failure advanced the theory. The synthesis of *unnatural* genetic systems taught us something about *natural* genetic systems. This drove the synthesis of more unnatural systems that replaced failure by success.

Without synthesis, this part of the first-generation theory for the double helix had remained largely unchallenged in the three decades since it was first adumbrated in 1953. It had appeared as "dogma" (Francis Crick's word) in textbooks and television series. Some of these said that "RNA is the same as DNA, except that each sugar has an additional –OH group" (a statement that, from any chemical perspective, is ignorant on its face). These facts all support the notion that, without the efforts of synthetic biology, this fascinating feature of the molecule behind Darwinian evolution would *never* have been recognized. Synthesis drove discovery and paradigm change in ways that analysis cannot. And this came about only because failure was analyzed and pursued.

3.2.3 Failure Changing the Phosphates

We also encountered failure when we attempted to replace the charged phosphates in the backbone of DNA by a linker that had about the same size as phosphate, but that lacked charges. The phosphate linkers were also viewed in the first-generation Watson–Crick theory as being largely incidental to the molecular recognition that is central to genetics and Darwinian evolution.

In fact, the repeating negative charges carried by the phosphate groups in the DNA backbone appeared to be downright undesirable. The repeating charges on the phosphate linkages prevented DNA from getting into cells. The charged phosphate linkers were sites of nuclease attack. The repulsion between two negatively charged backbones of two DNA strands seemed to weaken undesirably their association to form a double helix. DNA molecules without the negative charges in their backbone were expected to form *better* duplexes.

If, we thought, we could get rid of the charges without disrupting the rules for Watson–Crick pairing (A pairs with T, G pairs with C), then we might be able to create a new class of therapeutic molecules with an entirely new mechanism for biological activity. These were called "antisense drugs" [31]. The idea was simple. If we could synthesize an uncharged analog of DNA that could enter a cell by passive diffusion, it would survive degradation by nuclease attack. If the charges were indeed incidental to genetics, then this neutral synthetic DNA analog would still bind to complementary DNA molecules inside a cell following Watson–Crick rules. The antisense DNA analog would, therefore, target, with sequence specificity, only the unwanted DNA, perhaps from a virus or a mutated cancer gene. Antisense DNA might be a magic bullet for diseases associated with undesired DNA or RNA.

Following this theory, Zhen Huang, Christian Schneider, Clemens Richert, Andrew Roughton, and others in my group synthesized an uncharged unnatural DNA-like molecule that replaced the anionic phosphate diester linker in natural DNA and RNA with uncharged dimethylenesulfone linkers (Figure 3.5). This gave DNA and RNA analogs that have roughly the same geometry as the natural molecules [32, 33]. Indeed, Martin Egli solved a crystal structure of a short $G_{SO_2}C$ dinucleotide duplex. He found that the uncharged duplex was held together by G:C and C:G pairs in a mini-helix just like its RNA analog, whose crystal structure had been solved by Alex Rich two decades earlier [34].

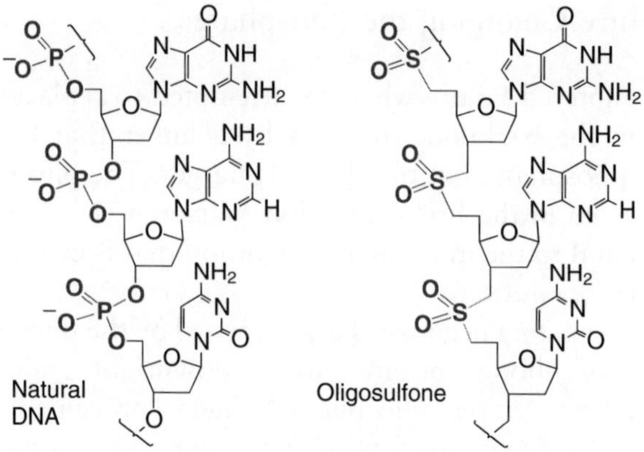

Figure 3.5 Replacing phosphates (–PO$_2$– units, each having a negative charge, left) in DNA by dimethylenesulfone linkers (the –SO$_2$– units, right, each lacking a negative charge) gave an uncharged analog of DNA. The uncharged analog of RNA was also synthesized

This appeared to validate the first-generation Watson–Crick theory for the double helix. It appeared that one *could* replace the charged phosphate linkers with uncharged linkers of approximately the same shape, and still form G:C and C:G pairs.

One theme already mentioned is that serious synthetic biologists do not neglect detailed analysis when a synthesis fails. The contranym theme is that one should extend the challenge when the synthesis appears to succeed. Success means that one has erred a bit on the safe side in selecting a challenge. To be consequential in driving discovery and paradigm change, if a theory seems to work, the challenge should be deepened until the theory fails.

Accordingly, we synthesized longer DNA and RNA analogs having more sulfone linkers. Instead of molecules with just one uncharged linker, we made molecules with two uncharged sulfone linkers to see how they worked. We then made molecules with three, five, and then seven uncharged sulfone linking bricks.

It was not long before the theory that we were using to guide the synthesis broke down. Longer oligosulfones folded on themselves [35]. Folding prevented them from pairing with *any* second strand, even one that was perfectly complementary in the Watson–Crick sense of the term. This failure led to a thought that should have been obvious, but was not in our culture (we, too, had been trained to view the DNA double helix as an unchallengeably elegant structure): pairing between two strands *requires* that neither strand fold on itself.

Another failure was then encountered. Different oligosulfones differing by only one nucleobase in their structure were found to display different levels of solubility, aggregation, folding, and chemical reactivity [36]. This prompted another thought that, in retrospect, should have been obvious. To support Darwinian evolution, a genetic molecule must have features that allow it to change its detailed structure, the details that encode genetic information. But the changes must be possible *without changing the overall properties of the system*. In particular, the changes in structure that correspond to changes in genetic information cannot change the rules by which the genetic molecules template the formation of their descendants. Changes do not do this in DNA and (in general) RNA. As we learned by synthesis, they do so in oligosulfones.

These results further drove the development of a second-generation model for the DNA double helix and the relation between its structure and Darwinian evolution. In this model, the phosphate linkers and the repeating backbone charge become quite important for four reasons.

First, and trivially, a polyanion is likely to be soluble in water. This was appreciated by Watson and Crick already in 1953, certainly more than Linus Pauling. Pauling had proposed an incorrect model for DNA where the phosphates did not point out into solvent, but rather (and paradoxically given their negative charges) interacted with each other [37]. When Watson and Crick first learned about the structure for DNA assemblies that Pauling was proposing, this feature immediately let them conclude that Pauling's model must be wrong.

Less trivially, the repeating charges in the backbone of natural polyanionic DNA repel each other. Within a strand, this repulsion helps keep DNA strands from folding on themselves. A polyanion is more likely to adopt an extended conformation suitable for templating than a neutral polymer, which is more likely to fold. As "not folding" is a property needed for a strand to bind to its complement, the repeating charges were proposed in the second-generation model to be important for the ability of DNA to support Darwinian evolution for this reason, as well as for solubility reasons.

The anion–anion repulsion between phosphates on two different strands is also important. When two strands approach each other, the repulsion forces interstrand interactions away from the backbone. This drives the contact between two strands to occur at the Watson–Crick

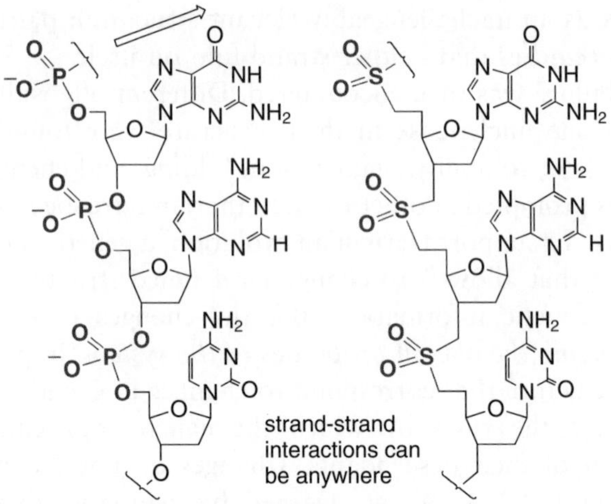

Figure 3.6 The repeating backbone anion drives the interaction between two strands as far from the backbone as possible. This guides strand–strand interactions, and forms the basis for Watson–Crick pairing rules

edge of the nucleobases (Figure 3.6). Without the polyanionic backbone, interstrand contacts can be anywhere [38]. Thus, the second-generation model views as naïve the assumption that this repulsion is bad. In fact, the repulsion moderates and controls the natural propensity of biomolecules to associate with other biomolecules, and directs in DNA that association to the part of the molecule where information is contained, the Watson–Crick edges of the nucleobases.

In the light cast by failure in our synthetic effort, the interstrand repulsion between two strands that both have repeating charges on their backbones is also seen to be important for pairing rules essential for Darwinian evolution. Without the repulsion from two backbones, both negatively charged, base pairing would not occur at the site where hydrogen bonding was needed. It would occur at other sites, including the Hoogsteen site, and not obey the simple rules required for genetics.

But the failure of the synthesis yielded a still more fundamental role for the repeating charge in a DNA molecule, one that suggested that repeating backbone charges were necessary for any biopolymer to support Darwinian evolution. Here, the argument is more subtle, and begins with the realization that replication alone is not sufficient for a genetic molecule to support Darwinian evolution. A Darwinian system must generate *inexact* replicates, descendants whose chemical structures are different from those of their parents. Further, these differences must then be replicable themselves. It does no good if the mutant has changed its biophysical properties so dramatically that the mutant genetic molecule precipitates, folds, or otherwise loses the ability to encode selectable information.

While self-replicating systems are well known in chemistry, those that generate inexact replicas with the inexactness itself being replicable are not [39]. As a rule, changing the structure of a molecule changes its physical behavior. Indeed, it is quite common in chemistry for *small* changes in molecular structure to lead to large changes in physical properties. This is certainly true in proteins, where a single amino acid replacement can cause the protein molecule to precipitate (the archetypal example of this is sickle cell hemoglobin). This means that inexact replicates need not retain the general physico-chemical properties of their ancestors; in particular, properties that are essential for replication.

This thought, again arising through the analysis of a failed synthesis, led us to realize that a repeating backbone charge might be universal for *all* genetic molecules that work in water, on Earth, Mars, and Titan, but also for extraterrestrial aliens whom we have not yet encountered. The polyanionic backbone dominates the physical properties of DNA. Replacing one nucleobase in the sequence of a DNA molecule by another, therefore, has only a second-order impact on the physical behavior of the molecule. This allows nucleobases to be replaced during Darwinian evolution without losing properties essential for replication.

This thought also puts in context the statement that DNA and RNA are "the same, except" for a replacement of an –H group by an –OH group on each of its biobricks. Such a change would have a major impact on the behavior of almost any other molecular system. It does not for DNA and RNA because their repeating backbone charges so dominate their overall behavior of these molecules that the changes expected through replacement of an –H by an –OH in each biobrick are swamped by the repeating charge. To this comment should be added the remark: "but only barely so." DNA and RNA still have many differences in their physical properties that can be attributed to the replacement of an –H by an –OH in each biobricks.

This is what engineering synthetic biologists are looking for. The repeating charge in the DNA backbone allows nucleotides to behave largely as *interchangeable parts*. It allows the whole to be the sum of its parts. It allows engineers, even those totally unfamiliar with structure theory, to design DNA molecules that pair with other DNA molecules according to simple rules. Because of this repeating backbone charge, and *only* because of this repeating backbone charge, is it possible to make "tiles" or biobricks from DNA, for example.

And *only* because of this repeating backbone charge can DNA and RNA support Darwinian evolution. The sequence ATCCGTTA behaves in most respects the same way as the sequence GCATGACA, even though these have very different molecular structures. This is because, in both cases, the molecules are polyanions. These differences hold the genetic information. Were it otherwise, we could not mutate ATCCGTTA to give GCATGACA, even if GCATGACA allowed us to survive and reproduce better.

For this reason, the second-generation model for DNA proposed that a repeating charge should be a universal structure feature of *any* genetic molecule that supports Darwinian evolution in water, regardless of where it is found on Earth [34]. Polycationic backbones are also predicted to be satisfactory under what has become called the "polyelectrolyte theory of the gene" [34]. Thus, if NASA missions *do* detect life in water on other planets, their genetics are likely to be based on polyanionic or polycationic backbones, even if their nucleobases and sugars differ from those found on Earth. This structural feature can be easily detected by simple instruments, some of which might eventually fly to Mars or Titan.

Again, it is hard to believe that these insights would have emerged without synthetic biology. After all, first-generation Watson–Crick theory had been in textbooks for three decades without recognizing the fundamental role of the repeating charge to the ability of DNA strands to bind their complements and support Darwinian evolution. Lacking that recognition, venture capitalists and other investors had bet billions of dollars on one particular antisense strategy, the one that required that molecular recognition remain in DNA analogs after the repeating charge was removed. Had they had the polyelectrolyte theory of the gene at their disposal, they would not have lost so much money. Synthesis drives discovery and paradigm changes in ways that analysis cannot.

3.3 BUILDING GENETICS FROM THE ATOM UP

One not trained in synthesis might not have expected that failure could be so rewarding. We did not generate flexible DNA. We did not generate antisense drugs. But we did show that our theory of the molecular structures behind the most fundamental of biological processes, replication and evolution, was inadequate. This led to a better theory.

3.3.1 Could Base Pairing behind Darwinian Evolution Be So Simple?

But what about the nucleobases, which had long been understood to be critical to the biological properties of DNA? And what about the simple rules that were proposed by Watson and Crick to account for genetics and Darwinian evolution: big pairs with small and hydrogen bond donors pair with hydrogen bond acceptors?

Could things be so simple? Again, if they were, then the synthetic biology paradigm (in the chemists' sense of the contranym) laid before us a grand challenge. On paper, if we shuffled the red hydrogen bond donor and hydrogen bond acceptor groups in the A:T and G:C pairs (Figure 3.7), treating them as interchangeable parts, we could write down eight new nucleobases that fit together to give four new base pairs having the same geometry as the A:T and G:C pairs (Figure 3.8). Photocopy the page from this book, cut out the nonstandard base pairs shown in Figure 3.8, and fit them together yourself as a modern James Watson would. As with the four standard nucleobases examined by Watson and Crick, the new nucleobases were predicted to pair with size complementarity (large with small) and hydrogen bond complementarity (hydrogen bond donors with acceptors), *if the theory behind the pairing were so simple.*

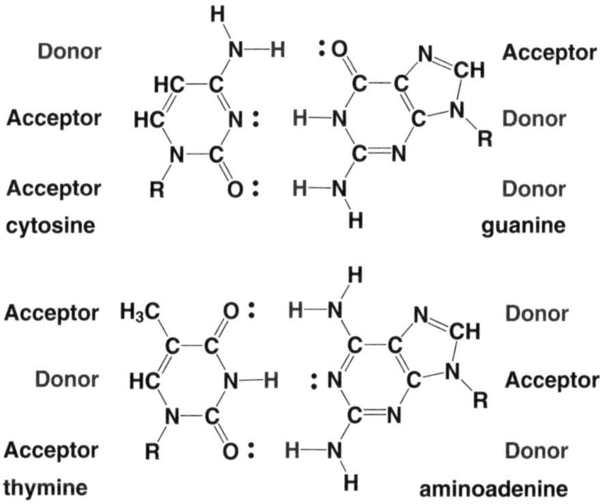

Figure 3.7 The two standard Watson–Crick pairs, idealized by replacing natural adenine (which lacks the bottom NH_2 group) with amino adenine

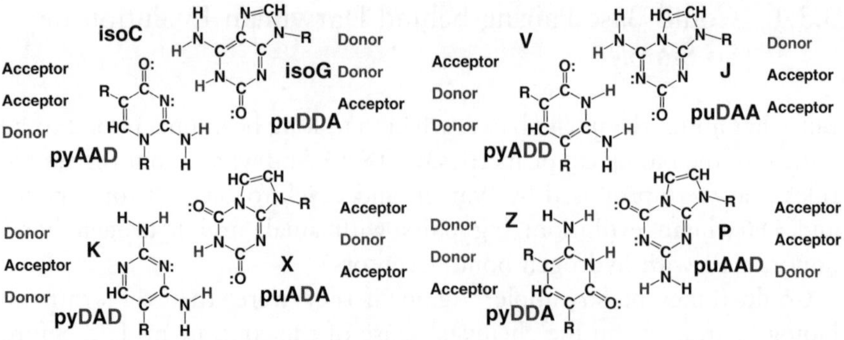

Figure 3.8 Shuffling hydrogen bond donor and acceptor groups in the standard nucleobase pairs generated eight additional heterocycles that, according to simple theory, should form four new, mutually independent, base pairs. This is called an "artificially expanded genetic information system" (AEGIS). Could molecular behavior at the center of genetics and Darwinian evolution be so simple? Synthesis was used to decide

As before, it was not enough to model the design on paper. Or even by computer. We needed to use synthetic technology from organic chemistry to *create* these new forms of matter, put them into DNA molecules, and see whether they worked as part of an artificially expanded genetic information system (AEGIS).

I will not leave you in suspense. Using the synthetic technology developed and enjoyed by chemists over the previous century, we were able to synthesize all of the synthetic components of our new artificial genetic alphabet. We were then able to put these synthetic nucleotides into synthetic DNA and RNA strands, and do all of the characterization of these that chemists do.

Once the synthetic task was complete, we observed that our artificial synthetic genetic system worked, and worked well. Artificial synthetic DNA sequences containing the eight new synthetic nucleotides formed double helices with their complementary synthetic DNA sequences. Complementation followed simple rules; just as A pairs with T and G pairs with C, P pairs with Z, V pairs with J, X pairs with K, and isoG pairs with isoC. The synthetic large nucleotides paired only with the correct synthetic small nucleotide. Our artificial synthetic DNA worked as well as natural DNA, at least in its ability to pair following simple rules.

An interesting irony is embedded in these results from synthetic biology. The base pairs were at the center of the Watson–Crick first-generation model for duplex structures; the phosphates and the sugars were not. Once synthetic biologists got their hands on this molecule, it was found that the base pairs were the easiest to change.

3.3.2 Synthetic Genetic Systems to Make Synthetic Protein Systems

Synthetic genetic systems can pair. But can they meet advanced challenges? Again, when the synthesis successfully meets the grand challenge originally laid out, discipline requires us to assume that we have not been sufficiently ambitious in the selection of the challenge. Accordingly, we must next make the challenge more difficult. For example, could the extra nucleotides be used in a natural translation system to increase the number of amino acids that could be incorporated into proteins by encoded ribosome-based protein synthesis?

Meeting this challenge required more synthesis, of transfer RNA molecules carrying AEGIS nucleotides in the anticodon loop charged with a nonstandard amino acid and of messenger RNA that contained the complementary nonstandard AEGIS nucleotides. Again, the synthesis based on simple molecules of ribosome-catalyzed protein synthesis was adequate as a guide; the challenge was met worked [40] (Figure 3.9). Expanding the number of biobricks in synthetic DNA could also expand the number of biobricks in encoded proteins. Further, the fact that the theory was adequate to meet this challenge constitutes support for the theory, under an inversion of the Feynman dictum ("What I can make, I understand").

It should be noted that, through this success with *unnatural* biology, something was learned about *natural* biology. As a control in one of the experiments with a messenger RNA molecule carrying an AEGIS base, we left out the charge-transfer RNA having a nonstandard nucleobase in its anticodon loop. We expected the synthesis of protein to stop at this point. Surprisingly, it did not. Instead, the ribosome paused, then skipped over the nonstandard codon via a frame shift, and continued translation. This does not happen with standard stop codons built from standard nucleobases. This contrast in the behavior between the synthetic and natural systems shed new light on the way in which natural genetic systems terminate protein synthesis [34].

3.3.3 Synthetic Genetics Supports Human Health Care

Pursuit of these "put-a-man-on-the-moon" challenges had taught us something. Base pairing *is* as simple from a molecular perspective as the

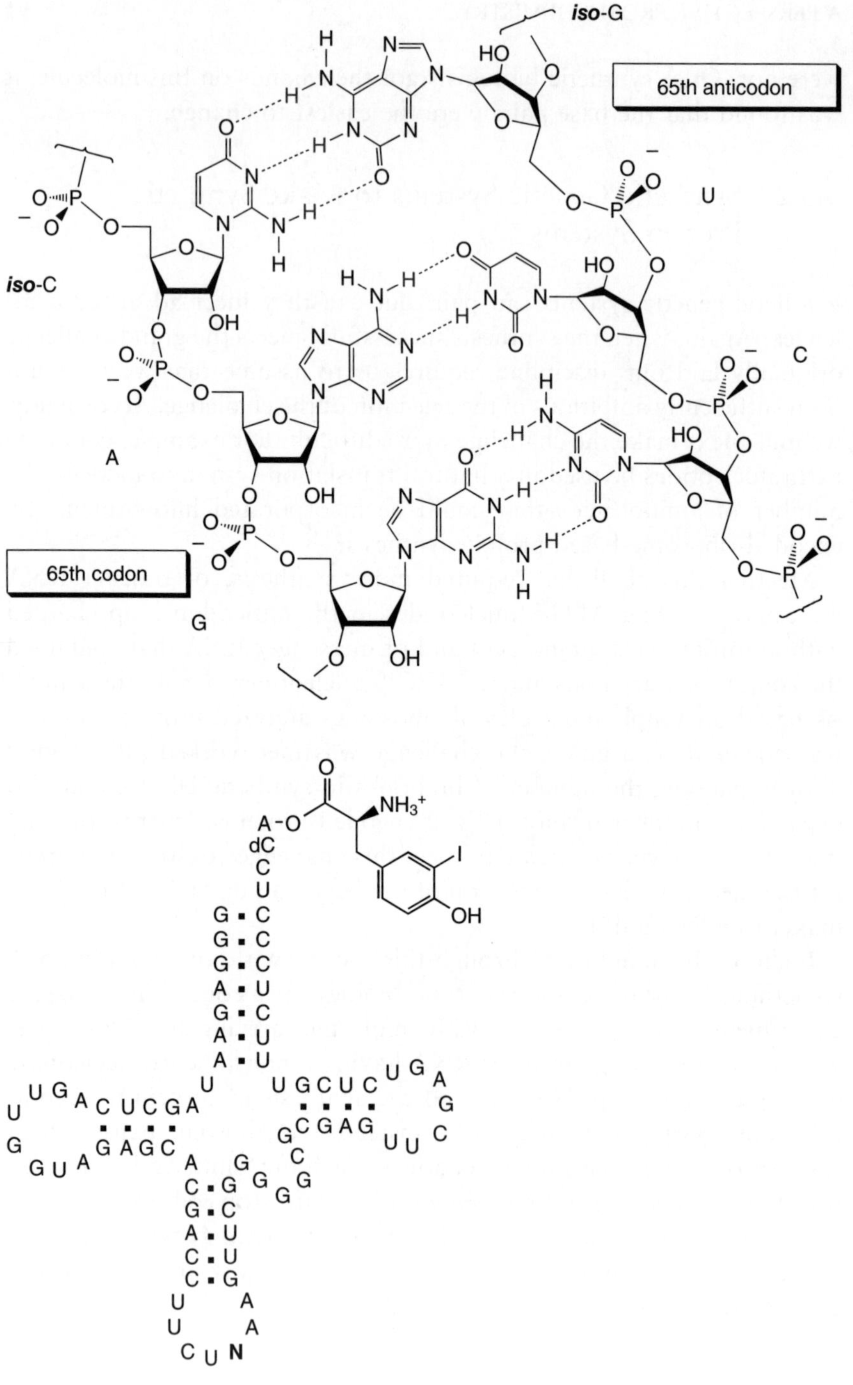

Figure 3.9 Putting a synthetic base into a messenger RNA, and providing a transfer RNA having the complementary nonstandard base in the anticodon loop (the "N") allowed the incorporation of a 21st amino acid (here, iodotyrosine) into a protein

first-generation theory proposed in 1953 by Watson and Crick implied. Simple theories used by Watson and Crick, together with the new polyelectrolyte theory of the gene, were sufficient to empower the design of a new genetic system that works as well as natural DNA. Thus, these simple theories delivered an understanding of the molecular behavior of natural DNA. They also provided the language sufficient to explain genetics. As genetics is a big part of Darwinian evolution, synthesis made a big contribution to our understanding of life, at least under our definition–theory of life as a chemical system capable of Darwinian evolution.

Of course, these successes required us to again set the bar higher. Perhaps the best demonstration of our better understanding of DNA is to use it to create new technology; mutating the Feynman dictum again, we might suggest: "If we understand it, we can *do* something with it."

Therefore, we set out to apply our synthetic genetic system in the clinic to support the care of human patients. The details are beyond the scope of this talk, but the general strategy is not. It might be worth a few words to explain how synthetic biology of this type has practical value.

Very often, diseases are caused by unwanted DNA. AIDS, for example, is caused by the human immunodeficiency virus (HIV), which delivers its own nucleic acid (RNA) into your body. A strep throat comes from unwanted bacteria carrying their unwanted DNA. Cancer comes from DNA from your own body that has mutated to give an unwanted sequence.

For such diseases, diagnosis involves detecting the unwanted DNA in a sample taken from a patient. But how can we find the unwanted DNA from the virus or the bacterium in that sample? After all, the unwanted DNA is present as just a few molecules in the sample; those few molecules are swamped by a background filled with considerable amounts of wanted DNA, the DNA from you the patient.

Accordingly, a general approach to detecting DNA involves two steps: (i) we must bind something to the unwanted DNA to form a bound complex; then (ii) we must move the bound complex to enrich and concentrate it at a spot where it can be detected.

Designing something to bind to unwanted DNA is easy if we know the sequence of the unwanted DNA. Following Watson–Crick pairing rules, we simply design a DNA strand that places a complementary A in a position where it can pair to each T in the unwanted DNA, a complementary T to pair with each A, a complementary C to pair with each G, and a complementary G to pair with each C. To illustrate with a trivial example, if the virus DNA sequence is TAAGCTTC, then the DNA sequence GAAGCTTA will bind to it, and bind to it selectively. If you have difficulty seeing this, remember that one of the sequences binds to the other in reverse order. This is, of course, the same idea as was pursued in the antisense industry.

To concentrate the bound complex at a spot in a detection architecture, it would be nice to do the same trick: place a DNA molecule with specific capture sequence at that spot and then place the complementary tag on the bound complex containing the unwanted DNA. The tag would drag the unwanted DNA to that spot, where it could be detected.

We make the tags from A, T, G, and C, and drag the tags to the detection spot using A:T and G:C pairing. Solving the "how to move DNA around" problem in this way encounters a problem in any real assay, however. This problem arises because biological samples that are actually examined in the clinic (your blood, for example) contain lots of DNA containing lots of A, T, G, and C. While the tag would be designed to have a different sequence than the sequence of the wanted DNA that is in your blood, it is difficult with A, T, G, and C to make a tag that is *very* different. For example, your DNA has just about every sequence 15 nucleotides long built from A, T, G, and C. These sequences will interfere with capture and concentration of unwanted DNA at a spot when the tag is built from the natural nucleotides A, T, G and C.

This problem can be solved by incorporating the extra synthetic nucleobases into the capture and concentration tags. This is exactly what was done by Mickey Urdea and Thomas Horn as they were developing at Chiron a system to detect HIV in the blood of AIDS patients. They used two of our synthetic nucleobases from the synthetic genetic alphabet (isoC and isoG, Figure 3.8) to move bound unwanted DNA to a place where it could be detected, exploiting pairing between two complementary components of the synthetic genetic system that do not pair with the natural A, T, G, and C. This left A, T, G, and C available to bind to the unwanted DNA directly.

Because neither isoG nor isoC is found in the wanted DNA from the human patient, the large amount of background DNA cannot interfere with the capture of the unwanted DNA attached to the probe. This reduces the level of "noise" in the system. As a result using our synthetic genetic system, a diagnostic tool can detect as few as a dozen molecules of unwanted DNA in a sample of patient blood even though that blood is full of wanted DNA from the patient. Together, this assay measured the level of the RNA from HIV (the causative agent of AIDS) in the blood of a patient, a measurement that allows the physician to adjust the treatment of the patient on a personal level to respond to the amount of virus that the patient has [41].

A similar diagnostic tool uses our synthetic genetic system to personalize the care of patients infected with hepatitis B and hepatitis C viruses. Still other applications of our synthetic genetic system are used in the analysis of cystic fibrosis, respiratory infections, influenza, and cancer. Today, our synthetic genetic systems help each year to personalize the care of 400 000 patients infected with HIV and hepatitis viruses. With the support of the National Human Genome Research Institute, we are developing tools that will allow synthetic genetics to sequence the genomes of patients rapidly and inexpensively. These tools will ultimately allow your physician to determine rapidly and inexpensively the genetic component of the malady that afflicts you.

The fact that synthetic genetic systems empower commercial activity as well academic research makes us still more confident of the theory that underlay the synthetic effort in the first place. The kind of confidence comes from making something entirely new that not only works in the laboratory, but also helps sick people; it is difficult to imagine a stronger way to obtain this confidence. In addition to driving discovery and paradigm change in ways that analysis cannot, synthetic biology allows us to generate multiple experimental approaches to decide whether our underlying view of reality is flawed.

3.3.4 The Next Challenge. Can Artificial Synthetic Genetic Systems Support Darwinian Evolution?

But why stop here? The next challenge in assembling a synthetic biology requires us to have our synthetic genetic system support Darwinian evolution. For this, we needed technology to copy synthetic DNA. Of course, copying alone would not be sufficient. The copies must occasionally be imperfect and the imperfections must themselves be copyable.

To copy our synthetic genetic system in pursuit of this goal, we turned to enzymes called DNA polymerases. Polymerases copy standard DNA strands by synthesizing new strands that pair A with T, T with A, G with C, and C with G [42]. The polymerases can then copy the copies, and then the copies of the copies. If done many times, this process is called the polymerase chain reaction, or PCR. PCR was developed by Kary Mullis, who was also awarded a Nobel Prize.

As we attempted to meet this grand challenge, we immediately encountered an unscripted problem. Natural polymerases have evolved for billions of years to accept natural genetic systems, not synthetic genetic systems. As we tried to use natural polymerases to copy our synthetic DNA, we found that our synthetic DNA differed from natural DNA too much. Natural polymerases, therefore, rejected our synthetic DNA as "foreign."

Fortunately, synthetic methods available from classical synthetic biology allow us to replace amino acids in the polymerases to get mutant polymerases. Several of these synthetic polymerases were found to accept our synthetic DNA. Michael Sismour and Zunyi Yang, working in my group, found combinations of polymerases (natural and synthetic) and synthetic genetic alphabets that worked together.

And so we went back to the laboratory to see whether DNA molecules built from synthetic nucleotides could be copied, whether the copies could be copied, and whether the copies of the copies could be copied. We also asked whether the polymerases would occasionally make mistakes (mutations), and whether those mistakes could themselves be copied.

To meet this challenge, we did accommodate a bit the preferences of DNA polymerases. These have evolved for billions of years on Earth to accept nucleobases that present electron density (the lobes in Figure 3.10) to the minor groove (down, in Figure 3.10) of the double helix. Many of our synthetic nucleobases do not do this, but two do: Z and P (Figure 3.10). These form a P:Z base pair that actually contributes to duplex stability more than the A:T and G:C pairs [43].

But would they work with natural enzymes? Again, we do not want to keep you in suspense. A six-letter synthetic genetic system built from four standard nucleotides and two synthetic nucleotides can be repeatedly copied (Figure 3.11).

Setting the bar higher, could they support Darwinian evolution? Here, careful experiments were done to determine whether they could participate in mutation processes. This work showed that Z and P could indeed mutate to C and G and, more surprisingly, that C and G could mutate to give Z and P. The details of the mutation process were studied. Sometimes Z is incorporated opposite G instead of C. Sometimes C is incorporated opposite P instead of Z. Sometimes P is incorporated opposite C instead of G. Sometimes G is incorporated opposite Z instead of P. This low level of mutation is just a few percent per copy. But once mutations are introduced into the children DNA, they themselves can be copied and, therefore, propagated to the next generation. Thus, the synthetic genetic system built from G, A, C, T, Z, and P (GACTZP) is capable of supporting Darwinian evolution.

A PERSPECTIVE FROM CHEMISTRY

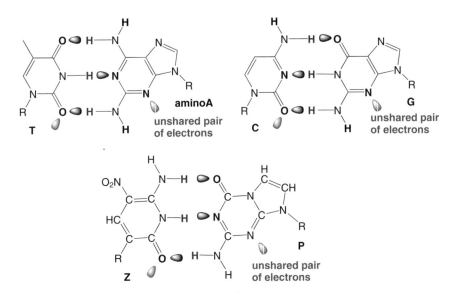

Figure 3.10 The Z and P pairs that have been incorporated into six-letter PCR, with mechanistic studies that show that this six-letter synthetic genetic system can support Darwinian evolution. Key to meeting this challenge was to make a small accommodation to the desire of natural DNA polymerases to have nucleobases that present electron density (the gray lobes) to the minor groove (down, in this structure) of the double helix. This is the case for T, A (shown here with an extra NH_2 unit), C, and G (top). It is also the case with the Z and P synthetic nucleobases (bottom)

3.3.5 Is This Synthetic Life?

A GACTZP synthetic six-letter genetic system that includes the "biobricks" G, A, C, T, Z, and P is clearly not homologous to the genetic system that we find naturally on Earth. It does, of course, share many structural features with natural genetic systems. Some of these we believe to be universal based on theories like the polyelectrolyte theory of the gene. The repeating backbone phosphates are not, according to that theory, dispensable.

But it can support Darwinian evolution. Is this artificial synthetic life? Our theory–definition holds that life is a self-sustaining chemical system capable of Darwinian evolution. The artificial genetic system that we have synthesized is certainly a chemical system capable of Darwinian evolution. It is not self-sustaining, however. For each round of evolution, a graduate student must add something by way of food; the system cannot go out to have lunch on its own.

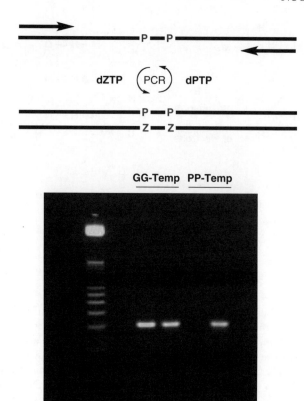

Figure 3.11 The polymerase chain reaction with a six-letter genetic alphabet, incorporating P and Z in addition to A, T, G, and C

Therefore, while our synthetic genetic system demonstrates that simple theory can empower/explain the molecular side of evolution, we are not yet at the point where we can use our synthetic genetic system as a "second example" of life. We are not ready to use our system to see whether it can spontaneously generate traits that we recognize from natural biology.

For this, we return to the need for bucks. Not surprisingly (and not inappropriately), funding is easier to find to research tools that help manage the medical care of patients infected with HIV and other viruses than to take synthetic biology to the next step. Because of the relevance of this work to these applications, work is proceeding. Further, the Defense Threat Reduction Agency (DTRA) is encouraging us to develop further the second-generation model of DNA.

Accordingly, we are attempting to meet the "put-a-man-on-the-moon" goal of obtaining a synthetic genetic system that can sustain to a greater degree its own access to Darwinian evolution.

Even should this be done, however, the community would not be unanimous in its view that a synthetic biology had been created. Various theories of biology constructively held by many in the community add criteria for a definition–theory of life. For example, those who subscribe to the cell theory of life will no doubt wait until the synthetic chemical system capable of Darwinian evolution is encapsulated in a cell. Those who subscribe to a metabolism theory of life might wait until the artificial synthetic genetic system also encodes enzymes that catalyze the transformation of organic compounds. Even those who subscribe to a Darwinian theory of life might insist that, before a synthetic biology is announced, the artificial system must evolve to a natural change in environment, not to one engineered in the laboratory.

Where one draws the line is, again, a matter of culture. Further, we expect that as these goals are pursued, the bar will be raised; again, if success is achieved, the discipline of a synthetic biologist requires the bar to be raised.

Again, this is not relevant to the value of the pursuit. The purpose of the synthetic effort is to force ourselves across uncharted territory where we must address unscripted questions. As we attempted to design a synthetic genetic system or a synthetic protein catalyst, we learned about genetics and catalysis in general, as well as the strengths and inadequacies of our theories that purported to understand these. In future pursuits of synthetic cells, metabolisms, and adaptation in a synthetic biological system, we cannot help but learn more about cells, metabolism, and adaptation in general, including these processes found naturally in life around us today on Earth. Illustrated by the four-wedge diagram in Figure 3.2, learning from synthesis will complement learning obtained from paleogenetics, exploration, and laboratory experiments attempting to understand the origin of life.

3.4 DOES SYNTHETIC BIOLOGY CARRY HAZARDS?

Provocative titles like "synthetic biology" and "artificial life" suggest a potential for hazard. They also conjure up images of Frankenstein. Is there any hazard associated with synthetic biology? If so, can we assess its magnitude?

As noted in Section 3.1, much of what is called "synthetic biology" today is congruent with the activities supported by the recombinant DNA technology that has been around for the 35 years since Waclaw Szybalski coined the term. There is no conceptual difference between how bacteria are constructed today to express genes from other places and how they were constructed in 1980; a straight line connects the synthetic biology of Szybalski to the current efforts of Venter, Smith, and their colleagues.

The hazards of this type of synthetic biology were discussed at a famous 1975 conference at the Asilomar conference site in Monterey, California. We now have a quarter century of experience with the processes used to mitigate any hazards that might exist from this type of synthetic biology. Placing a new name on an old research paradigm does not create a new hazard; much of the concern over the hazards of today's efforts of this type reflects simply the greater chance of success because of improved technology.

Those seeking to create artificial chemical systems to support Darwinian processes are, however, creating something new. We must consider the possibility that these artificial systems might escape from the laboratory. Does this possibility create a hazard?

Some general biological principles are relevant to assessing the potential for such hazards. For example, the more an artificial living system differs (at a chemical level) from a natural biological system, the less likely it is to survive outside of the laboratory. A living organism survives when it has access to the resources that it needs, and is more fit than competing organisms in recovering these resources from the environment where it lives. Thus, a completely synthetic life form having unnatural nucleotides in its DNA would have difficulty surviving if it were to escape from the laboratory. What would it eat? Where would it get its synthetic nucleotides?

Such principles also apply to less exotic examples of engineered life. Thirty years of experience with genetically altered organisms since Asilomar have shown that engineered organisms are less fit than their natural counterparts to survive outside of the laboratory. If they survive at all, they do so either under the nurturing of an attentive human or by ejecting their engineered features.

Thus, the most hazardous type of bioengineering is the type that is not engineering at all, but instead reproduces a known virulent agent in its exact form. The recent synthesis of smallpox virus or the

A PERSPECTIVE FROM CHEMISTRY

1918 influenza virus are perhaps the best examples of risky synthetic biology.

Further, we might consider the motivation of one actually wanting to do damage? Would one generate a genetically engineered *Escherichia coli*? Or place fuel and fertilizer in a rented truck and detonate it outside of the Federal Building in Oklahoma City? We know the answer to this question for one individual. We do not know it for all individuals. In most situations, however, it seems easier to do harm in nonbiotechnological ways than by engineering biohazards.

Any evaluation of hazard must be juxtaposed against the potential benefits that come from the understanding developed by synthetic biology. History provides a partial guide. In 1975, the city of Cambridge banned the classical form of synthetic biology within its six square miles to manage what was perceived as a danger. In retrospect, it is clear that had the ban been worldwide, the result would have been more than harmful. In the same decade that Cambridge banned recombinant DNA research, an ill-defined syndrome noted in patients having "acquired immune deficiency" was emerging around the planet as a major health problem. This syndrome came to be known as AIDS, and it was eventually learned that AIDS was caused by the HIV.

Without the technology that the city of Cambridge banned, we would have been hard pressed to learn what HIV was, let alone have compounds today that manage it. Today, classical synthetic biology and recombinant DNA technology allow us to manage new threats as they emerge, including SARS, bird influenza, and other infectious diseases. Indeed, it is these technologies that distinguish our ability to manage such threats today from how we would have managed them a century ago.

With these thoughts in mind, a Venn diagram can be proposed to assess risk in different types of synthetic biology (Figure 3.12). Activities within the red circle use standard terran biochemistry, more or less what Nature has developed on Earth over the past 4 billion years. Activities outside that circle concern activities with different biochemistry.

One circle contains systems that are capable of evolving. Those outside the circle cannot, and present no more hazard than a toxic chemical; regardless of its hazard, it is what it is, and cannot get any worse.

The second circle contains systems that are self-sustaining. They "live" without continuing human intervention. Those outside this circle require continuous feeding. Thus, these represent no more of a hazard than a pathogen that will die once released from the laboratory.

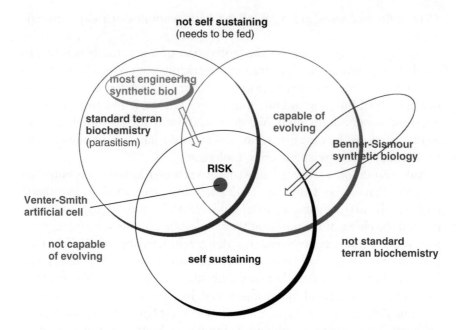

Figure 3.12 A Venn diagram illustrating the hazards of synthetic biology. One circle contains systems able to evolve. Those outside the circle cannot, and present no more hazard than a toxic chemical. Another circle contains systems that are self-sustaining. Those inside the circle "live" without continuing human intervention; those outside require continuous feeding, and are no more hazardous than a pathogen that dies when released from a laboratory. Systems within the third circle use standard terran molecular biology; those outside do not. The greatest chance for hazard comes from a system that is self-sustaining, uses standard biochemistry, and is capable of evolving, the intersection between the three circles

The greatest chance for hazard comes from a system that is self-sustaining, uses standard biochemistry, and is capable of evolving. This is, of course, the goal of the Venter–Smith artificial cell, which presents the same hazards as presented by natural nonpathogenic organisms: it might evolve into an organism that feeds on us. Those hazards, although not absent, are not large compared with those presented by the many natural nonpathogens that co-inhabit Earth with us.

3.5 CONCLUSIONS

Pursuit of the grandest challenge in contemporary synthetic biology, creating artificial life of our own, has already yielded fruits. Alien life with six letters in its genetic alphabet and more than 20 amino acids in its protein alphabet is possible. A repeating charge may be universal in the backbone of genetic biopolymers; a repeating dipole may be universal in the backbone of catalytic biopolymers. We have synthesized in the laboratory artificial chemical systems capable of Darwinian evolution.

This makes the next grand challenge still more ambitious. We would like a self-sustaining artificial chemical system capable of Darwinian evolution. To get it, the reliance of the current synthetic Darwinian systems on natural biology must be reduced. We have encountered unexpected problems as we attempt to do so. Some in the community are confident that, with a little more effort, we can surmount these problems; others of us are not so sure.

Achieving in the laboratory an artificial biology would expand our knowledge of life as universal more than anything else short of actually encountering alien life. Still better, it is more likely that synthetic biology will do this sooner than exploration will. If we had a simple form of designed life in our hands, we could ask key questions. How does it evolve? How does it create complexity? How does it manage the limitations of organic molecules related to Darwinian processes? And if we fail after we give the effort our best shot, it will directly challenge our simple (and possibly simplistic) definition–theory of life as being nothing more than a chemical system capable of Darwinian evolution.

ACKNOWLEDGMENTS

We are indebted to the Swiss Nationalfond, the National Human Genome Research Institute (R01HG004831, R01HG004647), and the National Institute of General Medical Sciences (R01GM086617) for support of the applied parts of this work, the NASA Astrobiology program for support of aspects of work relating to discovery of alien life (NNX08AO23G), and the DTRA (HDTRA1-08-1-0052) for supporting the basic research program. We are especially indebted to the encouragement of DTRA to develop the basic theory of DNA.

STATEMENT OF CONFLICTS OF INTEREST

Various authors are inventors on patents covering certain of these technologies.

REFERENCES

1. Hawking, S. (2009) *The Universe in a Nutshell*, Bantam Spectra, New York.
2. Szybalski, W. (1974) in *Control of Gene Expression* (eds A. Kohn and A. Shatkay), Plenum Press, New York, pp. 23–24, 404–405, 411–412, 415–417.
3. Adleman, L.M. (1994) Molecular computation of solutions to combinatorial problems. *Science*, **266**, 1021–1024.
4. Regis, E. (2009) *What is Life? Investigating the Nature of Life in the Age of Synthetic Biology*, Oxford University Press, Oxford.
5. Anonymous (2010) Genesis redux. *The Economist*, **395** (8683), 811–863, http://www.economist.com/displaystory.cfm?story_id=16163006 (accessed 23 November 2010).
6. Ulmer, K.M. (1983) Protein engineering. *Science*, **219**, 666–671.
7. Knowles, J.R. (1987) Tinkering with enzymes. *Science*, **236**, 1252–1258.
8. Benner, S.A. and Sismour, A.M. Synthetic biology. (2005) *Nature Reviews Genetics*, **6**, 533–543.
9. Benner, S.A. (2009) *Life, the Universe and the Scientific Method*, FfAME Press, Gainesville FL.
10. Benner, S.A. (2004) Understanding nucleic acids using synthetic chemistry. *Accounts of Chemical Research*, **37**, 784–797.
11. Lim, W.A. (2002) The modular logic of signaling proteins: building allosteric switches from simple binding domains. *Current Opinion in Structural Biology*, **12**, 61–68.
12. Moradian, A. and Benner, S.A. (1992) A biomimetic biotechnological process for converting starch to fructose: thermodynamic and evolutionary considerations in applied enzymology. *Journal of the American Chemical Society*, **114**, 6980–6987.
13. Lartigue, V., Vashee, S., Algire, M.A. et al. (2009) Creating bacterial strains from genomes that have been cloned and engineered in yeast. *Science*, **325**, 1693–1696.
14. Gibson, D.G., Glass, J.I., Lartigue, C. et al. (2010) Creation of a bacterial cell controlled by a chemically synthesized genome. *Science*, **328**, 52–56.
15. Medawar, P.B. (1967) *The Art of the Soluble*, Oxford University Press, Oxford.
16. Stackhouse, K.P., Nambiar, Stackhouse, J., Stauffer, D.M. et al. (1984) Total synthesis and cloning of a gene coding for the ribonuclease S protein. *Science*, **223**, 1299–1301.
17. Edge, M.D., Greene, A.R., Heathcliffe, G.R. et al. (1981) Total synthesis of a human leukocyte interferon gene. *Nature*, **292**, 756–762.
18. Benner, S.A. (1987) *Redesigning the Molecules of Life*, Springer, Berlin.
19. Benner, S.A. (2009) *The Life, the Universe and the Scientific Method*, FfAME Press, Gainesville, FL.

20. Kauffman, S.A. (1995) *At Home in the Universe*, Oxford University Press, Oxford.
21. Cleland, C.E. and Chyba, C.F. (2000) Defining 'life'. *Origins of Life and Evolution of Biospheres*, **32**, 387–393.
22. Joyce G.F. (1994) Foreword, in *Origins of Life: The Central Concepts* (eds D.W. Deamer and G.R. Fleischaker), Jones & Bartlett, Boston, MA, pp. xi–xii.
23. Benner, S.A., Ricardo, A., and Carrigan, M.A. (2004) Is there a common chemical model for life in the universe? *Current Opinion in Structural Biology*, **8**, 672–689.
24. Joyce, G.F., Schwartz, A.W., Miller, S.L., and Orgel, L.E. (1987) The case for an ancestral genetic system involving simple analogues of the nucleotides. *Proceedings of the National Academy of Sciences of the United States of America*, **84**, 4398–4402.
25. Cooper, G., Novelle, K., Belisle, W. et al. (2001) Carbonaceous meteorites as a source of sugar-related organic compounds for the early Earth. *Nature*, **414**, 879–883.
26. Imanishi, T. and Obika, S.J. (1999) Syntheses and properties of novel conformationally restrained nucleoside analogues. *Journal of Synthetic Organic Chemistry, Japan*, **57**, 969–980.
27. Wengel, J., Koshkin, A., Singh, S.K. et al. (1999) LNA (locked nucleic acid). *Nucleosides, Nucleotides and Nucleic Acids*, **18**, 1365–1370.
28. Eschenmoser, A. (1999) Chemical etiology of nucleic acid structure. *Science*, **284**, 2118–2124.
29. Augustyns, K. Van Aerschot, A., and Herdewijn, P. (1992) Synthesis of 1-(2,4-dideoxy-β-D-erythro-hexopyranosyl)thymine and its incorporation into oligonucleotides. *Bioorganic & Medicinal Chemistry Letters*, **2**, 945–948.
30. Renneberg, D. and Leumann, C.J. (2002) Watson–Crick base-pairing properties of tricyclo-DNA. *Journal of the American Chemical Society*, **124**, 5993–6002.
31. Burgess, K., Gibbs, R.A., Metzker, M.L., and Raghavachari, R. (1994) Synthesis of an oxyamide linked nucleotide dimer and incorporation into antisense oligonucleotide sequences. *Journal of the Chemical Society, Chemical Communications*, 915–916.
32. Huang, Z., Schneider, K.C., and Benner, S.A. (1991) Building blocks for oligonucleotide analogs with dimethylene sulfide, sulfoxide, and sulfone groups replacing phosphodiester linkages. *Journal of Organic Chemistry*, **56**, 3869–3882.
33. Huang, Z., Schneider, K.C., and Benner, S.A. (1993) Oligonucleotide analogs with dimethylenesulfide, -sulfoxide, and -sulfone groups replacing phosphodiester linkages. *Methods in Molecular Biology*, **20**, 315–353.
34. Roughton, A.L., Portmann, S., Benner, S.A., and Egli, M. (1995) Crystal structure of a dimethylene sulfone-linked ribodinucleotide analog. *Journal of the American Chemical Society*, **117**, 7249–7250.

35. Richert, C., Roughton, A.L., and Benner, S.A. (1996) Nonionic analogs of RNA with dimethylene sulfone bridges. *Journal of the American Chemical Society*, **118**, 4518–4531.
36. Schmidt, J.G., Eschgfaeller, B., Benner, S.A. (2003) A direct synthesis of nucleoside analogs homologated at the 3' and 5'- positions. *Helv. Chim. Acta*. **86**, 2937–2956.
37. Olby, R. (1994) *The Path to the Double Helix*, Dover, New York.
38. Steinbeck, C. and Richert, C. (1998) The role of ionic backbones in RNA structure: an unusually stable non-Watson–Crick duplex of a nonionic analog in an apolar medium. *Journal of the American Chemical Society*, **120**, 11576–11580.
39. Benner, S.A. and Hutter, D. (2002) Phosphates, DNA, and the search for nonterrean life: a second generation model for genetic molecules. *Bioorganic Chemistry*, **30**, 62–80.
40. Bain, J.D., Chamberlin, A.R., Switzer, C.Y., and Benner, S.A. (1992) Ribosome-mediated incorporation of a non-standard amino acid into a peptide through expansion of the genetic code. *Nature*, **356**, 537–539.
41. Elbeik, T., Surtihadi, J., Destree, M. *et al.* (2004) Multicenter evaluation of the performance characteristics of the Bayer VERSANT HCV RNA 3.0 assay (bDNA). *Journal of Clinical Microbiology*, **42**, 563–569.
42. Yang, Z., Sismour, A.M., Sheng, P. *et al.* (2007) Enzymatic incorporation of a third nucleobase pair. *Nucleic Acids Research*, **35**, 4238–4249.
43. Yang, Z., Hutter, D., Sheng, P. *et al.* (2006) Artificially expanded genetic information system: a new base pair with an alternative hydrogen bonding pattern. *Nucleic Acids Research*, **34**, 6095–6101.

4
Peptide Nucleic Acids (PNAs) as a Tool in Chemical Biology

Peter E. Nielsen[1,2]

[1]*University of Copenhagen, Department of Cellular and Molecular Medicine, Faculty of Health Sciences, The Panum Institute, Blegdamsvej 3c, Copenhagen N, DK-2200, Denmark*
[2]*University of Copenhagen, Department of Medicinal Chemistry, Faculty of Pharmaceutical Sciences, Universitetsparken 2, Copenhagen, DK-2100, Denmark*

4.1 Introduction 108
4.2 Chemistry 108
4.3 An Assay for Cellular Delivery Using PNA Antisense in pLuc–HeLa Cells 110
4.4 Duplex DNA Recognition 112
4.5 Targeted Gene Repair 113
4.6 Sequence Information Transfer 114
4.7 Concluding Remarks 115
    References 116

The fully synthetic nucleic acid mimic peptide nucleic acid (PNA) is a good example of exploiting organic chemistry and "small" molecules for addressing and exploring biological problems. PNA oligomerization is based on conventional Boc- or Fmoc-peptide chemistry, which makes conjugation chemistry very attractive. In particular, PNA–peptide conjugates are conveniently obtained by continuous solid-phase synthesis, but also conjugation to a large variety of small organic ligands (psoralen, fatty acids, cholesterol, acridine, ferrocene, phenanthroline, etc.) is readily available. PNA oligomers are potential RNA interference drug-discovery candidates relying on steric blocking of translation (initiation) or splicing upon sequence selectively binding to mRNA targets. Numerous cell culture studies, as well as a few *in vivo* mouse model studies, have illustrated this potential. Furthermore, it is argued that PNA may be a very versatile model system to explore novel cellular and *in vivo* delivery vehicles because it is inherently chemically and biologically "neutral"; therefore, in contrast to polyanionic nucleotides, for example, it has minimal influence on the properties of the carrier vehicle. PNA was originally designed as a sequence-specific triplex-forming ligand, and subsequently several duplex DNA binding modes have been described. Recent results on so-called duplex and double duplex invasion complexes are particularly interesting, and may provide novel approaches for targeted somatic gene repair. Finally, sequence information transfer processes are discussed, emphasizing the compatibility and easy orthogonality of PNA and peptide synthesis which makes PNA powerful sequence tags in peptide combinatorial library approaches.

4.1 INTRODUCTION

Although a few nucleobase amino acids and oligomers derived therefrom were described in the 1970s [1, 2], the aminoethylglycine PNA (peptide nucleic acid) (aegPNA) [3–5] (Figure 4.1) was the first PNA that exhibited good DNA and RNA sequence-directed recognition properties and, therefore, caught the interest in many fields of science, from pure chemistry, to (molecular) biology and medical drug discovery and (genetic) diagnostics, to nanotechnology and prebiotic chemistry. Thus, being a fully synthetic nucleic acid mimic, PNA is a good example of exploiting organic chemistry and "small" molecules for addressing and exploring biological problems and questions; that is, chemical biology.

4.2 CHEMISTRY

Although a wealth of derivatives, modifications, and analogues of the original noncyclic, achiral, and inherently uncharged aegPNA have been described during the past almost 20 years (a selection is presented in Figure 4.1), and many of these indeed do have interesting properties

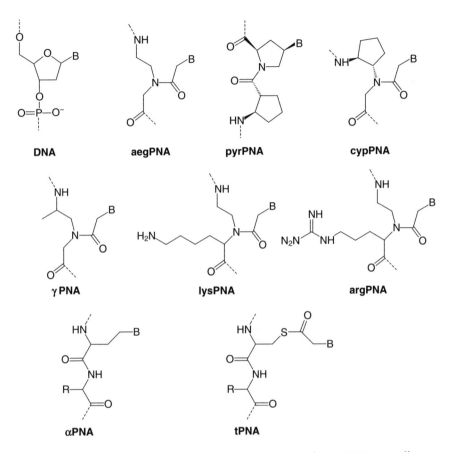

Figure 4.1 Chemical structures of DNA and the original (aeg)PNA as well as a series of PNA derivatives and analogues

and add new perspectives on PNA, most (molecular) biology applications are still based on the original aeg backbone, eventually incorporating dedicated modified nucleobases, such as pseudo-isocytosine, diaminopurine, thiouracil, and the G-clamp (*vide infra*). Also, the fact that PNA oligomerization is based on conventional Boc- or Fmoc-peptide chemistry makes conjugation chemistry very attractive, especially for PNA–peptide conjugates, which are conveniently obtained by continuous solid phase synthesis. Furthermore, conjugation to a large variety of small organic ligands (psoralen, fatty acids, cholesterol, acridine, ferrocene, phenanthroline, etc.) is readily available (e.g. [6–9]).

4.3 AN ASSAY FOR CELLULAR DELIVERY USING PNA ANTISENSE IN PLUC–HELA CELLS

PNA oligomers are potential RNA interference drug-discovery candidates relying on steric blocking of translation (initiation) upon sequence selectively binding to mRNA targets in the region of translation start or in the 5′ untranslated region, or on redirecting mRNA splicing by targeting exon–intron junctions in pre-mRNA [10, 11] (Figure 4.2). Numerous cell culture studies, as well as a few *in vivo* mouse studies, have illustrated this potential (e.g. [12–16]), but have also demonstrated that the major challenge for further progress is cellular delivery and, in particular, *in vivo* bioavailability, a challenge which is shared with other RNA interference and gene therapeutic technologies in general. In this context, PNA may be a very versatile model system to explore novel cellular and *in vivo* delivery vehicles because it is inherently chemically

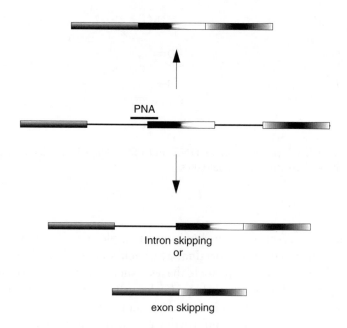

Figure 4.2 Schematic drawing of mRNA splicing modulation by antisense agents. Thin lines signify introns; while thick lines signify exons. Upon normal splicing (upwards) all introns are spliced out, but upon blockage of an intron–exon junction, either the intron is skipped and retained in the mRNA or the exon is skipped and an alternatively spliced mRNA results

and biologically "neutral" and, therefore, in contrast to polyanionic nucleotides, for example, has minimal influence on the properties of the carrier vehicle. Thus, the conclusions drawn from studies using PNA as the "active principle" may be more widely applicable for biotech drugs in general. Indeed, the pLuc–HeLa cell system has been used very successfully and extensively for discovery of novel delivery systems, as well as for characterizing existing ones (e.g. [17–21]). This system developed by Kole and co-workers [22] is based on a stably transfected luciferase gene in which an intron from a mutant thalassemia globin gene has been inserted. This intron is not correctly spliced out due to a mutation that activates an aberrant splice, thereby leaving a portion of the intron in the mRNA and consequently resulting in a nonfunctional luciferase enzyme. By blocking the aberrant splice site with a PNA (or other antisense) oligomer, normal splicing is restored and functional luciferase is expressed. The great advantage of this system is the positive readout, which gives very high sensitivity and accuracy and the option of using both an enzyme activity (luciferase) readout, which is compatible with high-throughput screening, and an RT-PCR readout quantifying the actual molecular biology targeting event of inhibiting a splice site in the pre-mRNA by measuring the relative ratio between correctly and incorrectly spliced mRNA. Thus, in principle, the assay can be used to determine the relative efficiency by which a delivery vehicle is able to deliver a PNA (or, by analogy, a chemically and physically comparable molecule) from the medium to the nucleus of the (HeLa) cell. A large variety of PNA conjugates, ranging from cell-penetrating peptides to lipids and various organic ligands, have been studied in this system [23] and also mediated by liposome and other carriers.

Obviously, the information extracted from the pLuc–HeLa system is limited by the fact that this is a specific (monolayer, cancer) cell type, and it is well established that the uptake of membrane nondiffusable compounds, such as oligonucleotides, PNAs, and peptides, varies very significantly between different cell types.

Interestingly, an analogous *in vivo* mouse model based on green fluorescent protein rather than luciferase was developed some years ago [16]. This model should also be very powerful in combination with antisense PNA constructs and complexes, but unfortunately it is not readily available and has only been used in a few studies. Both from a PNA point of view and from a more general gene therapeutics aspect, more comprehensive and comparative *in vivo* delivery, distribution, bioavailability, and also pharmacokinetics studies are highly warranted.

4.4 DUPLEX DNA RECOGNITION

PNA was originally designed as a sequence-specific duplex DNA-recognizing ligand and was discovered to bind duplex DNA by a novel mechanism: helix invasion [3, 24]. Subsequently, a range of other binding modes were discovered (apart from the original triplex invasion), including duplex invasion [25], double duplex invasion [26], and conventional major groove triplex binding [27] (Figure 4.3). Recent progress has demonstrated that 15-mer homopyrimidine PNAs conjugated to four lysines and/or a 9-aminoacridine, and in which all cytosines were replaced with pseudo-isoocytosine, bind to duplex DNA in a triplex mode with very high affinity (sub-nanomolar K_d) at physiologically relevant ionic conditions [27]. Because this binding does not require opening of the DNA helix, it is not sensitive to elevated ionic strength (as opposed to invasive binding modes) and may, therefore, have advantages for *in vivo* application despite the sequence constraints for homopurine targets. Duplex invasion binding has the advantage that it is based solely on Watson–Crick base pairing; therefore, a priori – in contrast to triplex (invasive) binding – does not have any target sequence constraints. However, because only one PNA strand is involved in the complex it is difficult to achieve sufficient free energy for effective

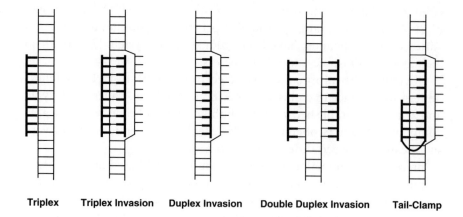

Triplex   Triplex Invasion   Duplex Invasion   Double Duplex Invasion   Tail-Clamp

Figure 4.3 Schematic drawing of dsDNA–PNA complexes so far discovered. Triplex and triplex invasion complexes require a homopurine (A/G) DNA target, as does the clamp of the tail–clamp complex, while duplex invasion and double duplex invasion in principle have no sequence constraints. However, the double duplex invasion complexes require use of (nucleobase- and/or backbone-induced) pseudo-complementary PNA oligomers

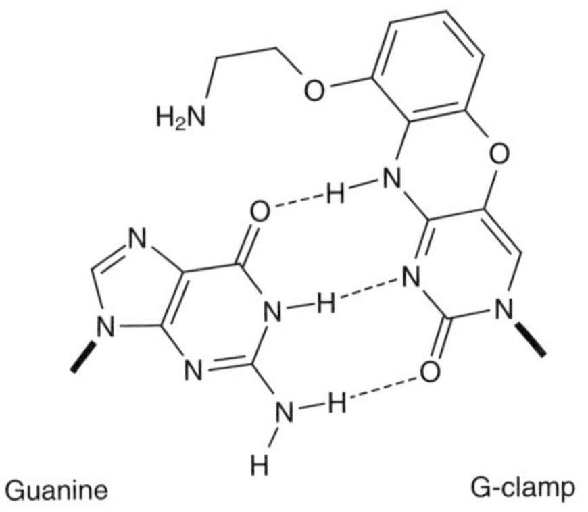

Figure 4.4 Guanine recognition by a helix-stabilizing G-clamp nucleobase

binding [25]. Using a combination of a γ-methyl-modified (Figure 4.1) PNA–DNA duplex-stabilizing backbone [28] and a highly stabilizing cytosine analogue, the G-clamp (Figure 4.4), it was recently found that such γ-PNAs are able to bind their sequence complementary target in duplex DNA efficiently via duplex invasion at micromolar concentrations [29]. Finally, it has been reported that PNA oligomers containing a cassette of D-lysine-backbone PNA (lysPNA, Figure 4.1) units enhance the efficiency of peudocomplementary PNAs in their ability to bind in a double duplex invasion mode, to a degree where only 20% pseudo-complementary diaminopurine–thiouracil base pairs are required [30]. These results promise that targeted chemical modification of the PNA backbone in combination with exploiting modified nucleobases (and various conjugation chemistry) may significantly expand the scope and efficiency of PNA targeting of duplex DNA, thereby increasing the possibilities of using these as sequence-specific genome-targeting agents in drug discovery, for instance within targeted gene repair.

4.5 TARGETED GENE REPAIR

Introduction of a single- or double-stranded DNA (donor) molecule which is homologous to a region of the cell's genome into this eukaryotic cell may, at a very low frequency (usually less than 0.1%), result in exchange of sequence information from the donor to the cell's genome. The mechanism is not very well understood, but is believed to involve homologous recombination and/or inscission repair. Most interestingly,

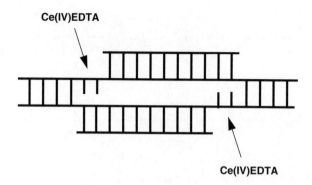

Figure 4.5 Schematic illustration of a staggered pcPNA dsDNA complex for Ce(IV)EDTA cleavage of the resulting single-stranded DNA stretches

the frequency can be significantly increased (up to one order of magnitude) if a DNA binding ligand, such as a triplex-forming oligonucleotide [31] or a PNA [32] is targeted proximal to the homologous region. If a double-strand DNA break is specifically introduced by an engineered zinc-finger nuclease which is genetically introduced to the cell, impressive repair efficiencies of up to 20% may be observed [33]. These results hold promise that it could be possible to develop medical treatment modalities allowing gene therapeutic correction of somatic mutations.

In this context, some recent results on Ce(IV)-induced double-stranded DNA cleavage of staggered PNA double-duplex invasion complexes [34] (Figure 4.5) are very interesting, as double-strand DNA breaks activate homologous recombination. Accordingly, it was shown that a DNA substrate cleaved with Ce(IV) as described above could activate targeted DNA repair approximately 50-fold over background [35]. Clearly, there is a very long way to clinically exploiting targeted gene repair, but the results obtained so far do inspire and warrant further exploration on this path.

4.6 SEQUENCE INFORMATION TRANSFER

Analogously to natural nucleic acids, PNAs contain sequence information that can be exploited and transmitted to other molecules, most directly to other nucleic acids and mimics. However, owing to their non-natural chemical composition, PNAs are not substrates for natural enzymes, such as polymerases; therefore, replication, transcription, or translational processes must be based on chemical rather than on enzymatic processes, and no efficient solutions apart from templated ligation reactions have yet been devised. PNA-templated DNA ligations and DNA-templated ligation reactions have been explored for possible mutation-specific diagnostic applications [36]. Furthermore, the peptide compatibility of PNA in combination with the sequence information that can be built in has been explored in PNA–peptide–PNA conjugates as a means to control peptide conformation (and thus activity) through conformational constraints imposed by PNA–PNA hybridization or via hybridization of such conjugates to cellular RNA or DNA molecules [37, 38].

The compatibility and easy orthogonality of PNA and peptide synthesis also makes it convenient to use PNA sequence tags in peptide combinatorial library approaches. This approach has been used with peptide libraries for profiling tyrosine kinases [39] and protease inhibitors [40, 41], and the peptide identities were deconvoluted through hybridization to dedicated DNA oligonucleotide microarrays. Because of their chemical and biological stability and relative "inertness," PNA sequence tags could find many applications in biology.

Very recently the principle of dynamic libraries was also exploited in a PNA context [42]. Specifically, thioester PNAs (tPNAs, Figure 4.1), which formally are derivatives of αPNAs [43] (Figure 4.1) built from a conventional α-amino acid peptide in which a nucleobase is attached to every second amino acid, were synthesized. However, in the tPNAs the nucleobase is attached via a labile thioester linkage which may engage in a (thiol-catalyzed) equilibrium with thioester nucleobase ligands free in solution. It was found that, in the presence of a DNA oligonucleotide to which the tPNA oligomer may hybridize, the tPNA oligomer would equilibrate into a sequence complementary to the DNA oligonucleotide, as this would produce the most stable tPNA–DNA duplex.

4.7 CONCLUDING REMARKS

Hopefully, the few examples presented above have illustrated the many areas of chemical biology where peptide nucleic acids and other DNA analogues and mimics can specifically modulate biological processes, provide tools for analyzing and understanding them, and in some cases even mimic them. Undoubtedly, new areas will be added in the future and new tools developed and novel principles discovered.

REFERENCES

1. De Koning, H. and Pandit, U.K. (1971) Unconventional nucleotide analogues VI. Synthesis of purinyl- and pyrimidinyl-peptides. *Recueil des Travaux Chimiques des Pays-Bas*, **91**, 1069–1080.
2. Buttrey, J.D., Jones, A.S., and Walker, R.T. (1975) Synthetic analogues of polynucleotides XIII. *Tetrahedron*, **31**, 73–75.
3. Nielsen, P.E., Egholm, M., Berg, R.H., and Buchardt, O. (1991) Sequence-selective recognition of DNA by strand displacement with a thymine-substituted polyamide. *Science*, **254**, 1497–1500.
4. Egholm, M., Buchardt, O., Christensen, L. *et al.* (1993) PNA hybridizes to complementary oligonucleotides obeying the Watson–Crick hydrogen-bonding rules. *Nature*, **365**, 566–568.
5. Nielsen, P.E. (1999) Peptide nucleic acid. A molecule with two identities. *Accounts of Chemical Research*, **32**, 624–630.
6. Murtola, M., Ossipov, D., Sandbrink, J., and Stromberg, R. (2007) RNA cleavage by 2,9-diamino-1,10-phenanthroline PNA conjugates. *Nucleosides, Nucleotides and Nucleic Acids*, **26**, 1479–1483.
7. Aoki, H. and Tao, H. (2007) Label- and marker-free gene detection based on hybridization-induced conformational flexibility changes in a ferrocene-PNA conjugate probe. *Analyst*, **132**, 784–791.
8. Kim, K.H., Fan, X.J., and Nielsen, P.E. (2007) Efficient sequence-directed psoralen targeting using pseudocomplementary peptide nucleic acids. *Bioconjugate Chemistry*, **18**, 567–572.
9. Bentin, T. and Nielsen, P.E. (2003) Superior duplex DNA strand invasion by acridine conjugated peptide nucleic acids. *Journal of the American Chemical Society*, **125**, 6378–6379.
10. Karras, J.G., Maier, M.A., Lu, T. *et al.* (2001) Peptide nucleic acids are potent modulators of endogenous pre-mRNA splicing of the murine interleukin-5 receptor-alpha chain. *Biochemistry*, **40**, 7853–7859.
11. Siwkowski, A.M., Malik, L., Esau, C.C. *et al.* (2004) Identification and functional validation of PNAs that inhibit murine CD40 expression by redirection of splicing. *Nucleic Acids Research*, **32**, 2695–2706.
12. Pooga, M., Soomets, U., Hällbrink, M. *et al.* (1998) Cell penetrating PNA constructs regulate galanin receptor levels and modify pain transmission in vivo. *Nature Biotechnology*, **16**, 857–861.
13. Hamilton, S.E., Simmons, C.G., Kathiriya, I.S., and Corey, D.R. (1999) Cellular delivery of peptide nucleic acids and inhibition of human telomerase. *Chemistry & Biology*, **6**, 343–351.
14. Shiraishi, T. and Nielsen, P.E. (2004) Down-regulation of MDM2 and activation of P53 in human cancer cells by antisense 9-aminoacridine-PNA (peptide nucleic acid) conjugates. *Nucleic Acids Research*, **32**, 4893–4902.
15. Fabani, M.M. and Gait, M.J. (2008) MiR-122 targeting with LNA/2'-O-methyl oligonucleotide mixmers, peptide nucleic acids (PNA), and PNA-peptide conjugates. *RNA*, **14**, 336–346.

16. Sazani, P., Gemignani, F., Kang, S.-H. et al. (2002) Systemically delivered antisense oligomers upregulate gene expression in mouse tissues. *Nature Biotechnology*, **20**, 1228–1233.
17. Bendifallah, N., Rasmussen, F.W., Zachar, V. et al. (2006) Evaluation of cell-penetrating peptides (CPPs) as vehicles for intracellular delivery of antisense peptide nucleic acid (PNA). *Bioconjugate Chemistry*, **17**, 750–758.
18. Shiraishi, T., Pankratova, S., and Nielsen, P.E. (2005) Calcium ions effectively enhance the effect of antisense peptide nucleic acids conjugated to cationic tat and oligoarginine peptides. *Chemistry & Biology*, **12**, 923–929.
19. Shiraishi, T., Hamzavi, R., and Nielsen, P.E. (2008) Subnanomolar antisense activity of phosphonate–peptide nucleic acid (PNA) conjugates delivered by cationic lipids to HeLa cells. *Nucleic Acids Research*, **36**, 4424–4432.
20. Koppelhus, U., Shiraishi, T., Zachar, V. et al. (2008) Improved cellular activity of antisense peptide nucleic acids by conjugation to a cationic peptide–lipid (catlip) domain. *Bioconjugate Chemistry*, **19**, 1526–1534.
21. Abes, S., Turner, J.J., Ivanova, G.D. et al. (2007) Efficient splicing correction by PNA conjugation to an R6-penetratin delivery peptide. *Nucleic Acids Research*, **35**, 4495–4502.
22. Kang, S.H., Cho, M.J., and Kole, R. (1998) Up-regulation of luciferase gene expression with antisense oligonucleotides: implications and applications in functional assay development. *Biochemistry*, **37**, 6235–6239.
23. Shiraishi, T. and Nielsen, P.E. (2009) Cellular bioavailability of peptide nucleic acids (PNAs) conjugated to cell penetrating peptides, in *Delivery Technologies for Biopharmaceuticals, Peptides, Proteins, Nucleic Acids and Vaccines* (eds L. Jørgensen and H. Mørck Nielsen), John Wiley & Sons, Copenhagen, Denmark, Chapter 16, pp. 305–338.
24. Nielsen, P.E., Egholm, M., and Buchardt, O. (1994) Evidence for (PNA)2/DNA triplex structure upon binding of PNA to dsDNA by strand displacement. *Journal of Molecular Recognition*, **7**, 165–170.
25. Nielsen, P.E. and Christensen, L. (1996) Strand displacement binding of a duplex-forming homopurine PNA to a homopyrimidine duplex DNA target. *Journal of the American Chemical Society*, **118**, 2287–2288.
26. Lohse, J., Dahl, O., and Nielsen, P.E. (1999) Double duplex invasion by peptide nucleic acid: a general principle for sequence-specific targeting of double-stranded DNA. *Proceedings of the National Academy of Sciences of the United States of America*, **96**, 11804–11808.
27. Hansen, M.E., Bentin, T., and Nielsen, P.E. (2009) High-affinity triplex targeting of double stranded DNA using chemically modified peptide nucleic acid oligomers. *Nucleic Acids Research*, **37**, 4498–4507.
28. He, G., Rapireddy, S., Bahal, R. et al. (2009) Strand invasion of extended, mixed-sequence B-DNA by GammaPNAs. *Journal of the American Chemical Society*, **131**, 12088–12090.

29. Chenna, V., Rapireddy, S., Sahu, B. et al. (2008) A simple cytosine to G-clamp nucleobase substitution enables chiral G-PNAs to invade mixed-sequence double-helical B-form DNA. *Chembiochem*, **9** (15), 2388–2391.
30. Ishizuka, T., Yoshida, J., Yamamoto, Y. et al. (2008) Chiral introduction of positive charges to PNA for double-duplex invasion to versatile sequences. *Nucleic Acids Research*, **36**, 1464–1471.
31. Knauert, M.P., Lloyd, J.A., Rogers, F.A. et al. (2005) Distance and affinity dependence of triplex-induced recombination. *Biochemistry*, **44**, 3856–3864.
32. Lonkar, P., Kim, K.H., Kuan, J.Y. et al. (2009) Targeted correction of a thalassemia-associated beta-globin mutation induced by pseudo-complementary peptide nucleic acids. *Nucleic Acids Research*, **37**, 3635–3644.
33. Urnov, F.D., Miller, J.C., Lee, Y.L. et al. (2005) Highly efficient endogenous human gene correction using designed zinc-finger nucleases. *Nature*, **435**, 646–651.
34. Miyajima, Y., Ishizuka, T., Yamamoto, Y. et al. (2009) Origin of high fidelity in target-sequence recognition by PNA–Ce(IV)/EDTA combinations as site-selective DNA cutters. *Journal of the American Chemical Society*, **131**, 2657–2662.
35. Katada, H., Chen, H.J., Shigi, N., and Komiyama, M. (2009) Homologous recombination in human cells using artificial restriction DNA cutter. *Chemical Communications*, 6545–6547.
36. Grossmann, T.N. and Seitz, O. (2009) Nucleic acid templated reactions: consequences of probe reactivity and readout strategy for amplified signaling and sequence selectivity. *Chemistry*, **15**, 6723–6730.
37. Röglin, L., Ahmadian, M.R., and Seitz, O. (2007) DNA-controlled reversible switching of peptide conformation and bioactivity. *Angewandte Chemie, International Edition in English*, **46**, 2704–2707.
38. Röglin, L., Altenbrunn, F., and Seitz, O. (2009) DNA and RNA-controlled switching of protein kinase activity. *Chembiochem*, **10**, 758–765.
39. Winssinger, N., Damoiseaux, R., Tully, D.C. et al. (2004) PNA-encoded protease substrate microarrays. *Chemistry & Biology*, **11**, 1351–1360.
40. Urbina, H.D., Debaene, F., Jost, B. et al. (2006) Self-assembled small-molecule microarrays for protease screening and profiling. *Chembiochem*, **7**, 1790–1797.
41. Pouchain, D., Diaz-Mochon, J.J., Bialy, L., and Bradley, M. (2007) A 10 000 member PNA-encoded peptide library for profiling tyrosine kinases. *ACS Chemical Biology*, **2**, 810–818.
43. Ura, Y., Beierle, J.M., Leman, L.J. et al. (2009) Self-assembling sequence-adaptive peptide nucleic acids. *Science*, **325**, 73–77.
42. Howarth, N.M. and Wakelin, L.P.G. (1997) α-PNA: a novel peptide nucleic acid analog of DNA. *Journal of Organic Chemistry*, **62**, 5441–5450.

Part Two
Peptides and Proteins

5

High Solubility of Random-Sequence Proteins Consisting of Five Kinds of Primitive Amino Acids

Nobuhide Doi, Koichi Kakukawa, Yuko Oishi, and Hiroshi Yanagawa

Keio University, Department of Biosciences and Informatics, 3-14-1 Hiyoshi, Kohoku-ku, Yokohama 223-8522, Japan

5.1 Introduction 122
5.2 Materials and Methods 123
    5.2.1 Construction of a DNA Library of Random-Sequence Proteins 123
    5.2.2 Cloning, Expression and Purification of the Random-Sequence Proteins 124
    5.2.3 CD and Fluorescence Measurements 124
    5.2.4 Size-Exclusion Chromatography 125
5.3 Results 125
    5.3.1 Design of DNAs Encoding Random-Sequence Proteins with Primitive Alphabets 125

Chemical Synthetic Biology, First Edition. Edited by Pier Luigi Luisi and Cristiano Chiarabelli.
© 2011 John Wiley & Sons, Ltd. Published 2011 by John Wiley & Sons, Ltd.
Reprinted in full with permission from Oxford University Press. Copyright 2005.

5.3.2 High Solubility of the Random-Sequence VADEG Proteins 126
5.3.3 Structural Characterization of the VADEG Proteins 127
5.4 Discussion 131
5.4.1 Why are the Random-Sequence VADEG Proteins Highly Soluble? 132
5.4.2 Origin and Early Evolution of Proteins through Functional Selection 132
Acknowledgments 134
References 134

Searching for functional proteins among random-sequence libraries is a major challenge of protein engineering; the difficulties include the poor solubility of many random-sequence proteins. A library in which most of the polypeptides are soluble and stable would, therefore, be of great benefit. Although modern proteins consist of 20 amino acids, it has been suggested that early proteins evolved from a reduced alphabet. Here, we have constructed a library of random-sequence proteins consisting of only five amino acids, Ala, Gly, Val, Asp, and Glu, which are believed to have been the most abundant in the prebiotic environment. Expression and characterization of arbitrarily chosen proteins in the library indicated that five-alphabet random-sequence proteins have higher solubility than do 20-alphabet random-sequence proteins with a similar level of hydrophobicity. The results support the reduced-alphabet hypothesis of the primordial genetic code and should also be helpful in constructing optimized protein libraries for evolutionary protein engineering.

5.1 INTRODUCTION

The question of what proportion of random-sequence proteins exhibit folded structure or functional activity in a given sequence space is important for understanding how natural proteins have evolved and how novel proteins can be engineered (for reviews, see Refs [1–4]). Since 1990, several researchers have reported the construction and screening of large libraries of random-sequence proteins and the biophysical characterization of such proteins [5–12]. However, experimental studies using random-sequence protein libraries face the difficulty of identifying novel proteins with native-like structures and desired functions. In the only successful example so far reported, Keefe and Szostak isolated novel ATP-binding proteins from a random-sequence library, but the biophysical characterization of these proteins was difficult owing to poor solubility [12], although the X-ray crystal structure of one of them was recently solved [13]. Generally, random-sequence proteins have a strong tendency to form aggregates [5]. This is unfavorable for functional selection and further improvement of their solubility [14], and folding stability [15] by directed evolution is required. Hence, in the laboratory it would be profitable to start from a library in which most proteins are soluble and stable [16], in order to evolve novel proteins, as would also have been the case in the prebiotic soup.

Although modern proteins consist of 20 amino acids, it has been proposed that the origin and early evolution of protein synthesis involved a reduced alphabet that was gradually extended through co-evolution of the genetic code and the primordial biochemical system for amino acid synthesis [17–19]. If this is so, the properties of random-sequence proteins with a reduced alphabet may be different from those of the 20-alphabet random-sequence proteins previously reported. Davidson and co-workers constructed and characterized random-sequence proteins consisted of only Gln, Leu, and Arg [20, 21]. These QLR proteins showed remarkable helical structures, but their solubility was fairly low. In a computational study using inverse folding techniques, Babajide *et al.* demonstrated that native-like folded structures of proteins tested were maintained with restricted alphabets containing primitive amino acids such as Ala and Gly, but were not maintained with a non-primitive QLR alphabet [22]. In this chapter we describe the first attempt to construct and characterize random-sequence proteins using a restricted set of primitive amino acids.

5.2 MATERIALS AND METHODS

5.2.1 Construction of a DNA Library of Random-Sequence Proteins

The 88-bp DNA [GGTAGATCTGGAAGACTGTGG (GNW)$_{15}$TGGG CGAGACCGCTCGAGGTTC] consisting of 15 consecutive random codons (GNW, in which N = T:C:A:G = 15:30:30:25 and W = T:A = 40:60) was synthesized at FASMAC (Kanagawa, Japan) and amplified by polymerase chain reaction (PCR) using primers GNW-F (5'-GGTAG ATCTGGAAGACTGTGG-3') and GNW-R (5'-GAACCTCGAGCGG TCTCGC-3') with *Ex Taq* DNA polymerase (Takara Shuzo). The PCR product was purified by phenol–chloroform extraction and ethanol precipitation and separated into two equal aliquots that were digested with either *Bbs*I or *Bsa*I. The resulting fragments were purified by 2% agarose gel electrophoresis and Recochip (Takara Shuzo), ligated with T4 DNA ligase (Toyobo) and amplified by PCR. Repeating this procedure three times yielded a final library with eight (=2^3) contiguous random regions.

5.2.2 Cloning, Expression and Purification of the Random-Sequence Proteins

The DNA library was cloned, randomly selected, and sequenced with an ABI PRISM 3100 (Applied Biosystems). The random-sequence region of the eight in-frame genes was digested with *Bgl*II and *Xho*I and then subcloned into a derivative of a pET vector (Novagen) containing the N-terminal T7·tag sequence and the C-terminal His$_6$ tag sequence. *Escherichia coli* BL21-CodonPlus(DE3) cells (Stratagene) transfected with individual recombinant plasmids were grown in LB broth containing 100 μg ml^{-1} ampicillin and 40 μg ml^{-1} chloramphenicol at 37 °C. When the culture achieved an optical density of 0.6–0.7 at 600 nm, isopropylthio-β-D-galactoside was added to a final concentration of 0.1 mM. After an additional 3 h of incubation, the cells were harvested by centrifugation and lysed in a BugBuster (Novagen) containing a protease inhibitor cocktail (Sigma). The centrifuged supernatants were used as the soluble fractions. The pellets were resuspended in a buffer containing 8 M urea and the centrifuged supernatants were used as insoluble fractions. These fractions were analyzed by 16.5% Tricine SDS–PAGE [23]. The proteins were detected with CBB (Coomassie brilliant blue R250) staining and Western blotting with anti-T7·tag antibody. The soluble fractions were loaded on the affinity column of nickel–NTA agarose resin (Qiagen) and the recombinant proteins were eluted with an imidazole gradient. The protein molar concentration was determined from the UV absorption at 280 nm and the molar absorption coefficient was calculated from $\varepsilon = 5690 \, \text{M}^{-1} \text{cm}^{-1}$ for Trp [24].

5.2.3 CD and Fluorescence Measurements

Circular dichroism (CD) spectra of purified proteins in the absence and presence of 1–5 M GdnHCl (guanidine hydrochloride) were measured on a J-820 spectropolarimeter (JASCO) at 25 °C. The protein concentration was 3 μM and the light pathlength used was 1 mm. The results were expressed as mean residue molar ellipticity [θ].

Fluorescence measurements were performed at 25 °C on a Shimadzu RF-1500 spectrofluorometer. The emission spectra of Trp residues of 1 μM proteins were measured at an excitation wavelength of 280 nm and the fluorescence spectra of 50 μM ANS (1-anilinonaphthalene-8-sulfonic acid) (Molecular Probes) in the absence and presence of 1 μM protein were measured with excitation at 371 nm.

5.2.4 Size-Exclusion Chromatography

Gel-filtration experiments on purified proteins were performed using a Shodex KW-803 column (Showa Denko) on a Vision Workstation (Applied Biosystems). The column was calibrated with a low molecular weight gel filtration calibration kit (Amersham Pharmacia Biotech). The Stokes radii of purified proteins and the control proteins (bovine serum albumin (BSA), ovalbumin, chymotrypsinogen, and ribonuclease) were calculated from their elution volumes as described previously [25].

5.3 RESULTS

5.3.1 Design of DNAs Encoding Random-Sequence Proteins with Primitive Alphabets

The most abundant amino acids in the prebiotic environment as inferred from the results of spark-discharge experiments were Ala, Gly, Asp, and Val [26, 27], whereas those deduced from analysis of the Murchison meteorite were Gly, Ala, Glu, and Val [28]. Interestingly, codons for all these amino acids have guanosine (G) at the first nucleotide (Figure 5.1), and thus codons GNC and GNN, where N denotes U, C, A, or G, were proposed to have formed the early genetic code [27, 29]. We chose the five amino acids Ala, Gly, Val, Asp, and Glu as a primitive alphabet.

Using the strategy shown in Figure 5.2, we constructed a DNA library encoding polypeptides of more than 100 residues of random mixtures of the five amino acids. The semi-random codon GNW (N = T, C, A, or G; W = T or A) encodes either Val, Ala, Asp/Glu, and Gly with probabilities determined by the percentages of T, C, A, and G in the base mixture at the second nucleotide position of each randomized codon. The frequency of each amino acid was set to reflect the putative prebiotic abundance of Ala and Gly [26–28]. The GNW codons contained only A and T at the third nucleotide position in order to reduce the total GC content of the DNA sequences to no more than 50%.

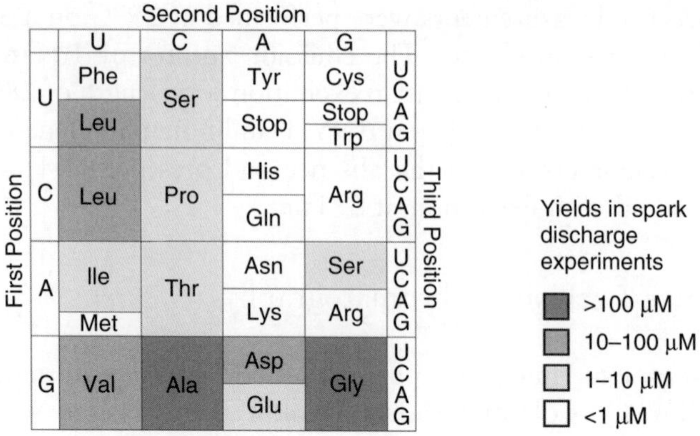

	Second Position				
		U	C	A	G
First Position	U	Phe / Leu	Ser	Tyr / Stop	Cys / Stop / Trp
	C	Leu	Pro	His / Gln	Arg
	A	Ile / Met	Thr	Asn / Lys	Ser / Arg
	G	Val	Ala	Asp / Glu	Gly

Yields in spark discharge experiments:
- ■ >100 μM
- ▨ 10–100 μM
- ▫ 1–10 μM
- □ <1 μM

Figure 5.1 The universal genetic code and the abundance of amino acids in simulated prebiotic synthesis. Typical yields in spark-discharge experiments were: Ala, 790 μM; Gly, 440 μM; Asp, 34 μM; Val, 19.5 μM; Leu, 11.3 μM; Glu, 7.7 μM; Ser, 5.0 μM; Ile, 4.8 μM; Thr, 1.6 μM; Pro, 1.5 μM; and others (data from Ref. [27])

The DNA library was cloned into *E. coli* and the sequences of arbitrarily chosen clones were analyzed. As shown in Figure 5.3, the amino acid sequences were designed to encode an N-terminal T7·tag to allow Western blot analysis, a C-terminal His$_6$ tag to allow affinity purification and several fixed Trp residues at the random cassette junctions to allow fluorescence studies and UV measurements for protein quantitation. Consequently, 92–94% of each polypeptide sequence consisted of random combinations of the five amino acids (Table 5.1). The length of each sequence varied (Figure 5.3), perhaps owing to unexpected recombination in a repetitive GNW region during PCR and/or unexpected deletion in the synthetic DNA cassettes (single base deletions were observed in 17% of the DNA cassette sequences). However, all random cassettes were different from each other, as we expected.

5.3.2 High Solubility of the Random-Sequence VADEG Proteins

As shown in Figure 5.4, we examined the solubility of eight of the random-sequence VADEG proteins and found that all of them were expressed in the soluble fraction. In the case of G4, the protein was also detected in the insoluble fraction. In previous studies, only five of twenty-five 20-alphabet random-sequence proteins [7] and only two of

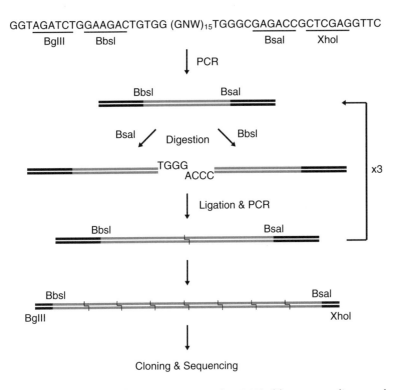

Figure 5.2 The strategy for construction of a DNA library encoding random-sequence proteins of >100 amino acid residues. The synthesized 88 bp DNA (top) consisting of 15 consecutive random codons, GNW, was amplified by PCR, purified and separated into two equal aliquots that were digested with either *Bbs*I or *Bsa*I type IIS restriction endonuclease. The resulting fragments were purified, ligated with T4 DNA ligase, and amplified by PCR. Repeating this procedure three times yielded a final library with eight (=2^3) contiguous random regions (i.e. 120 random codons). The *Bgl*II–*Xho*I sites were used for subcloning of genes. This strategy based on type IIS restriction sites is similar to that described previously [11], although "preselection" was not performed because the GNW repeats contain no stop codon

eleven random-sequence QLR proteins [21] were found to be soluble. The VADEG proteins presented here thus seem to possess remarkably high solubility.

5.3.3 Structural Characterization of the VADEG Proteins

Previous reports indicated that 20-alphabet random-sequence proteins did not show any marked secondary structure [9, 10], whereas the

```
     1         10
     MASMTGGQQMGRSGRLW
       T7·tag

          20        30        40        50        60        70        80
G1   AVVDGGVDWAAVAGGVADDDGVGEWVGADVGAAGDDDGDGWVDAGAAVGEAAVGAAWADVGAD
G2   GEADAGDVAGAGVAAWAAGAGVAGGDAVVAEWEAVGVAGEAAAEGAGWGAVGAVVGEGGDGAG
G3   AEGGVAGAEAGAAAVWGDEDDGAVVGGVVEGWGGGVAAAGAAGAWAAGVEGEVAEEGGVDWAE
G4   AAGAAAEDVWAVVDAVVGEVGVGAAWAVGVGGEVGEAGGGVWEAVAVAAAEGDGAGEWGDVDG
G5   AVAGDDGGAVGAEGVWDVEAVAAAGDAVEWAGAVGAGGGGEEGGEWVAAGAVEDAAADAWEGA
G6   ADEVGDGVAGDGAEVWADAGVVEGVDAGAAEWAVAAADEAVGGGGGAWEVGAEDVGEAAGEVA
G7   DAAEAVDVGGGDWAEGAEAAEGVAGGDGWAEVAGVDVADGGEVWGEVEGDAVEGDAAAGWAGE
G8   AAAGGGGGWAAEADVAGAVEGGEGEWVGGGADAGEADEVGGVEAAVAVAVWVDDAGGAAAGEA

          90        100       110       120       130       140
G1   GAVEAAAEAGWEGAGDGVAGVGGWGGGADAVEGVAAGWVGVVEGAAGGAEVAD---------
G2   WAGAEDGAGADAAAWVDGGGVGAAAGGADAWDAAVAGAADAGDDGAWVEVGDGVAGEVVGVG-
G3   AVAVGAGGDVGGAEWEDDGAAVAAAADEAAWVVAGDAAGGEVDWEVDGDEAAEGVGEAA----
G4   GADGDAAEVVWAEAVVEEAEVEEVADWDAGEAVDAGAGAAAVV--------------------
G5   EAVADVEVAAGGWGGVAGVAGVAAAEGVWVGDGDADGAGAGGGAWEADVAEEAVEVVGAG---
G6   WVVGAADEAVGAVGDDWAGGGVAVADDGEGVDWAGAGAGEGGAVGGAWGGVAADAGVVAEAAE
G7   VEVVAVDAAGGADWGAVGDVDEVVAAVGVWEAVGAADVAAVAGEWAVGGAAVEGGGEVDA---
G8   DDDWGGGDGAGAAAEVDAAWAVAAGEEVAVEVGDE----------------------------

     150
     WARPLEHHHHHH
      6xHis tag
```

Figure 5.3 Predicted amino acid sequences of arbitrarily chosen random-sequence VADEG proteins. The N-terminal T7·tag sequence and the C-terminal His$_6$ tag sequence are also shown

Table 5.1 Properties of the random-sequence proteins

Protein	Composition (%)[a]					Monomer M_r	Solubility[b]	$[\theta]_{222}$[c]	ANS binding[d]
	V	A	D	E	G				
G1	18	27	14	7	27	13 781	S	−2634	1.2
G2	15	32	10	7	28	14 234	S	−783	1.1
G3	15	30	10	14	24	14 526	S	−1107	1.3
G4	21	29	9	14	19	13 308	S and I	−2127	1.5
G5	17	30	9	12	26	14 465	S	−1207	1.1
G6	17	28	11	11	26	14 766	S	−102	1.1
G7	20	26	11	13	22	14 823	S	−744	1.3
G8	14	30	11	13	25	12 244	S	−935	1.3

[a]Does not contain the N- and C-terminal tail sequences.
[b]Determined by CBB staining (Figure 5.4). S, soluble; I, insoluble.
[c]The mean residue ellipticity at 222 nm (see also Figure 5.5).
[d]Ratio of ANS fluorescence in the presence and absence of purified proteins (see also Figure 5.6).

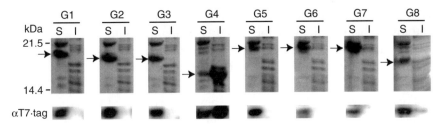

Figure 5.4 Expression and solubility of the random-sequence proteins. The soluble (lanes S) and insoluble (lanes I) fractions of overexpressed proteins were analyzed by 16.5% Tricine SDS–PAGE [23]. The arrows indicate the positions of random-sequence proteins. The proteins were detected by CBB staining (top) and Western blotting with anti-T7·tag antibody (bottom)

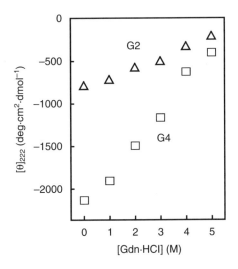

Figure 5.5 The mean residue ellipticity values of VADEG proteins (triangles, G2; squares, G4) at 222 nm as a function of GdnHCl concentration. CD spectra of eight purified proteins at a concentration of 3 μM were measured on a J-820 spectropolarimeter (JASCO). The other six proteins (Table 5.1) revealed a similar level of ellipticity to that of G2 and G4

three-alphabet QLR proteins had remarkably high levels of helical structure (fractional helicity 32–70%) and it was suggested that proteins with a limited range of amino acids have a greater tendency to form secondary structure [21]. However, the five-alphabet VADEG proteins revealed no marked secondary structure as analyzed by means of CD measurements (Figure 5.5 and Table 5.1), in spite of the abundance of Glu and

Ala, which are known to be strong helix formers [30]. One reason for the low helical content of the VADEG proteins would be the presence of the strong helix breaker, Gly, within the polypeptides. The highly structured QLR proteins contain the strong helix former Leu and have no helix breaker in their sequences.

For tertiary structure analysis, the emission spectra of the Trp residues doped in the random-sequence proteins in advance were measured at an excitation wavelength of 280 nm. The emission maxima ranged from 348 to 352 nm in an aqueous buffer, suggesting that almost all the Trp residues are exposed to the solvent [31]. However, the intensity at the emission maximum of the G8 random-sequence protein was decreased by half in buffer containing 5 M urea (data not shown). Hence, some of the Trp side chains of the G8 protein appear to be located in a hydrophobic environment under native conditions and to undergo denaturation in 5 M urea. The presence of hydrophobic clusters in the random-sequence polypeptide was also supported by the results of ANS binding experiments. The fluorescence emission spectrum of ANS is known to be enhanced when the dye binds to hydrophobic regions of proteins [32]. As shown in Figure 5.6, ANS fluorescence increased in the

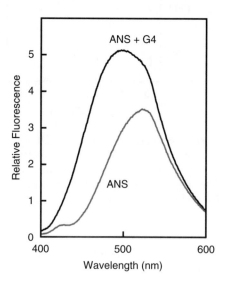

Figure 5.6 Fluorescence spectra of 50 μM ANS (excitation at 371 nm) in the absence (gray line) and presence of 1 μM G4 protein (black line). Fluorescence measurements were performed on a Shimadzu RF-1500 spectrofluorometer. For the other seven proteins, see also Table 5.1

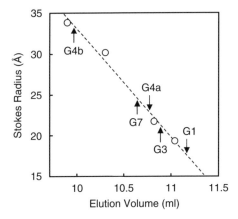

Figure 5.7 Stokes radius plot for VADEG proteins. The elution volume was determined by gel filtration of purified proteins using a Shodex KW-803 column (Showa Denko) on a Vision Workstation (Applied Biosystems). The circles indicate the elution volume of four control proteins (BSA, ovalbumin, chymotrypsinogen, and ribonuclease) plotted against their known Stokes radii and the line shows an empirical equation relating Stokes radius to elution position [25]. The arrows indicate the elution volumes of VADEG proteins. The G4 protein purified from the soluble fraction showed two peaks (G4a and b), whereas the other seven proteins each gave a single peak between the positions of G1 and G7

presence of the G4 random-sequence protein. A relatively small increase was also observed for the G1, G3, G7, and G8 proteins (Table 5.1).

Next, the oligomeric state of the VADEG proteins was analyzed by means of gel filtration experiments. Each VADEG protein, except for the G4 protein, was eluted as a single peak and the Stokes radius calculated from the elution position of each VADEG protein was slightly larger than that deduced from data for monomeric, globular proteins with similar molecular weights (Figure 5.7). This result suggests that the VADEG proteins with high solubility tended to exist as monomers with slightly extended shape, whereas the QLR proteins [21] and the 20-alphabet random-sequence proteins [10] with poor solubility tended to form multimeric structures.

5.4 DISCUSSION

We constructed a library of random-sequence proteins consisting of five amino acids (Ala, Gly, Val, Asp, and Glu) which are believed to have been abundant under prebiotic conditions. We found that the random-sequence VADEG proteins in the library have higher solubility than that of previously reported 20-alphabet random-sequence proteins. Structural analyses of some of our five-alphabet random-sequence proteins indicated the presence of flexible monomeric structures. The high solubility of these random-sequence proteins with a primitive alphabet supports the reduced-alphabet hypothesis of the primordial genetic code and also implies that such protein libraries may be favorable tools for evolutionary protein engineering.

5.4.1 Why are the Random-Sequence VADEG Proteins Highly Soluble?

The hydrophobicity of the VADEG proteins calculated based on the index [33] was in the range −0.2 to 0.2, which is slightly higher (i.e. more hydrophobic) than that of the previous random-sequence proteins with poor solubility: −2.2 to −0.2 for the QLR proteins [21] and −1.0 to −0.2 for the 20-alphabet random-sequence proteins [14]. Therefore, the content of hydrophobic residues of the VADEG proteins cannot explain their high solubility. As suggested by Wilkinson and Harrison, protein solubility may not always correlate with hydrophobicity but it would also be affected by net charge and the fraction of turn-forming residues (Asp, Asn, Pro, Gly, Ser) [34], both of which are relatively high in the VADEG proteins. Since these proteins contain the negatively charged amino acids Asp and Glu, but no positively charged amino acid, they are expected to be highly charged at neutral pH. Further, the abundant Gly has no side chain and tends to favor high solubility. Indeed, the G4 random-sequence protein that was expressed as both soluble and insoluble forms has relatively low contents of Asp and Gly compared with the other seven proteins, which were only expressed in the soluble fraction (Table 5.1).

5.4.2 Origin and Early Evolution of Proteins through Functional Selection

Do the unfolded structures of the VADEG proteins, as well as the 20-alphabet random-sequence proteins [9, 10], mean that these proteins have no function? Although the functions of modern proteins are closely related to their tertiary structures, numerous examples of proteins that are unstructured in solution, but which become structured on binding to the target molecule, have now been recognized [35]. The unfolded structure of a polypeptide has functional advantages, including the ability to bind to several different targets through induced folding [35]. Similarly, it appears likely that the flexible structures of primitive polypeptides would have provided a variety of functions to be optimized during molecular evolution [36].

Keefe and Szostak estimated the frequency of functional proteins in a library of 20-alphabet random-sequence proteins as 1 in 10^{11} by using mRNA display technology [12]. It would be interesting to study whether a library of simple-alphabet random-sequence proteins with higher solubility contains functional proteins in the reduced sequence space. Such studies are in progress by using *in vitro* display technologies developed in our laboratory [37–40]. The acidic residues, Asp and Glu, can bind to metals such as Mg^{2+} and Ca^{2+}, so primitive enzymes with the VADEG alphabet might act as metalloenzymes, like ribozymes in the RNA world. Recent studies *in vitro* [41–43] and *in silico* [44–47] provide insight into the extent to which native protein structure and function can be achieved with reduced alphabets, such as IKEAG [41]. In order to acquire artificial proteins with various functions by laboratory evolution, it might be effective to dope basic amino acids such as Lys and Arg into a set of reduced-alphabet random-sequence libraries.

As an alternative approach to laboratory protein evolution and selection, Hecht *et al.* constructed a "binary code" library by designing a binary pattern of polar and nonpolar amino acids that would favor proteins containing abundant secondary structure [48, 49]. The quality of the library is high; almost all of the proteins in the library are soluble and have a well-ordered structure, although their structural diversity is limited to a predesigned unique fold, such as a four-helix bundle structure. Libraries of modest quality and diversity containing polypeptides with various folds and abundant secondary structure, but no well-ordered structure, can be constructed by randomly combining naturally occurring polypeptide segments such as repetitive peptide motifs [50, 51], secondary structure units [52, 53], and random cDNA fragments [16]. In general, there would be a trade-off between quality and diversity of combinatorial polypeptide libraries. However, a random-sequence library with a reduced alphabet can achieve relatively high quality (e.g. high solubility) with enormous diversity.

ACKNOWLEDGMENTS

We thank Dr Toru Tsuji for helpful discussions. This research was supported in part by the Industrial Technology Research Grant Program in '04 from the NEDO of Japan and by a Grant-in-Aid for Scientific Research and a Special Coordination Fund grant from the MEXT of Japan.

REFERENCES

1. Doi, N. and Yanagawa, H. (1998) Screening of conformationally constrained random polypeptide libraries displayed on a protein scaffold. *Cellular and Molecular Life Sciences*, **54**, 394–404.
2. Kauffman, S. and Ellington, A.D. (1999) Thinking combinatorially. *Current Opinion in Chemical Biology*, **3**, 256–259.
3. Saven, J.G. (2002) Combinatorial protein design. *Current Opinion in Structural Biology*, **12**, 453–458.
4. Watters, A.L. and Baker, D. (2004) Searching for folded proteins *in vitro* and *in silico*. *European Journal of Biochemistry*, **271**, 1615–1622.
5. Mandecki, W. (1990) A method for construction of long randomized open reading frames and polypeptides. *Protein Engineering*, **3**, 221–226.
6. LaBean, T.H., Kauffman, S.A., and Butt, T.R. (1995) Libraries of random-sequence polypeptides produced with high yield as carboxy-terminal fusions with ubiquitin. *Molecular Diversity*, **1**, 29–38.
7. Prijambada, I.D., Yomo, T., Tanaka, F. et al. (1996) Solubility of artificial proteins with random sequences. *FEBS Letters*, **382**, 21–25.
8. Doi, N., Itaya, M., Yomo, T. et al. (1997) Insertion of foreign random sequences of 120 amino acid residues into an active enzyme. *FEBS Letters*, **402**, 177–180.
9. Doi, N., Yomo, T., Itaya, M., and Yanagawa, H. (1998) Characterization of random-sequence proteins displayed on the surface of *Escherichia coli* RNase HI. *FEBS Letters*, **427**, 51–54.
10. Yamauchi, A., Yomo, T., Tanaka, F. et al. (1998) Characterization of soluble artificial proteins with random sequences. *FEBS Letters*, **421**, 147–151.
11. Cho, G., Keefe, A.D., Liu, R. et al. (2000) Constructing high complexity synthetic libraries of long ORFs using *in vitro* selection. *Journal of Molecular Biology*, **297**, 309–319.
12. Keefe, A.D. and Szostak, J.W. (2001) Functional proteins from a random-sequence library. *Nature*, **410**, 715–718.
13. Lo Surdo, P., Walsh, M.A., and Sollazzo, M. (2004) A novel ADP- and zinc-binding fold from function-directed in vitro evolution. *Nature Structural & Molecular Biology*, **11**, 382–383.

14. Ito, Y., Kawama, T., Urabe, I., and Yomo, T. (2004) Evolution of an arbitrary sequence in solubility. *Journal of Molecular Evolution*, **58**, 196–202.
15. Chaput, J.C. and Szostak, J.W. (2004) Evolutionary optimization of a nonbiological ATP binding protein for improved folding stability. *Chemistry & Biology*, **11**, 865–874.
16. Fischer, N., Riechmann, L., and Winter, G. (2004) A native-like artificial protein from antisense DNA. *Protein Engineering, Design and Selection*, **17**, 13–20.
17. Crick, F.H.C. (1968) Origin of genetic code. *Journal of Molecular Biology*, **38**, 367–379.
18. Wong, J.T. (1975) A co-evolution theory of the genetic code. *Proceedings of the National Academy of Sciences of the United States of America*, **72**, 1909–1912.
19. Brooks, D.J., Fresco, J.R., Lesk, A.M., and Singh, M. (2002) Evolution of amino acid frequencies in proteins over deep time: inferred order of introduction of amino acids into the genetic code. *Molecular Biology and Evolution*, **19**, 1645–1655.
20. Davidson, A.R. and Sauer, R.T. (1994) Folded proteins occur frequently in libraries of random amino acid sequences. *Proceedings of the National Academy of Sciences of the United States of America*, **91**, 2146–2150.
21. Davidson, A.R., Lumb, K.J., and Sauer, R.T. (1995) Cooperatively folded proteins in random sequence libraries. *Nature Structural & Molecular Biology*, **2**, 856–864.
22. Babajide, A., Hofacker, I.L., Sippl, M.J., and Stadler, P.F. (1997) Neutral networks in protein space: a computational study based on knowledge-based potentials of mean force. *Folding and Design*, **2**, 261–269.
23. Schägger, H. and von Jagow, G. (1987) Tricine sodium dodecyl-sulfate polyacrylamide-gel electrophoresis for the separation of proteins in the range from 1-kDa to 100-kDa. *Analytical Biochemistry*, **166**, 368–379.
24. Gill, S.C. and von Hippel, P.H. (1989) Calculation of protein extinction coefficients from amino-acid sequence data. *Analytical Biochemistry*, **182**, 319–326.
25. Uversky, V.N. (1993) Use of fast protein size-exclusion liquid chromatography to study the unfolding of proteins which denature through the molten globule. *Biochemistry*, **32**, 13288–13298.
26. Miller, S.L. (1953) A production of amino acids under possible primitive Earth conditions. *Science*, **117**, 528–529.
27. Eigen, M. (1978) Hypercycle – principle of natural self-organization. C. Realistic hypercycle. *Naturwissenschaften*, **65**, 341–369.
28. Kvenvolden, K., Lawless, J., Pering, K. *et al.* (1970) Evidence for extraterrestrial amino-acids and hydrocarbons in the Murchison meteorite. *Nature*, **228**, 923–926.

29. Kuhn, H. and Waser, J. (1994) On the origin of the genetic-code. *FEBS Letters*, **352**, 259–264.
30. Chou, P.Y. and Fasman, G.D. (1978) Empirical predictions of protein conformation. *Annual Review of Biochemistry*, **47**, 251–276.
31. Teale, F.W. (1960) Ultraviolet fluorescence of proteins in neutral solution. *Biochemical Journal*, **76**, 381–388.
32. Stryer, L. (1965) The interaction of a naphthalene dye with apomyoglobin and apohemoglobin. A fluorescent probe of non-polar binding sites. *Journal of Molecular Biology*, **13**, 482–495.
33. Kyte, J. and Doolittle, R.F. (1982) A simple method for displaying the hydropathic character of a protein. *Journal of Molecular Biology*, **157**, 105–132.
34. Wilkinson, D.L. and Harrison, R.G. (1991) Predicting the solubility of recombinant proteins in *Escherichia coli*. *Biotechnology*, **9**, 443–448.
35. Wright, P.E. and Dyson, H.J. (1999) Intrinsically unstructured proteins: re-assessing the protein structure-function paradigm. *Journal of Molecular Biology*, **293**, 321–331.
36. James, L.C. and Tawfik, D.S. (2003) Conformational diversity and protein evolution – a 60-year-old hypothesis revisited. *Trends in Biochemical Sciences*, **28**, 361–368.
37. Nemoto, N., Miyamoto-Sato, E., Husimi, Y., and Yanagawa, H. (1997) In vitro virus: bonding of mRNA bearing puromycin at the 3'-terminal end to the C-terminal end of its encoded protein on the ribosome *in vitro*. *FEBS Letters*, **414**, 405–408.
38. Doi, N. and Yanagawa, H. (1999) STABLE: protein-DNA fusion system for screening of combinatorial protein libraries *in vitro*. *FEBS Letters*, **457**, 227–230.
39. Yonezawa, M., Doi, N., Kawahashi, Y. et al. (2003) DNA display for in vitro selection of diverse peptide libraries. *Nucleic Acids Research*, **31**, e118.
40. Horisawa, K., Tateyama, S., Ishizaka, M. et al. (2004) *In vitro* selection of Jun-associated proteins using mRNA display. *Nucleic Acids Research*, **32**, e169.
41. Riddle, D.S., Santiago, J.V., Bray-Hall, S.T. et al. (1997) Functional rapidly folding proteins from simplified amino acid sequences. *Nature Structural & Molecular Biology*, **4**, 805–809.
42. Silverman, J.A., Balakrishnan, R., and Harbury, P.B. (2001) Reverse engineering the $(\beta/\alpha)_8$ barrel fold. *Proceedings of the National Academy of Sciences of the United States of America*, **98**, 3092–3097.
43. Akanuma, S., Kigawa, T., and Yokoyama, S. (2002) Combinatorial mutagenesis to restrict amino acid usage in an enzyme to a reduced set. *Proceedings of the National Academy of Sciences of the United States of America*, **99**, 13549–13553.
44. Chan, H.S. (1999) Folding alphabets. *Nature Structural & Molecular Biology*, **6**, 994–996.

45. Wang, J. and Wang, W. (1999) A computational approach to simplifying the protein folding alphabet. *Nature Structural & Molecular Biology*, **6**, 1033–1038.
46. Murphy, L.R., Wallqvist, A., and Levy, R.M. (2000) Simplified amino acid alphabets for protein fold recognition and implications for folding. *Protein Engineering*, **13**, 149–152.
47. Fan, K. and Wang, W. (2003) What is the minimum number of letters required to fold a protein? *Journal of Molecular Biology*, **328**, 921–926.
48. Kamtekar, S., Schiffer, J.M., Xiong, H. *et al.* (1993) Protein design by binary patterning of polar and nonpolar amino acids. *Science*, **262**, 1680–1685.
49. Wei, Y. and Hecht, M.H. (2004) Enzyme-like proteins from an unselected library of designed amino acid sequences. *Protein Engineering, Design and Selection*, **17**, 67–75.
50. Shiba, K., Takahashi, Y., and Noda, T. (1997) Creation of libraries with long ORFs by polymerization of a microgene. *Proceedings of the National Academy of Sciences of the United States of America*, **94**, 3805–3810.
51. Shiba, K., Shirai, T., Honma, T., and Noda, T. (2003) Translated products of tandem microgene repeats exhibit diverse properties also seen in natural proteins. *Protein Engineering*, **16**, 57–63.
52. Tsuji, T., Yoshida, K., Satoh, A. *et al.* (1999) Foldability of barnase mutants obtained by permutation of modules or secondary structure units. *Journal of Molecular Biology*, **286**, 1581–1596.
53. Tsuji, T., Onimaru, M., and Yanagawa, H. (2001) Random multi-recombinant PCR for the construction of combinatorial protein libraries. *Nucleic Acids Research*, **29**, e97.

# 6

Experimental Approach for Early Evolution of Protein Function

Hitoshi Toyota[1], Yuuki Hayashi[2], Asao Yamauchi[1], Takuyo Aita[3], and Tetsuya Yomo[2,4,5]

[1]*Osaka University, Department of Biotechnology, Graduate School of Engineering, 2-1 Yamadaoka, Suita, Osaka 565-0871, Japan*
[2]*Osaka University, Department of Bioinformatic Engineering, Graduate School of Information Science and Technology, 2-1 Yamadaoka, Suita, Osaka 565-0871, Japan*
[3]*Saitama University, Department of Functional Materials Science, 255 Shimo-okubo, Saitama 338-8570, Japan*
[4]*Osaka University, Graduate School of Frontier Science, 2-1 Yamadaoka, Suita, Osaka 565-0871, Japan*
[5]*ERATO, JST, 2-1 Yamadaoka, Suita, Osaka 565-0871, Japan*

6.1 Introduction 140
    6.1.1 How does Functional Protein Arise? 140
    6.1.2 Fitness Landscape in Protein-Sequence Space 141
    6.1.3 Early Evolution of Functional Protein 142
6.2 Experimental Evolution Starting from Random Sequences 143
    6.2.1 Frequency of Evolvable Sequences in Protein Sequence Space 143
    6.2.2 Structure of Fitness Landscape in Protein-Sequence Space 147

Chemical Synthetic Biology, First Edition. Edited by Pier Luigi Luisi and Cristiano Chiarabelli.
© 2011 John Wiley & Sons, Ltd. Published 2011 by John Wiley & Sons, Ltd.

6.3 Discussion 149
  6.3.1 Evolution of Protein Features Other than Functionality 149
  6.3.2 Evolution of Natural Protein 150
  6.3.3 Early Evolution of Life 151
  References 151

It is not yet understood how functional proteins arise, and in particular whether the original polypeptide could possess the capability of evolution (evolvability) so as to undergo a kind of Darwinian evolution. The notion of fitness landscape in protein-sequence space is introduced, whereby the distribution of fitness over the sequence space is represented by the concept of fitness landscape. One can define accordingly whether a given sequence is evolvable toward higher fitness. The chapter focuses on the early evolution of protein functions. How frequently and to what extent can a random sequence evolve function? These questions are equivalent to how many positions over the fitness landscape can have a route to higher position and how high the route can continuously rise without being trapped into local maxima.

Through experimental evolution started from random sequences, we examined unseen regions of protein-sequence space. We prepared random sequences of a definite length, about 140 residues composed of the 20 naturally occurring amino acids. To get the least diversity required for early evolution of protein function, we arbitrarily chose less than 10 of the synthesized random sequences as the initial material for some independent experimental evolutions based on different functions, esterase activity, DNA binding, and phage infectivity. The three experimental evolutions were successful, indicating that a substantial fraction in all possible sequences is evolvable. The analysis of the experimental evolution with various diversity sizes indicated that up to 40% in the altitude from the bottom in the fitness landscape is so smooth that the evolution is hardly trapped in the local maxima, whereas over 50% the landscape rapidly increases the ruggedness, requiring a huge diversity for further evolution. Based on the experimental data, together with other information, we terminate the chapter with an overall discussion on the early evolution of protein function and its implication on the origin of life.

6.1 INTRODUCTION

6.1.1 How does Functional Protein Arise?

Natural proteins have evolved by iterative processes of sequence mutation and natural selection over long periods. The resulting modern proteins possess highly efficient biological functions, such as molecular recognition or catalysis. It is well known that a novel functional protein can be created by utilizing the sequence of a preexisting functional protein as a scaffold through molecular mechanisms such as gene duplication or recombination [1, 2]. On the other hand, it remains unclear how the first functional protein could arise without relying on a preexisting one. A random polypeptide might have synthesized [3, 4], or small functional polypeptides might have assembled into a large polypeptide [5–7], and then have evolved in some way or other. In any case, as such a rudimentary polypeptide probably had low functional activity at the beginning, it must have possessed potentiality for Darwinian evolution – that is, evolvability toward higher functional activity by the processes of mutation and selection – to become an efficient functional protein. It is not known, however, what protein sequences possess such evolvability.

6.1.2 Fitness Landscape in Protein-Sequence Space

The evolvability of protein depends on the structure of a fitness landscape [8] in protein-sequence space [9] (Figure 6.1). The protein-sequence space is a high-dimensional space, in which points represent each of all

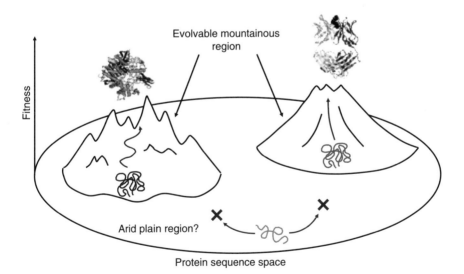

Figure 6.1 An imaginary schematic fitness landscape in protein sequence space. Evolutionary processes (arrows) by mutation and selection can be regarded as an ascending trajectory on the surface of the fitness landscape. The trajectory, that is evolutional dynamics, is affected by the structure of the fitness landscape in protein sequence space

possible sequences and those differing by one residue substitution are arranged neighboring each other. For example, the sequence space of a 100-residue protein, representative of small natural proteins, consists of 20^{100} (i.e. about 10^{130}, far beyond astronomical numbers) sequences. Each sequence has a fitness for an evolutionary process (for example, the functional activity for a functional selection) under certain conditions. The distribution of a fitness over the sequence space is represented by the concept of fitness landscape, where the sequence space is mapped on a Cartesian plane and the height in the vertical direction to the plane at a point corresponds to the fitness value of a sequence located at the point. An evolutionary process of protein by mutation and selection is observed as a trajectory of a point on the surface of a fitness landscape. Hence, a sequence is evolvable toward higher fitness when an ascending slope extending from the sequence can be founded within the scope of mutational steps in the landscape. The structures of the fitness landscape around the vicinity of several natural protein sequences have been examined and the evolvabilities of the sequences have been revealed [10–16]. However, since those sequences cover only tiny fractions of the immense sequence space, landscape structures of almost all regions remain unknown.

6.1.3 Early Evolution of Functional Protein

To reveal the early evolution of protein, we consider the following two questions.

First, what is the frequency of sequences evolvable toward higher functionality in protein sequence space? The frequency of evolvable sequences in the sequence space determines the number of sequences which should be explored for starting evolution of the first functional protein. If only the natural protein sequences and their vicinity (probably about 10^{12} sequences [17]) constitute evolvable mountainous regions and the rest are arid plain regions in a fitness landscape, starting evolution of the first protein would be improbable (e.g. 1 in 10^{118}, assuming that all the proteins are 100 residues in length). Hence, sufficient numbers of functional sequences and evolvable sequences toward them will exist other than naturally occurring ones among all possible sequences. Indeed, Keefe and Szostak selected ATP-binding proteins from a large library of 6×10^{12} random sequences sampled uniformly from the sequence space of about 80-residue protein, and estimated the frequency of the sequences with a certain level of efficient funcional activity to be roughly 1 in 10^{11} [18]. This result suggests that mountainous regions of a fitness landscape certainly exist in unseen regions of protein-sequence space. Here, we examine the frequency of evolvable sequences in unseen regions of protein-sequence space by experimental evolution starting from random sequences. For clarity, we chose a definite small number of sequences and explore around them by cycles of random point mutation and functional selection. The results of experimental evolution suggest that any arbitrary sequence would be easily evolvable for any requirements of (at least simple) functionality.

Second, how and to what extent does an evolvable protein sequence evolve? The evolutionary dynamics of protein depend on the ruggedness of the fitness landscape. If a landscape is smooth, protein evolution proceeds steadily toward higher fitness with sufficient search. On the other hand, if a landscape is rugged, evolution may eventually stop at a local optimum, where all neighbors within the scope of a mutational step have lower fitness than the optimum. Here, we estimated the ruggedness of a fitness landscape by analyzing the results of experimental evolution above. The results of the analysis suggest that the structure of the fitness landscape varies from smooth to rugged as the fitness rises.

6.2 EXPERIMENTAL EVOLUTION STARTING FROM RANDOM SEQUENCES

To elucidate the evolvability of proteins in unseen regions of protein-sequence space, we set out to investigate arbitrarily sampled points in the space (that is, random proteins) and explore around them by experimental evolution using the cycles of random point mutation and functional selection for several common biological functions. For simplicity, the random proteins of a definite length, about 140 residues, composed of the 20 naturally occurring amino acids were used [19]. The proteins consist of a variable part of about 120 residues between N- and C-terminal fixed sequence tags of about 10 residues each. Hence, the size of the sequence space, which we consider, is 20^{120}, or about 10^{156}.

6.2.1 Frequency of Evolvable Sequences in Protein Sequence Space

Yamauchi *et al.* carried out experimental evolution by functional selection based on affinity to a transition state analog (TSA) for an esterase

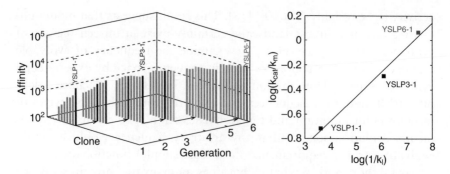

Figure 6.2 (A) The affinity is expressed as the number of eluted phage bound to the TSA. The longest black bar in each generation indicates the clone with highest affinity in the daughter clones to be the parent clone for the next generation. The black bars pointed by arrows indicate the affinity ranking of the parent clone among its daughter clones in next generation. (B) Plot of TSA $1/K_I$ values versus k_{cat}/K_m for YSLP1-1, YSLP3-1 and YSLP6-1 proteins, derived from the selected phage clones at 1st, 3rd and 6th generations (Reprinted with permission from [20]. Copyright 2002 Oxford University Press)

activity using a phage display system, and demonstrated that a random protein is evolvable within a small library [20]. First, 10 arbitrary sequences were chosen from the sequence space and the TSA-binding activity of the each protein was examined. Notably, experimentally detectable differences of the activity were observed among the random proteins (Figure 6.2). Next, the sequence of the protein exhibiting the highest activity among the 10 sequences was selected and then mutated to produce 10 descendant sequences for the next generation. The mutant proteins also showed differences of the activity even though only two or three amino acid residues were changed from the parent sequence (Figure 6.2). For the generations by the same processes of mutation and selection, the TSA-binding activity of the selected proteins was gradually increased (Figure 6.2). The esterase activity of the selected proteins was also increased along with the generations in accordance with the principle of transition-state stabilization [21] (Figure 6.2). From the results of the experiment that exploring such a small number of sequences was sufficient for evolution of a functional protein, it is conceivable that any arbitrary sequence might be evolvable. This conjecture is verified by the following experiments.

Hayashi *et al.* carried out another experimental evolution by functional selection for infectivity of bacteriophage fd to *Escherichia coli* starting from only one arbitrary sequence of a random protein and demonstrated that the one arbitrary sequence is evolvable [22]. The

EXPERIMENTAL APPROACH FOR EARLY EVOLUTION 145

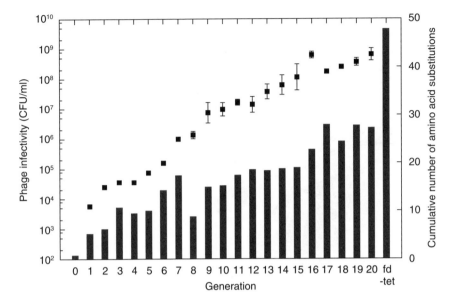

Figure 6.3 Time series of phage infectivity and cumulative number of the substituted amino acids in *in vitro* evolution experiment. Each bar represents the infectivity of the selected phage in each generation. Each filled square represents the average number of substituted amino acids of the selected clones from the initial sequence in each generation (Reprinted with permission from [25]. The Public Library of Science (PLoS))

infectivity of the phage, in which the functional domain of the g3p coat protein mediating infection to *E. coli* was replaced with a random protein, was measured by the number of infected *E. coli* cells by the phage. For the generations by the basically similar processes of mutation and selection with the above experiment for esterase activity, the infectivity of the phage carrying the selected proteins was gradually increased (Figure 6.3). The following experiment demonstrated that the success of the evolution was not by coincidence of a fortuitous choice of the phage infectivity selection for the starting sequence.

Toyota *et al.* carried out experimental evolution by functional selection for DNA-binding activity starting from the same sequence of a random protein used as the starting sequence in the above experiment for phage infectivity using a phage display system and demonstrated that the evolvable sequence for a function of phage infectivity was also evolvable for another distinct function of DNA-binding activity [23]. For the generations by the basically similar processes of mutation and selection with the above experiments, the DNA-binding activity was gradually increased (Figure 6.4). From these results – that the one

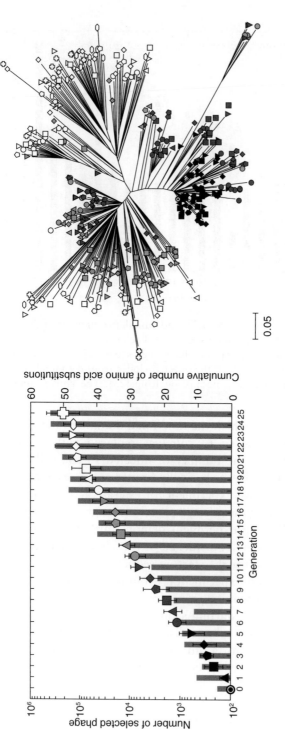

Figure 6.4 (A) Number of the selected phage bound to the immobilized DNA and cumulative number of amino acid substitutions in *in vitro* evolution experiment. Each bar and symbol represents the number of the selected phage and the cumulative average number of substituted amino acids in the selected sequence from the initial sequence in each generation, respectively. (B) Phylogenetic tree constructed by the Neighbor-Joining method using the distances among all the deduced amino acid sequences determined. The symbols indicates the protein sequences of each generation in the figure (A). The details are shown in his original paper [23] (Reprinted with permission from [23]. Copyright 2008 Oxford University Press)

arbitrary sequence of a random protein was able to evolve for the two distinct functions of phage infectivity and DNA-binding activity – since the functions were also taken arbitrarily, it is reasonable to assume that the one sequence would be evolvable for any (at least simple) functions rather than that the sequence is specific to the two functions.

Taken together, these results of experimental evolution suggest that any sequence of protein would be easily evolvable for any requirements of (at least simple) functionality. This means that evolvable mountainous regions of a fitness landscape for any function would exist densely over protein-sequence space. This inference is further supported by sequence analysis of the evolving proteins in experimental evolution. Toyota *et al.* showed by phylogenetic analysis of the evolving sequences obtained from experimental evolution for DNA-binding activity that multiple paths exist extending from one sequence for functional protein evolution [23] (Figure 6.4).

6.2.2 Structure of Fitness Landscape in Protein-Sequence Space

It is suggested from experimental evolution so far that evolvable mountainous regions would be abundant in a fitness landscape for a protein function. What is the structure of a fitness landscape? The structure of a fitness landscape affects the dynamics of protein evolution. If a landscape is smooth, evolution proceeds steadily toward higher fitness with sufficient search. On the other hand, if a landscape is rugged, evolution may eventually stop at a local optimum, where all neighbors within the scope of a mutational step have lower fitness than the optimum. The ruggedness of a fitness landscape is determined by the degree of nonadditivity, or nonindependence, effects of amino acid substitutions on the fitness [24]. If the effects of substitutions are additive, since positive effects can be accumulated, smooth ascending slopes toward higher fitness always exist. On the other hand, if the effects of substitutions are nonadditive, since accumulated positive effects by then can be affected by subsequent substitutions, local optima occur when further accumulation of positive effects by substitutions becomes difficult.

Hayashi *et al.* estimated the ruggedness of a fitness landscape by the analysis of the results of experimental evolution [25, 26]. The details of the analysis are described in the original paper. Briefly, the ruggedness was estimated by the changes of functional activity of evolving proteins along with generations. The results of the analysis indicated that one

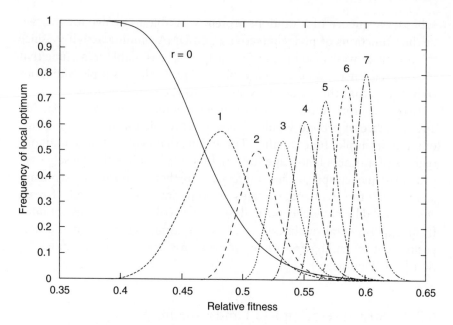

Figure 6.5 Emergence of local optima on the fitness landscape. Each Gaussian-like curve represents the fraction of local optima in the r-th order (Reprinted with permission from [25]. The Public Library of Science (PLoS))

residue substitution affects about 20% of the other residues of a protein on average. The frequency of local optima at a given fitness was calculated with this value representing the degree of nonadditivity effects, as shown in Figure 6.5. Here, relative fitnesses of zero and one correspond to the mean and the highest fitness among all possible sequences respectively; hence, it turns out that the starting sequence of an arbitrarily chosen random protein has a relative fitness of around zero. A local optimum of the rth order means the optimum where all neighbors within r substitutions have a lower fitness than the optimum and at least one neighbor with $r + 1$ substitutions have higher fitness than or equal fitness to the optimum, so the r represents a basin size of the optimum. The landscape is smooth up to a relative fitness of 0.4 and becomes more rugged with local optima having larger basin size as the fitness becomes higher. For example, during the 8th to the 13th generations under the conditions of 10^3 sequence diversity in a mutant library, the fitness was rapidly increased and reached a steady state, although only about 30% residues of the evolved proteins are substituted from the starting random protein (Figure 6.3). This is due to the nonaddtivity effects of substitutions, or the ruggedness of the fitness landscape.

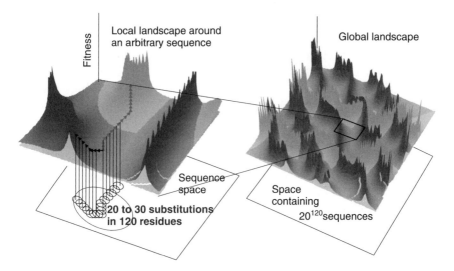

Figure 6.6 A schematic fitness landscape based on the results of our evolution experiments. The start and end points of each arrow represents the parent sequence and the selected sequence in its daughter clones, respectively. Each small circle indicates the library consisting of its daughter clones. The two series of arrows represent divergent evolution trajectories from a single common sequence

6.3 DISCUSSION

The studies of experimental evolution starting from arbitrarily chosen random proteins suggest that any sequence of protein would be easily evolvable for any requirements of (at least simple) functionality. This means that evolvable mountainous regions of a fitness landscape for any function would exist densely over protein-sequence space (Figure 6.6). The structure of the fitness landscape is estimated to be smooth up to middle regions of relative fitness (between the mean and the highest fitness among all possible sequences) and to become more rugged with local optima having larger basin size as the fitness becomes higher. To evolve toward even higher fitnesses over local optima, an evolutionary strategy, such as high mutation rate or recombination, other than used in the experiments would be needed.

6.3.1 Evolution of Protein Features Other than Functionality

Characteristic features other than functionality, such as solubility, foldability into regular structures, or stability, are also often observed in natural proteins. How do these features arise? These features of a protein affect its own functionality and, therefore, may be subjected to evolution. About 80% of random sequences of protein composed of the 20 naturally occurring amino acids are expected to be insoluble [19]. Ito *et al.* demonstrated, however, that an insoluble random protein can evolve toward a soluble protein by processes of mutation and selection [27]. Average random proteins would not form regular structures like natural proteins [28]. On the other hand, an ATP-binding protein selected from a random-sequence library by Keefe and Szostak [18] has a regular structure, treble clef finger, commonly occurring in natural proteins [29, 30]. This indicates that there evidently exist sequences forming regular structures other than natural sequences in protein-sequence space. A theoretical proposition using toy models indicates a possibility that protein structures can evolve gradually through functional selection [31–33]. This proposition is supported by the results of experimental evolution described above. Toyota *et al.* [23] demonstrated that polyproline II-like structure, which is not necessarily composed of proline residues and commonly observed in natural proteins [34, 35], emerged in evolved proteins obtained in experimental evolution starting from random proteins with unordered structures.

6.3.2 Evolution of Natural Protein

To what extent of fitness does natural protein have evolvability? There is substantial evidence that natural proteins allow further optimization of their native functionalities by directed evolution in the laboratory [36]. Hence, it is clear that natural proteins are not completely optimized for their native functionality. It is likely that natural proteins may have evolved under constraints of multiple selection criteria. As an example, Arnold *et al.* demonstrated that thermostability and functionality of natural proteins are not inversely correlated; that is, it is possible to optimize each different property independently by evolution [37]. In actual processes of evolution in a biological context, cell population dynamics plays an important role. In experimental evolution of a cell population, cells carrying mutated genes of one vital enzyme, glutamine synthetase, were cultured in a chemostat [38]. Under low population density conditions, only cells carrying mutant enzymes with higher activity survived. In contrast, under high population density conditions, cells carrying mutant enzymes with lower activity and those with higher activity coexisted due to cellular interactions via the environment in which the enzyme products were leaked from cells. This result suggests a case that protein functionality is not necessarily highly optimized.

6.3.3 Early Evolution of Life

On considering the problems of the origin of life, it is remarkable that the early evolution of functional protein could start even within only a small fraction of sequence diversity of random protein and could proceed relatively easily up to a certain level of functionality. It is estimated that the minimal number of functional proteins required to form a primitive living cell is about 10^2–10^3 [39–41]. Suppose roughly that 100 of the functional proteins have mutually unrelated sequences and originate independently. It is estimated that the minimal number of functional proteins required to form a primitive living cell is about 10^2–10^3 [39–41]. Suppose roughly that 100 of the functional proteins have mutually unrelated sequences and originate independently. If the frequency of evolvable sequences for each functional protein in protein sequence space is as low as 1 in 10^{10}, since the probability to capsule 100 evolvable sequences simultaneously in a cell is $(1/10^{10})^{100}$, it is unlikely to emerge a primitive living cell. In contrast, if any sequence of protein possesses evolvability for any functionality as shown in this article, then a primitive living cell might have emerged with a set of functional proteins having low activities and then evolved even under inefficient and slow dynamics. After all, protein would have come to be used in modern living organisms due to its flexible nature on functional evolution.

REFERENCES

1. Ohno, S. (1970) *Evolution by Gene Duplication*, Springer-Verlag.
2. Nei, M. (1975) *Molecular Population Genetics and Evolution*, North-Holland.
3. Paecht-Horowitz, M., Berger, J., and Katchalsky, A. (1970) Prebiotic synthesis of polypeptides by heterogeneous polycondensation of amino-acid adenylates. *Nature*, **228** (5272), 636–639.
4. Ferris, J.P., Hill, A.R., Jr, Liu, R., and Orgel, L.E. (1996) Synthesis of long prebiotic oligomers on mineral surfaces. *Nature*, **381** (6577), 59–61.
5. Doolittle, W.F. (1978) Genes in pieces: were they ever together? *Nature*, **272** (5654), 581–582.
6. Blake, C.C. (1978) Do genes-in-pieces imply proteins-in-pieces? *Nature*, **273** (5660), 267.

7. Gilbert, W. (1987) The exon theory of genes. *Cold Spring Harbor Symposia on Quantitative Biology*, **52**, 901–905.
8. Wright, S. (1931) Evolution in mendelian populations. *Genetics*, **16** (2), 97–159.
9. Smith, J.M. (1970) Natural selection and the concept of a protein space. *Nature*, **225** (5232), 563–564.
10. Kauffman, S.A. and E.D. Weinberger (1989) The NK model of rugged fitness landscapes and its application to maturation of the immune response. *Journal of Theoretical Biology*, **141** (2), 211–245.
11. Wells, J.A. (1990) Additivity of mutational effects in proteins. *Biochemistry*, **29** (37), 8509–8517.
12. Wagner, C.R., Huang, Z., Singleton, S.F., and Benkovic, S.J. (1995) Molecular basis for nonadditive mutational effects in *Escherichia coli* dihydrofolate reductase. *Biochemistry*, **34** (48), 15671–15680.
13. Matsuura, T., Yomo, T., Trakulnaleamsai, S. et al. (1998) Nonadditivity of mutational effects on the properties of catalase I and its application to efficient directed evolution. *Protein Engineering*, **11** (9), 789–795.
14. Lunzer, M., Miller, S.P., Felsheim, R., and Dean, A.M. (2005) The biochemical architecture of an ancient adaptive landscape. *Science*, **310** (5747), 499–501.
15. Weinreich, D.M., Delaney, N.F., Depristo, M.A., and Hartl, D.L. (2006) Darwinian evolution can follow only very few mutational paths to fitter proteins. *Science*, **312** (5770), 111–114.
16. Amitai, G., Gupta, R.D., and Tawfik, D.S. (2007) Latent evolutionary potentials under the neutral mutational drift of an enzyme. *HFSP Journal*, **1** (1), 67–78.
17. Choi, I.G. and Kim, S.H. (2006) Evolution of protein structural classes and protein sequence families. *Proceedings of the National Academy of Sciences of the United States of America*, **103** (38), 14056–14061.
18. Keefe, A.D. and Szostak, J.W. (2001) Functional proteins from a random-sequence library. *Nature*, **410** (6829), 715–718.
19. Prijambada, I.D., Yomo, T., Tanaka, F. et al. (1996) Solubility of artificial proteins with random sequences. *FEBS Letters*, **382** (1-2), 21–25.
20. Yamauchi, A., Nakashima, T., Tokuriki, N. et al. (2002) Evolvability of random polypeptides through functional selection within a small library. *Protein Engineering*, **15** (7), 619–626.
21. Pauling, L. (1946) Molecular architecture and biological reactions. *Chemical Engineering News*, **24** (10), 1375–1377.
22. Hayashi, Y., Sakata, H., Makino, Y. et al. (2003) Can an arbitrary sequence evolve towards acquiring a biological function? *Journal of Molecular Evolution*, **56** (2), 162–168.
23. Toyota, H., Hosokawa, M., Urabe, I., and Yomo, T. (2008) Emergence of polyproline II-like structure at early stages of experimental evolution from random polypeptides. *Molecular Biology and Evolution*, **25** (6), 1113–1119.
24. Kauffman, S.A. (1993) *The Origin of Order*, Oxford University Press.

25. Hayashi, Y., Aita, T., Toyota, H. et al. (2006) Experimental rugged fitness landscape in protein sequence space. *PLoS ONE*, **1**, e96.
26. Aita, T., Hayashi, Y., Toyota, H. et al. (2007) Extracting characteristic properties of fitness landscape from in vitro molecular evolution: a case study on infectivity of FD phage to *E. coli. Journal of Theoretical Biology*, **246**, 538–550.
27. Ito, Y., Kawama, T., Urabe, I., and Yomo, T. (2004) Evolution of an arbitrary sequence in solubility. *Journal of Molecular Evolution*, **58** (2), 196–202.
28. Yamauchi, A., Yomo, T., Tanaka, F. et al. (1998) Characterization of soluble artificial proteins with random sequences. *FEBS Letters*, **421** (2), 147–151.
29. Surdo, P.L., Walsh, M.A., and Sollazzo, M. (2004) A novel ADP- and zinc-binding fold from function-directed *in vitro* evolution. *Nature Structural & Molecular Biology*, **11** (4), 382–383.
30. Krishna, S.S. and Grishin, N.V. (2004) Structurally analogous proteins do exist! *Structure*, **12** (7), 1125–1127.
31. Saito, S., Sasai, M., and Yomo, T. (1997) Evolution of the folding ability of proteins through functional selection. *Proceedings of the National Academy of Sciences of the United States of America*, **94** (21), 11324–11328.
32. Yomo, T., Saito, S., and Sasai, M. (1999) Gradual development of protein-like global structures through functional selection. *Nature Structural Biology*, **6** (8), 743–746.
33. Nagao, C., Terada, T.P., Yomo, T., and Sasai, M. (2005) Correlation between evolutionary structural development and protein folding. *Proceedings of the National Academy of Sciences of the United States of America*, **102** (52), 18950–18955.
34. Adzhubei, A.A. and Sternberg, M.J. (1993) Left-handed polyproline II helices commonly occur in globular proteins. *Journal of Molecular Biology*, **229** (2), 472–493.
35. Bochicchio, B. and Tamburro, A.M. (2002) Polyproline II structure in proteins: identification by chiroptical spectroscopies, stability, and functions. *Chirality*, **14** (10), 782–792.
36. Kaur, J. and Sharma, R. (2006) Directed evolution: an approach to engineer enzymes. *Critical Reviews in Biotechnology*, **26** (3), 165–199.
37. Arnold, F.H., Wintrode, P.L., Miyazaki, K., and Gershenson, A. (2001) How enzymes adapt: lessons from directed evolution. *Trends in Biochemical Sciences*, **26** (2), 100–106.
38. Kashiwagi, A., Noumachi, W., Katsuno, M. et al. (2001) Plasticity of fitness and diversification process during an experimental molecular evolution. *Journal of Molecular Evolution*, **52** (6), 502–509.
39. Deamer, D. (2005) A giant step towards artificial life? *Trends in Biotechnology*, **23** (7), 336–338.
40. Luisi, P.L., Ferri, F., and Stano, P. (2006) Approaches to semi-synthetic minimal cells: a review. *Naturwissenschaften*, **93** (1), 1–13.
41. Forster, A.C. and Church, G.M. (2006) Towards synthesis of a minimal cell. *Molecular Systems Biology*, **2**, 45.

7
Searching for *de novo* Totally Random Amino Acid Sequences

Cristiano Chiarabelli, Cecilia Portela Pallares, and Anna Quintarelli
University of Roma 3, Biology Department, Viale G. Marconi 446, 00146 Roma, Italy

7.1 Introduction 156
7.2 Strategies for Peptide Libraries Production 157
7.3 Combinatorial Chemistry 159
   7.3.1 Preparation of the Library 160
   7.3.2 Screening of Library Components 161
   7.3.3 Identification of Active Compounds 162
7.4 A Model for Chemical Evolution of Macromolecular Sequences 163
7.5 Never-Born Proteins Project 167
   References 173

Random protein space has been explored several times to isolate new functional proteins. Generally, such investigations, defined as directed molecular evolution, have been carried out starting from selected extant protein scaffolds and randomizing either restricted regions or different parts of the entire sequence in a work of protein redesign. Alternatively, using recombination techniques, protein

---

Chemical Synthetic Biology, First Edition. Edited by Pier Luigi Luisi and Cristiano Chiarabelli.
© 2011 John Wiley & Sons, Ltd. Published 2011 by John Wiley & Sons, Ltd.

fragments have been mixed to obtain novel combinations. We can define these approaches as "directed randomizations," in the sense that randomization is performed in order to achieve certain desired properties.

7.1 INTRODUCTION

The rationale of a "total randomization" approach is completely different, as there is no bias towards any given structural or functional property and it leads almost necessarily to novel proteins that are not present in nature.

In this chapter we present different approaches, chemical and biological, to create novel protein sequences in a completely random fashion. The final aim, in the field of chemical synthetic biology, is to obtain large random libraries of chemical structures alternative to those present in nature.

The sequencing of several organism genomes and recent progress in the field of bioinformatics and computational biology have laid the foundations for synthetic biology, a new discipline at the interface between biology and engineering, which aims to "(i) design and construct new biological parts, devices, and systems and (ii) re-design existing, natural biological systems for useful purposes" (http://www.syntheticbiology.org).

Synthetic biology will certainly have an impact on basic knowledge, in that the implementation of existing genes, proteins, and metabolic pathways in non-native settings will help to shed light on their function and dynamic behavior. Within this framework, synthetic biology will fill the gap between description and understanding of biological systems, clarifying fundamental principles of biological organization.

On the other hand, synthetic biology will revolutionize technology and production paradigms in the twenty-first century and foster the development of new technological tools to produce innovative medicines (red biotechnology), generate new sources of energy and chemical processes (white biotechnology), and in prevention and opening new gateways in environmental risk research (green biotechnology).

To date, research in the field of synthetic biology has mainly focused on engineering extant life forms in order to introduce novel, desirable tracts. In fact, over the past few years, synthetic biologists have generated remarkable systems. These remarkable achievements have largely been "one-offs," since each one is a special case, and though they must be regarded as milestones in the respective field, they do not provide a comprehensive and coherent engineering roadmap for the next step [1].

Indeed, the rationality behind this approach assumes that single enzymes or metabolic pathways can be "cut and pasted" among different organisms, meanwhile retaining their functionality. Traditional synthetic biology approaches very often tend to rely on a strong top-down bioengineering programming approach which does not account for emergent behavior that is, by definition, not derivable from the single components of the system.

An alternative concept in synthetic biology is the so-called chemical synthetic biology, which is the term to define that part of the field that, instead of assuming an engineering approach based on genome manipulation, is oriented towards the synthesis of chemical structures alternative to those present in nature, such as proteins, nucleic acids, vesicular forms, and others. The idea of investigating biological structures alternative to extant ones is not new; actually, it was pursued even before the advent of synthetic biology, and a lot of examples are reported in this book.

A very interesting field is the one related to protein libraries; in particular, random proteins. The synthesis of libraries of peptides with random amino acid sequences has become a relatively widely exploited area of research. The approaches to create such libraries are chemical or biological.

7.2 STRATEGIES FOR PEPTIDE LIBRARIES PRODUCTION

Two global strategies for constructing new peptide sequences are random and rational design. In the former approach, peptide sequences are implemented without restrictions, while in the latter strategy they are designed, residue by residue, to yield a sequence with a desired structure. Although rational design has shown great promise in creating novel proteins and can prove successful for designing proteins with desired structures, it does not typically explore extensive regions of sequence space. In the random approach the peptides obtained can be selected from a large population; for example, a combinatorial library not relying on prior structural information. Therefore, new bioactive peptides can be obtained by generating libraries containing the highest possible number of different sequences and then fishing for the desired function, or by rational design of an appropriate structure, hopefully with the desired function. These two opposite approaches can be combined at different steps of the selection procedure and the relative contributions of rational and irrational selection depend on available knowledge about protein and peptide structure-function and may change at different steps of the process [2].

The combinatorial peptide library approach is mainly based on two methods: phage display and synthetic peptide libraries. Both strategies have been proven powerful tools having specific benefits and limitations. Libraries of peptides displayed on biological surfaces, such as bacteriophage particles, have become a widely studied laboratory procedure for the identification of specific ligand molecules in research and drug discovery. The technology is based on an *in vitro* selection procedure in which a peptide gene is genetically fused to a bacteriophage coat protein, resulting in display of the peptide on the surface of the viral particle. From the mid 1980s, when the first phage peptide library was constructed, random peptide libraries displayed on phage have been used for a variety of biological applications, including ligands to target receptors, specific ligands for DNA sequences, enzyme inhibitors, peptides that mimic carbohydrate structures, protein–protein interfaces, and receptor binding sites. Phage-display libraries have led to selection of peptides against cancer cells and against cancer-associated proteins [2].

The great advantage of phage display libraries is the very large number of different peptide sequences (up to 10^{10}) that can be displayed on viral surfaces. Identification of a specific peptide sequence from a selected phage is possible because phage display libraries enable pairing of phenotype with genotype. But while the generation, screening, and deconvolution through affinity selection of highly complex phage display libraries is straightforward and well established, as the peptides are presented multivalently and covalently attached to the phage, this may result in an increase of their apparent affinity to the target molecule. When tested individually (that is, monomers not attached to the phage), the affinity of the peptides may be lower [3]. Of course, only L-amino acids can be encoded and the construction of phage peptide libraries containing non-natural amino acids is not possible.

Compared with recombinant peptide and protein libraries, the generation of synthetic peptide libraries tends to be more laborious and the selection of individual peptides typically requires rather elaborate deconvolution strategies. This drawback, however, is offset by the far higher chemical diversity that can be presented by synthetic peptide libraries, which is achieved by incorporating nonproteinogenic amino acids as well as other types of building blocks. Since these nonproteinogenic amino acids are not recognized by common proteases, chemically modified peptides have a higher propensity for metabolic stability and, consequently, bioavailability [3].

7.3 COMBINATORIAL CHEMISTRY

Combinatorial chemistry began in the mid 1980s and has complemented the classical approaches of finding new drugs based on rational drug design and on screening of natural extracts. It provides a way to explore chemical space through the preparation of large collections, or libraries, of different compounds. Combinatorial chemistry was first applied to prepare libraries of oligomers, mainly peptides and oligonucleotides, although it was rapidly adapted to make libraries of small organic molecules of potential pharmaceutical interest. It is now used to create chemical diversity in the synthesis of new exploratory compounds, as well as to optimize hits by making targeted libraries for structure–activity relationship studies. Recent advances in virtual combinatorial approaches, which permit the *in-silico* design, construction, and even screening of libraries of up to millions of compounds, strengthen the combinatorial approach as a tool to accelerate and cheapen early-stage drug discovery. Combinatorial chemistry is not a new chemistry; rather, it is intended as a different way of using pre-existing synthetic routes or methods and has the aim of increasing productivity in terms of numbers of new highly diverse molecules that must cover the chemical space allowed by a predetermined set of starting building blocks. The higher the number of building blocks and the number of transforming synthetic steps, the higher will be the number of attainable molecules. However, the higher the number of attainable molecules, the higher should be the handling capacity of so many reagents and products. Therefore, automation in combinatorial chemistry plays a very critical role [4, 5]. Although combinatorial libraries can be synthesized in solution, they are most often prepared on solid phase, for which purification of the intermediates is not required, as excess solvents or reagents are simply removed by filtration. Automation in chemical synthesis has been essentially developed around the solid-phase method introduced by Merrifield for the synthesis of peptides [6], which after its introduction has been continuously improved, making it nowadays one of the most robust and well-established synthetic methods. For this reason, the basic concepts of combinatorial chemistry, such as compound libraries, molecular repertoires, chemical diversity, and library complexity, have been developed using peptides [7]. The high purity levels achieved with short peptides and the built-in code represented by their own sequences have promoted the employment of large, but rationally encoded mixtures instead of single compounds, leading to the generation and manipulation of libraries composed of hundreds of thousands and even millions of different peptides.

All combinatorial library methods involve three main steps: preparation of the library, screening of the library components, and determination of the chemical structures of active compounds.

7.3.1 Preparation of the Library

There are different strategies for library synthesis on solid phase. The first is the *parallel synthesis*, whereby peptide libraries are synthesized in individual reaction chambers; therefore, each product is pure, separated, and well determined [8]. This approach is highly amenable to miniaturization and automation. As each vessel contains a single known compound, identification of the active compounds in biological screening is direct. However, this method generates only a small set of peptides library.

Another strategy is the *split and mix synthesis* method [9]. This provides equimolar mixtures of random and large peptide libraries. Resin beads or any other solid support are used as micro-reactors onto which subsequent steps of amino acid coupling and redistribution are performed in order to increase the number of new molecules exponentially at each step. It consists of three steps: splitting, coupling, and mixing. First, resin beads are split into a number of aliquots that equals the number of building blocks to be utilized in the synthesis (splitting). Then, to each resin aliquot one building block is coupled and the reactions are driven to completion (coupling). Following this step, beads are randomly and thoroughly mixed (mixing) and then re-split into the same number of aliquots, achieving a set of homogeneous and equimolar collections of compounds. Repetitive execution of these basic steps n times will thus produce a rapid increase of newly generated molecules, while bead number remains constant. Importantly, since resin beads encounter one reactant at a time, no more than one compound per bead can be generated, therefore the total number of final compounds is limited by the number of initial beads. Furthermore, as reactions are driven to completion, all peptides are generated in equal ratios in the mixtures.

By this procedure, couplings are carried out on spatially separated samples, adding single amino acids to each separate resin aliquot. As a result, both the number of peptides and their equimolar ratio are preserved in each aliquot. By choosing suitable sets of acid-resistant linkers and side-chain protections, the compounds can be left on the beads, freed of the protecting groups. In this way, peptides can be tested still attached to beads, exploiting the high local concentration of compounds on the beads' surfaces and the "one bead, one compound" (OBOC) concept. The use of suitable linkers also allows the controlled released of molecules, thus permitting testing in solution. A major drawback of OBOC libraries is the limited set of peptides; indeed, the complexity is dictated by the number of beads [4].

Another approach that can be used is the *reagent mixture* or *pre-mix synthesis*. It is more convenient than the split and mix method for the synthesis of large peptide libraries. The amino acids chosen for the library are pre-mixed and then coupled to a single resin batch. Before coupling the last amino acid, the resin is split into a number of aliquots that equals the number of building blocks and to each aliquot one single amino acid is separately coupled. This procedure leads to the generation of sub-libraries that are labeled by the N-terminal residue. Every single bead contains all the ensemble of peptides of the sub-library, as any single bead encounters the reagents under the same conditions. This method is faster than the mix and split, producing a wide chemical diversity. The distribution of any single peptide within the final mixtures is strictly dependent on the relative kinetic rates of individual protected amino acids. To compensate for the different reaction rates of each amino acid reagent, isokinetic ratios are calculated and the corresponding amounts of each amino acid reagent are employed in the coupling reactions. To avoid the use of "weighted" mixtures, alternative methods can be applied; for example, the use of large excesses of mixtures.

7.3.2 Screening of Library Components

Once synthesized, libraries are required to be submitted to a screening procedure to determine active components for a given target. A reliable high-throughput assay is essential to screen a combinatorial library successfully. Both solid-phase and solution-phase assays have been developed [9]. In the solid-phase assays, target–ligand interactions can be detected by incubating immobilized peptide libraries (covalently attached to the solid support) with the target, which can be visualized by taking advantage of suitable groups, such as enzymes, fluorescent dyes, radionuclides, or biotin. The peptide-bound target can be detected by colorimetric assays, fluorescent microscopy, or flow cytometry. Solid-phase-linked peptide libraries can also be screened using whole cells or microorganisms and the cell–ligand interactions can be visualized directly by microscope. The major drawback in which selection of bioactive ligand is performed on support-linked peptides is that the activity of the selected sequences, once these are synthesized as soluble compounds, may be different from that of the immobilized peptides. On the other hand, when performing high-throughput screening of free peptides in solution, automation, miniaturization, and very sensitive detection methodologies are required. Different assays, such as enzyme-linked immunosorbent assay (ELISA), cell-based cytotoxic assay, antimicrobial assay, affinity chromatography, and radiometric and fluorescence-based assays can be used [2]. In some instances, it may be advantageous to combine solution-phase assays with on-bead assays to screen a specific target. Positive beads isolated by this approach are more likely to be true positives.

7.3.3 Identification of Active Compounds

As we have mentioned before, when performing parallel synthesis, identification of the active compounds of the library is direct, as each vessel contains a single known compound.

In the case of the split and mix synthesis, identification of the peptide sequence of the selected bioactive bead is the main drawback of the method. Micro-sequencing and mass spectrometry, performed directly on the beads, are possible approaches for peptide identification, together with chemical encoding, a method based on tagged peptides in which the tag enables the peptide sequence to be retraced [2].

In reagent mixture libraries, identification of bioactive peptide sequences is obtained by deconvolution. Deconvolution methods evaluate the contribution of each residue to the desired biological activity. There are two main approaches: an iterative process [7, 10] and positional scanning [11]. In the first approach, a progressive elucidation of the active sequence is carried out by identifying one amino acid at a time for each position. The process involves subsequent cycles of screening and resynthesis of active pools, thus producing a progressive activity enrichment and a contextual simplification of the peptide pools. The chemical synthesis of the next sub-library is dictated by the results from the previous one; thus, only a subpopulation of the library components is resynthesized, arriving at the last assay where single molecules, differing only for the last amino acid, are prepared in parallel and tested. In the positional scanning approach, all sub-libraries needed to determine the final active sequence are prepared at the beginning of the screening and, in this case, the library is replicated a number of times equal to the sequence length. Each variable position is tested independently and at the same time, using a set of sub-libraries with known residues for every individual position. Whereas in the iterative process the sequence is obtained step by step, in the case of the positional scanning the most active sequence is reconstructed at the end of the process by selecting the most active pools on each known position. The two approaches are not mutually exclusive, and the choice of one approach or the other is often dictated by the synthesis capacity and also on the structures to be synthesized [4].

7.4 A MODEL FOR CHEMICAL EVOLUTION OF MACROMOLECULAR SEQUENCES

An interesting approach for the chemical evolution of macromolecules has been developed in the group of Professor Luisi [12]. A classic synthetic procedure based on the Merrifield solid-phase synthesis of peptides has been utilized for the synthesis of randomly produced peptides as well as for their stepwise fragment condensation.

The idea is based on the assumption that short peptide fragments can produce chain elongation by successive fragment condensation. In particular, the fragment condensation can be induced by the catalytic action of peptides present in the same library of shorter sequences. The idea is not new; and it has been already reported that relatively simple peptides may be endowed with proteolytic activity. For example, histidine-containing dipeptides appear capable of cleaving peptide and nucleic acid bonds [13, 14], and even Gly–Gly [15] appears to possess some catalytic activity. This catalytic power, as in the case of the Ser–His peptide, is present also in longer peptides, and thus the hypothesis that the generation of simple peptides may produce proteolytic catalysis is reasonable. In our case, we need the reverse reaction of course, namely the synthesis of peptide bonds catalyzed by such peptides. Peptide synthesis by the reverse proteolytic reaction is well known. Detailed papers and extensive review articles have been presented in the past by Jakubke [16] and by others [17, 18]; and within the field of the origin of life, scenarios of alternate dry and wet environment have been proposed as conditions for bond -formation [19].

Initially, it should be noted that the random condensation of a large library of co-oligopeptides gives rise, in principle, to an astronomic number of longer chains. For example, starting from a library of 100 different peptides, an ideal successive-step fragment condensation would give rise to approximately 10^4 different compounds, which, in the following fragment condensation step, would become of the order of 10^8 different hetero-polypeptides. To arrive, instead, at a selection of multiple copies of only one or a few chain configurations, some natural screening criteria must be introduced. The selection criteria adopted in the work is one that is assumed to simulate natural chemical evolution: namely, selection is operated by the contingency of the environmental conditions, such as pH, solubility, temperature, salinity, etc. Dependent upon these contingent conditions, the largest majority of products may be eliminated (e.g. by lack of solubility), and only a few chain products may "survive" at each step in solution, then undergoing further elongation.

A random library of decapeptides has been generated by a computer procedure; four of these peptides randomly chosen have been synthesized and condensed with four other randomly conceived decapeptides, thus producing 16 totally random icosapeptides. Successive fragment condensation proceeded only with those compounds which remained soluble under the given initial conditions, and so on for the following steps, until only one co-oligopeptide sequence of over 40 residues was obtained. Under our simulated conditions, in a few steps, a *de novo* protein was produced, namely, one without any significant homology with extant proteins of similar length. Furthermore, this small protein (44 residues long) showed to assume a stable three-dimensional folding.

First, two parent 40-residue peptides, P_1 and P_2, were designed randomly but with the constraint that the relative abundance of the 20 amino acids used in their construction maintained a $1:1:1$ relationship. This, in turn, meant that the products, on average, would also have this relative composition. Other frequencies could have been imposed (e.g. the natural abundance in modern proteins) or a restricted set of amino acids used, but this was considered to be unwarranted in the absence of a convincing a priori information.

The sequences of the parent 40-residue peptides are:

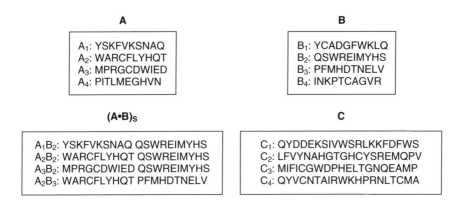

Figure 7.1 The peptide sequences used in the study

P_1: YSKFVKSNAQ WARCFLYHQT MPRGCDWIED PITLMEGHVN
P_2: YCADGFWKLQ QSWREIMYHS PFMHDTNELV INKPTCAGVR

A matrix, **A** × **B**, of 16 × 20-residue peptides was constructed by the systematic combination of two small libraries **A** and **B** each comprising four 10-residue peptide sequences (Figure 7.1).

The individual peptides A_i ($i = 1 - 4$) and B_j ($j = 1 - 4$) each represented 10-residue sequences obtained from the two above-mentioned parent 40-residue peptides. The 16 × 20-residue sequences arrived at in this way were synthesized by the solid-phase method.

The peptide products were subjected to selection on the basis of their solubility in water under well-defined conditions. It was found that A_1B_2, A_2B_2, and A_3B_2 were completely soluble in aqueous 100 mM Tris buffer in the pH range 5.2–8.6; A_1B_3 and A_3B_3 were insoluble, whereas A_2B_3 was totally soluble, counter to prediction. Peptides insoluble in Tris buffer remained insoluble irrespective of salt concentration, whereas water-soluble peptides remained in solution. The subsets (**A** × **B**)$_s$ that fulfilled the criterion of being soluble in water were then subjected to chain elongation by combination with a further small set of 20-residue sequences **C**, giving rise to the new library **C** × (**A** × **B**)$_s$ consisting of 16 peptides which are 40-residues long.

None of the latter was soluble in aqueous buffer, but two of them, $A_1B_2C_1$ and $A_2B_2C_1$, turned out to be soluble in 6 M guanidinium chloride (GuCl). The addition of a polar N-terminal extension to them (DDEE) resulted in the 44-residue sequences DDEE-$A_1B_2C_1$ and DDEE-$A_2B_2C_1$. Of these two samples, only the latter was soluble in water and was studied in more detail. The whole sequence of this peptide is

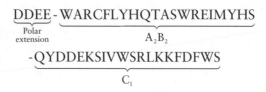

It is important to mention that this sequence has no homologies or similarities with proteins of comparable length (and up to 100 residues) in the data bank. As such, it can be considered a protein that is not extant on Earth – a truly *de novo* protein – or, as colloquially said in this new area of peptide chemistry, one of the never-born proteins (NBPs) [20].

The three-dimensional structure of the DDEE-$A_2B_2C_1$ peptide has been modeled using the ROKKY protein structure-suite [21, 22], which allows ab-initio protein structure prediction using a fragment-assembly-simulated annealing protocol. The lowest energy model obtained from this procedure is shown in Figure 7.2. In agreement with spectroscopic data, a mainly α-helical fold is predicted for the peptide. The structure is made up of a long, bent α-helix, packed together with a shorter helix, and residues in helical conformation account for 60% of the total number of residues. The structure appears to be compact, with all the charged and polar residues located on the solvent-accessible surface of the molecule. Moreover, a well-packed hydrophobic core is formed at the interface of the two α-helices by residues which span the entire sequence length (Phe9, Leu10, Trp17, Ile20, Met21, Tyr26, Val33, Trp34, Leu37, Phe40, Phe42).

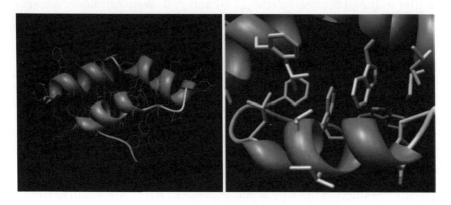

Figure 7.2 Model of the three-dimensional structure of the DDEE-$A_2B_2C_1$ peptide. Left: global view; right: detailed view of the hydrophobic core

In conclusion, the approach proposed can be conceptually generalized to a primordial mechanism that appears capable of producing a specific macromolecular sequence from an initial oligopeptide, with a step-by-step elongation, which is determined by the contingency of the environmental pressure – be that pH, temperature, salinity, solubility, aggregation, or other physical factors. This is also an attempt to address the question of the design, synthesis, and standardization of completely *de novo* proteins to be exploited as novel functional scaffolds for synthetic biology.

Another study carried out in the group of Professor Luisi that lies within the framework of chemical synthetic biology, which aims at exploring the sequence space for novel biological biochemical structures that do not exist in nature, is the previously mentioned project on NBPs [20]. In particular, the project is characterized by its biological nature, since the approach is based on phage display technology, not on chemical synthesis.

7.5 NEVER-BORN PROTEINS PROJECT

The rationale behind this project relies on the observation that the number of natural proteins on our Earth, although apparently large, is only an infinitesimal fraction of the possible ones. This means that there is an astronomically large number of proteins that have never been sampled by natural evolution on Earth: the NBPs that await for human exploration and exploitation.

Speaking about proteins, there are thought to be roughly 10^{13} proteins of all sizes in extant organisms. This apparently huge number represents less than noise when compared with the number of all theoretically different proteins. The discrepancy between the actual collection of proteins and all possible ones becomes clear if one considers that the number of all possible 50-residue peptides that can be synthesized with the standard 20 amino acids is 20^{50}, namely 10^{65}. Moreover, the number of theoretically possible proteins increases with length, so that the related sequence space is beyond contemplation [23]. This means that there is an astronomically large number of proteins that have never been subjected to the long pathway of natural evolution on Earth: the "NBPs.

Following the construction of a purely random DNA library of 150 bases, totally new proteins have been generated. These random DNA sequences of 150 bp contain the codons for a tri-peptide substrate for thrombin: proline–arginine–glycine (PRG). A large number of these

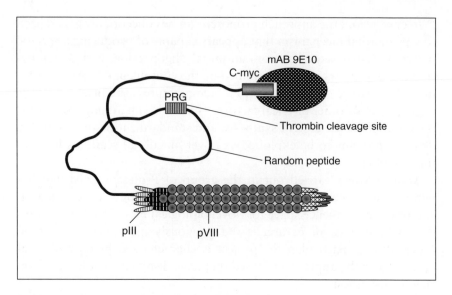

Figure 7.3 Schematic drawing of the selection system. The random peptide library is bound to the minor coat protein (pIII) of M13K07. In the middle of the random sequences is present a tripeptide PRG (proline–arginine–glycine) substrate for thrombin. The c-myc tag serves to bind the specific antibody 9E10 during the selection of the thrombin-resistant sequences

sequences (approximately 10^{12}) have been inserted into phagemid vectors in order to express the corresponding polypeptide chains by phage display techniques using *Escherichia coli* as host cells. Expressed proteins were then produced and displayed fused to protein III (Figure 7.3). Those phage which successfully expressed the NBP were detected by the ELISA panning technique and the selection of resistant clones was performed using proteolytic attack by thrombin (Figure 7.3).

The DNA library cloned in the phagemid system was obtained using two groups of oligonucleotides (forward and reverse) with codon schemes NNK and NNM respectively, where N is an equimolar representation of all four bases and where K and M represent only G or T for K and C or A for M. These schemes use 32 codons to encode all 20 amino acids and one stop codon (TAG), yielding an acceptably low frequency of stop codons when used to encode short polypeptides. The oligonucleotides consist of random nucleotides encoding 23 amino acids in the forward group and 24 in reverse, flanked by 11–18 fixed residues, which are necessary for annealing and cloning (Figure 7.4a). After annealing and incubation with Klenow, the DNA library was

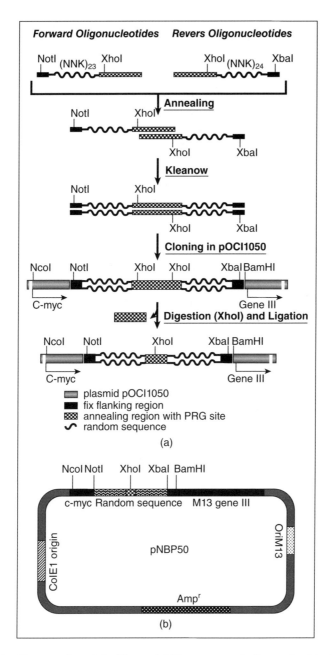

Figure 7.4 Construction of the library. (a) Two groups of oligonucleotides (forward and reverse) with codon schemes NNK and NNM respectively have been used to construct the library. After annealing and incubation with Klenow, the DNA library has been cloned in pOCI1050 vectors using the unique *Not*I, *Xba*I sites, subsequently the bases in excess in the annealing region have been removed by cleavage with *Xho*I restriction enzymes. (b) The resulting phagemid vector pNBP50 has been constructed in such a way that it allowed in-frame expression of the pelB signal sequence, the c-myc tag-sequence, the library, and the C-terminal part of gene III. It contains ColE1 origin, OriM13, and a gene to confer the ampicillin resistance

cloned in pOCI1050 vectors (Figure 7.4a). The final result is shown in Figure 7.4b.

Following the creation of a library, the next step was to prepare separate phage particle pools expressing random peptides–gene III fusion (Figure 7.3) and to select them with thrombin. The different phage groups were incubated separately with a thrombin solution chosen on the basis of preliminary studies.

Interestingly, the distribution of peptides tested displayed on phage, divided into categories according to thrombin digestion, shows that a large majority is either completely digested or resistant (Figure 7.5). Various digestion categories represent different levels of resistance against thrombin. For example, the clones in class 0–10 present a percentage of total protein digestion between 0 and 10, namely they are practically not affected by thrombin.

A preliminary structural analysis was carried out on two resistant sequences (named NBP1 and NBP127) belonging to the 0–10 digestion category (Figure 7.5).

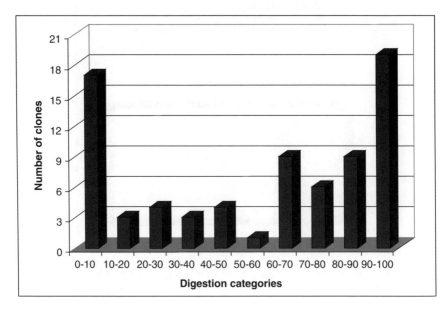

Figure 7.5 Distribution of the peptide library with respect to thrombin digestion. The peptides tested are divided into categories according to the different levels of resistance against thrombin digestion shown by each peptide sequence; for example, the clones in class 0–10 display a digestion between 0 and 10%

DE NOVO TOTALLY RANDOM AMINO ACID SEQUENCES

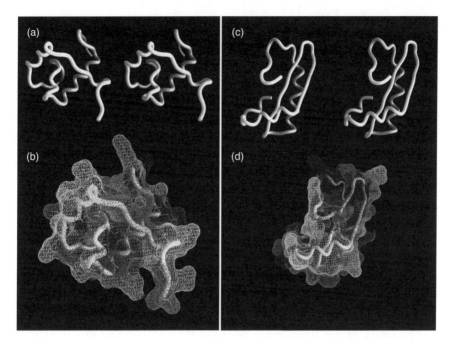

Figure 7.6 Schematic representation of the three-dimensional structure of protein NBP1 and NBP127. (a) Stereo view of the protein NBP1 backbone; (b) mesh representation of the molecular surface of protein NBP1; (c) stereo view of the protein NBP127 backbone; (d) mesh representation of the molecular surface of protein NBP127. Apolar, polar, and positively/negatively charged surface areas are colored in green, white, and blue/red respectively

Analysis of the three-dimensional model of protein NBP1 revealed a compact globular structure containing three α-helices (Figure 7.6a). Both polar and charged amino acids are located on the protein surface and hydrophobic residues are positioned inside the molecule forming a "canonical" hydrophobic core (Figure 7.6b). Two of the three cysteine residues of the protein (Cys19 and Cys28) are located within the hydrophobic core, which are in principle closed sufficiently to allow the formation of a disulfide bond. The single Trp residue present in the amino acid sequence is located on the protein surface, making it an ideal probe to test the accuracy of the model by fluorescence analysis.

Analysis of the three-dimensional model of protein NBP127 reveals an elongated bean-shaped structure containing three α-helices and two β-strands (Figure 7.6c). Also in this case, polar and charged amino acids are located on the protein surface, whereas nonpolar residues form a hydrophobic core partly exposed to the solvent in correspondence with a cleft of the protein surface (Figure 7.6d). Interestingly, four out of the five cysteine residues of the protein could form disulfide bridges (the first between Cys20 and Cys44 and the second between Cys36 and Cys46). One of the two Trp residues is located on the protein surface and the second is embedded in the protein structure.

One of the most surprising results of this investigation is the high degree of folded sequences observed in the NBP library. We are aware that our criterion of folding is an operational one, and valid only as a first general qualitative screening, but we tend to accept the validity of the picture coming out from this investigation and the suggestion that protein folding is, indeed, a general feature of this protein library. Generalizations have to be taken with great care, however; note that our protein library has been built without any preconceived structural constraint, except for the sequence length. In this sense, the distribution of foldability obtained in our study – namely the fact that a large part of the clones is very rapidly cleaved by thrombin, and a significant part (around 22%) is not – may suggest an important general feature of *de novo* protein structures.

The study of the structure of proteins that do no exist in nature will permit us to establish whether additional principles of structure stability may be present in the folding of proteins. Alternatively, it could be that what we know now about the structure of proteins (the elements of secondary, tertiary, and quaternary structure) is sufficient to describe all NBPs. If novel structural elements are to be found, these could be used as a scaffold for a new kind of structural chemistry of proteins, which can be useful also from the biotechnological standpoint. At any rate, the comparison between the structural and thermodynamic features of NBPs with extant ones will contribute to tackle the basic question "By which principles have the extant proteins been selected out by evolution?" Is the selection due to the fact that "our" proteins possess particular structural (e.g. folding, hydrodynamics) and thermodynamic properties (e.g. stability)? In this case, the NBPs should not display (not to the same extent) those particular properties.

The NBPs may display catalytic functions, and it will be interesting to see whether some of the clones display hydrolytic properties – which is most likely. Less likely, but interesting, is the question of whether they display more sophisticated enzymatic properties like those of "our" enzymes. Even more challenging will be the question of whether the NBPs may display totally new catalytic properties; for example, against recalcitrant substrates which are not recognized by our enzymatic proteins.

From the applied science standpoint, NBPs will represent completely novel scaffolds for synthetic biology applications. It is well known that enzymes find several applications in biotechnology and chemical industry as specific catalysts or in general as active therapeutic molecules. The search for catalytic properties in the NBPs can, therefore, be relevant for applied science. Notice that NBPs will be completely orthogonal (insulated) to extant organisms and, therefore, may be implemented into synthetic biology chassis, possibly reducing cross-talk and parasitic effects. In addition, NBPs (unlike natural polymers) are not the results of any evolutionary pathway and can, therefore, be regarded as virgin polymers whose functionality can be engineered without constrictions to meet the user's requirements. In this regard, NBPs may pave the way for novel design paradigms in synthetic biology and applications in metabolic engineering and fine-chemical production, as well as in complex system engineering.

REFERENCES

1. Marguet, P., Balagadde, F., Tan, C., and You, L. (2007) Biology by design: reduction and synthesis of cellular components and behaviour. *Journal of the Royal Society Interface*, **4**, 607–623.
2. Falciani, C., Lozzi, L., Pini, A., and Bracci, L. (2005) Bioactive peptides from libraries. *Chemistry & Biology*, **12**, 417–426.
3. Eichler, J. (2008) Peptides as protein binding site mimetics. *Current Opinion in Chemical Biology*, **12**, 707–713.
4. Marasco, D., Perretta, G., Sabatella, M., and Ruvo, M. (2008) Past and future perspectives of synthetic peptide libraries. *Current Protein and Peptide Science*, **9**, 447–467.
5. Messeguer, A. and Cortés, N. (2007) Combinatorial chemistry in cancer research. *Clinical and Translational Oncology*, **9**, 83–92.
6. Merrifield, R.B. (1963) Solid phase peptide synthesis. I. The synthesis of a tetrapeptide. *Journal of the American Chemical Society*, **85**, 2149–2154.
7. Houghten, R.A., Pinilla, C., Blondelle, S.E. et al. (1991) Generation and use of synthetic peptide combinatorial libraries for basic research and drug discovery. *Nature*, **354**, 84–86.
8. Shin, D.S., Kim, D.H., Chung, W.J., and Lee, Y.S. (2005) Combinatorial solid phase peptide synthesis and bioassays. *Journal of Biochemistry and Molecular Biology*, **38**, 517–525.
9. Lam, K.S., Lebl, M., and Krchnák, V. (1997) The 'one-bead–one compound' combinatorial library method. *Chemical Reviews*, **97**, 411–448.

10. Lam, K.S., Salmon, S.E., Hersch, E.M. *et al.* (1991) A new type of synthetic peptide library for identifying ligand-binding activity. *Nature*, **354**, 82–84.
11. Pinilla, C., Appel, J.R., Blanc, P., and Houghten, R.A. (1992) Rapid identification of high affinity peptide ligands using positional scanning synthetic peptide combinatorial libraries. *Biotechniques*, **13**, 901–905.
12. Chessari, C., Thomas, R., Polticelli, F., and Luisi, P.L. (2006) The production of *de novo* folded proteins by a stepwise chain elongation: a model for prebiotic chemical evolution of macromolecular sequences. *Chemistry & Biodiversity*, **3**, 1202–1210.
13. Li, Y., Zhao, Y., Hatfield, S. *et al.* (2000) Dipeptide seryl–histidine and related oligopeptides cleave DNA, protein, and a carboxyl ester. *Bioorganic & Medicinal Chemistry*, **8**, 2675–2680.
14. Shen, C., Mills, T., and Oro, J. (1990) Prebiotic synthesis of histidyl-histidine. *Journal of Molecular Evolution*, **31**, 175–179.
15. Plankensteiner, K., Righi, A., and Rode, B.M. (2002) Glycine and diglycine as possible catalytic factors in the prebiotic evolution of peptides. *Origins of Life and Evolution of the Biosphere*, **32**, 225–236.
16. Jakubke, H.D. (1995) Hydrolysis and formation of peptides, in *Enzyme Catalysis in Organic Synthesis: A Comprehensive Handbook*, vol. I (eds K. Drauz and H. Waldmann), Wiley–VCH Verlag GmbH, Weinheim, pp. 431–458.
17. Jost, R., Brambilla, E., Monti, J.C., and Luisi, P.L. (1980) Papain catalyzed oligomerization of α-amino acids. Synthesis and characterization of water-insoluble oligomers of L-methionine. *Helvetica Chimica Acta*, **63**, 375–384.
18. Pellegrini, A. and Luisi, P.L. (1978) Pepsin-catalyzed peptide synthesis. *Biopolymers*, **17**, 2573–2580.
19. Commeyras, A., Taillades, J., Collet, H. *et al.* (2004) Dynamic co-evolution of peptides and chemical energetics, a gateway to the emergence of homochirality and the catalytic activity of peptides. *Origins of Life and Evolution of the Biosphere*, **34**, 35–55.
20. Chiarabelli, C., Vrijbloed, J.W., de Lucrezia, D. *et al.* (2006) Investigation of *de novo* totally random biosequences. Part II. On the folding frequency in a totally random library of *de novo* proteins obtained by phage display. *Chemistry & Biodiversity*, **3**, 840–859.
21. Ginalski, K., Elofsson, A., Fischer, D., and Rychlewski, L. (2003) 3D-Jury: a simple approach to improve protein structure predictions. *Bioinformatics*, **22**, 1015–1018.
22. van Gunsteren, W.F. and Berendsen, H.J.C. (1990) Computer simulation of molecular dynamics: methodology, applications, and perspectives in chemistry. *Angewandte Chemie*, **29**, 992–1023.
23. Thomas, R.M., Vrijbloed, J.W., and Luisi, P.L. (2001) Toward random polypeptide synthesis. *Chimia*, **55**, 114–118.

Part Three
Complex Systems

8

Synthetic Genetic Codes as the Basis of Synthetic Life

J. Tze-Fei Wong and Hong Xue

Hong Kong University of Science and Technology, Fok Ying Tung Graduate School and Department of Biochemistry, Hong Kong, China

8.1 Introduction 178
8.2 Natural Code Expansion 179
    8.2.1 Phase 2 Expansion 179
    8.2.2 Proteome-Wide and Site-Specific Expansions 180
8.3 Natural Code Turnover 181
8.4 The Deep Freeze 184
8.5 Synthetic Code Expansion 186
    8.5.1 Proteome-Wide Expansion 186
    8.5.2 Site-specific Expansion 188
8.6 Synthetic Code Turnover 190
8.7 Usefulness of Synthetic Codes 191
8.8 Discussion 192
    Acknowledgments 194
    References 194

The genetic code likely originated from the post-transcriptional additions of peptidyl side chains to ribozymes to augment their catalytic activities, much as post-translational modifications augment protein activities today. Since the universal genetic code contains many amino acids that were not available from the primordial environment, yet excludes others that were readily produced by prebiotic synthesis, the primitive code must have evolved and expanded to arrive at its present-day ensemble of 23 standard amino acids. Evidence strongly suggests that the code started with the primordially available Phase 1 amino acids, and co-evolved with the early amino acid biosynthetic pathways, which brought novel Phase 2 amino acids into the code, and that the code underwent not only expansion but also turnover with respect to its amino acid composition. We then discuss how rising noise/benefit ratios accompanying the further introduction of novel amino acids shut off evolution, and caused the encoded amino acid ensemble to enter into a 2.5 billion year "deep freeze." In recent years, however, proteome-wide and site-specific expansions of the universal genetic code, as well as its turnover, have been accomplished through mutations or gene insertions, thereby thawing the code out of its prolonged deep freeze. These experimental genetic code mutants represent forms of *synthetic life* that differ from all earthly life with respect to their basic building blocks, which is a more fundamental departure than that called upon in synthetic biology where the same earthly standard building blocks are always employed despite the scrambling and *de novo* design of genes.

8.1 INTRODUCTION

Proteins are the workers of the cell responsible for catalysis, structure, transport, signaling, and immunity. To perform these tasks, the protein language has to be information rich, both quantitatively and qualitatively. The maximum information content I_{max} of a protein sequence is given by [1]

$$I_{max} = N \log_2(M)$$

Quantitatively, any protein molecule 100 amino acid (aa) residues long and containing 23 kinds of amino acid will have an I_{max} of $100 \log_2(23)$ regardless of what the 23 amino acids happen to be. Qualitatively, however, the performance of the protein language depends on the choice of its amino acid alphabet. An all-alkyl protein language, for instance, may consist of 23 kinds of aa-letter with side chains of methyl-, ethyl-, ... $CH_3(CH_2)_{22}$ groups. The polypeptides in such a potential language will lag in terms of chemical versatility far behind present-day proteins constructed with the 23 amino acids in the genetic code: 20 regular amino acids that receive proteome-wide encoding by their own codon(s) and three extra ones that receive in each instance only a shared codon for entry into specific sites in the proteome, namely selenocysteine (Sec) sharing the nonsense codon UGA with the chain termination (ter) signal,

UUU	Phe	UCU	Ser	UAU	Tyr	UGU	Cys
UUC	Phe	UCC	Ser	UAC	Tyr	UGC	Cys
UUA	Leu	UCA	Ser	UAA	ter	UGA	ter/Sec
UUG	Leu	UCG	Ser	UAG	ter/Pyl	UGG	Trp
CUU	Leu	CCU	Pro	CAU	His	CGU	Arg
CUC	Leu	CCC	Pro	CAC	His	CGC	Arg
CUA	Leu	CCA	Pro	CAA	Gln	CGA	Arg
CUG	Leu	CCG	Pro	CAG	Gln	CGG	Arg
AUU	Ile	ACU	Thr	AAU	Asn	AGU	Ser
AUC	Ile	ACC	Thr	AAC	Asn	AGC	Ser
AUA	Ile	ACA	Thr	AAA	Lys	AGA	Arg
AUG	Met/fMet	ACG	Thr	AAG	Lys	AGG	Arg
GUU	Val	GCU	Ala	GAU	Asp	GGU	Gly
GUC	Val	GCC	Ala	GAC	Asp	GGC	Gly
GUA	Val	GCA	Ala	GAA	Glu	GGA	Gly
GUG	Val	GCG	Ala	GAG	Glu	GGG	Gly

Figure 8.1 The 23 amino acid genetic code

pyrrolysine (Pyl) sharing the nonsense codon UAG with ter, and formylmethionine (fMet), which enters only into chain initiation sites in bacterial proteins encoded by AUG, the regular codon of Met (Figure 8.1). The composition of the aa-alphabet, therefore, is important for proteins.

8.2 NATURAL CODE EXPANSION

Since many more than 23 L-amino acids are known but left out of the genetic code, the factors that separated noncanonical, or unnatural, amino acids from the canonical 23 during the formation of the present-day code need to be examined.

8.2.1 Phase 2 Expansion

Cellular functions require biocatalysts, either enzymes today or ribozymes in an earlier RNA world [2]. The transition between these biocatalyst regimes depended on the birth of the genetic code, which likely arose from the post-transcriptional additions of peptidyl side chains to ribozymes to augment their catalytic activities, much as post-translational modifications (PTMs) augment protein activities today. This eventually led to the adoption of polypeptides as biocatalysts on account of their superior catalytic proficiency, and abandonment by

RNAs of most of their ribozyme functions to focus on their polypeptide encoding role [3–5].

To start off the genetic code, prebiotic-type syntheses are known to yield amino acids, and meteorites also bring amino acids from space. Thus, abundance in the prebiotic environment could be a foremost factor favoring the inclusion of an amino acid in the code, as in the case of Gly, Ala, Asp, and Glu, which are found on the Murchison meteorite, and also formed by electric discharge, thermal synthesis, ultraviolet irradiation, and high-energy particle irradiation from primitive gases [6]. However, this cannot be the case with Gln and Asn, which are thermally so unstable that their concentrations in the primitive ocean could not exceed 3.7 pM and 24 nM respectively [7], the ultraviolet-unstable Cys, Met, Trp, His, Tyr, and Phe [8, 9], or Sec, which is produced inside the cell, not as free Sec but only as Sec-tRNA compound [10]. Accordingly, the code could only begin with the prebiotically abundant amino acids. To arrive at the present-day universal code, sooner or later it had to expand to bring in the prebiotically unavailable ones. Evidence indicates that the encoded amino acid ensemble coevolved with the amino acid biosynthetic pathways which generated novel amino acids to add to the code either as free compounds, or via *pretran synthesis*, whereby a (product aa)-tRNA conjugate is synthesized from a (precursor aa)-tRNA conjugate through pre-translational modification [11, 12].

Based on this coevolution mechanism, a distinction was drawn between the prebiotically derived Phase 1 amino acids Gly, Ala, Ser, Asp, Glu, Val, Leu, Ile, Pro, and Thr and the biosynthesis-derived Phase 2 amino acids Phe, Tyr, Arg, His, Asn, Gln, Lys, Trp, Cys, Met, Sec, Pyl, and fMet [13, 14]. The findings that all the Phase 1 but none of the Phase 2 amino acids are produced by high-energy particle irradiation of primitive gases [15, 16] or found on meteorites [17] have provided exceptional confirmation for this proposed Phase 1–Phase 2 partition of the encoded amino acids [3, 13]. Accordingly, only those amino acids that were either abundant in the prebiotic environment or produced by primordial amino acid biosynthesis stood to be recruited into the code. Being at the right place at the right time was essential with respect to joining the code.

8.2.2 Proteome-Wide and Site-Specific Expansions

To add a new Phase 2 amino acid to the code requires (i) biosynthesis of the new amino acid and (ii) securing its attachment to a tRNA so that it may enter ribosomal protein synthesis guided by the anticodon on the tRNA. These requirements can be fulfilled by two different pathways. In the aaRS pathway, the amino acid is formed and a cognate aaRS is evolved to attach it to tRNA. In the pretran synthesis pathway, a (Phase 1 amino acid)–tRNA conjugate is converted to a (Phase 2 amino acid)–tRNA conjugate, giving the Phase 2 amino acid at once the anticodon on the tRNA [10, 12, 18], as in:

Met–tRNA → fMet–tRNA
Glu–tRNA → Gln–tRNA
Asp–tRNA → Asn–tRNA
PhosphoSer–tRNA → Cys–tRNA
PhosphoSer–tRNA → Sec–tRNA

In some organisms, the original pretran synthesis of Gln–tRNA, Asn–tRNA or Cys–tRNA from Glu–tRNA, Asp–tRNA or phosphoSer–tRNA has since been replaced by an evolved GlnRS, AsnRS, or CysRS respectively. Some of the remaining Phase 2 amino acids also might have entered the code initially through pretran synthesis, only switching subsequently to reliance on aaRS. So far, no SecRS or fMetRS has been encountered in any organism, and Sec and fMet yet enter proteins solely via pretran synthesis today [19]. Interestingly, in the ciliate *Euplotes crassus*, Sec still shares the UGA codon, but with Cys instead of ter [20].

PylS, the PylRS gene, and *pylT*, the gene for the UAG-decoding Pyl-acceptor tRNA, are located on a gene cassette together with Pyl-synthesis genes in the archaeon *Methanosarcina acetivorans*. Insertion of this cassette into *Escherichia coli* induces the appearance of Pyl within *E. coli*, which is normally devoid of Pyl. Because the PylRS and Pyl-acceptor tRNA introduced into *E. coli* react only with each other and not with any host tRNA or aaRS, the presence of this mutually preoccupied aaRS–tRNA pair does not perturb other aaRS–tRNA interactions inside *E. coli* [21]. This usage of a mutually preoccupied aaRS–tRNA pair to achieve natural site-specific code expansion parallels the use of these kinds of aaRS–tRNA pairs in synthetic site-specific code expansion.

8.3 NATURAL CODE TURNOVER

Amino acids such as α-amino-*n*-butyric acid, α-aminoisobutyric acid, and norvaline are produced by electric discharge under prebiotic-like

Table 8.1 Relative abundances of amino acids in Murchsion meteorite and from electric discharge synthesis [22][a]

Amino Acid	Murchison meteorite	Electric discharge
Glycine	++++	++++
Alanine	++++	++++
α-Amino-*n*-butyric acid	+++	++++
α-Aminoisobutyric acid	++++	++
Valine	+++	++
Norvaline	+++	+++
Isovaline	++	++
Aspartic acid	+++	+++
Glutamic acid	+++	++
β-Alanine	++	++
β-Amino-*n*-butyric acid	+	+
β-Aminoisobutyric acid	+	+
γ-Aminobutyric acid	+	++
Sarcosine	++	+++
N-Ethylglycine	++	+++
N-Methylalanine	++	++

[a]Mole ratio relative to Gly (=100): +, 0.05–0.5; ++, 0.5–5; +++, 5–50; ++++, >50.

conditions and found on the Murchison meteorite (Table 8.1), but are excluded from the genetic code. Their exclusion suggests that the natural code underwent not only expansion to bring in Phase 2 amino acids, but also underwent turnovers causing the displacement of amino acids that were included at one time in the primitive codes. The historical occurrence of code turnovers is supported by the evidence that at least some of the amino acids in the extant code are endowed with superior functional fitness compared with other prebiotically available building blocks absent nowadays from the code [6]:

(a) Use of hydroxy-acids instead of amino acids as monomers generates polyesters, which are more labile toward basic hydrolysis than polypeptides.
(b) Although polypeptides conceivably might be replaced by polymers of alternating dicarboxylic acids and diamines, polymerization of a single class of subunits is inherently simpler than co-polymerization of two (or more) different classes of subunits.
(c) Prebiotic syntheses give higher yields of α-amino acids than β- or γ-amino acids. Compared with poly-α-amino acids, poly-β- or

γ-amino acids also do not form stable secondary structures well on account of higher entropy from an increased number of freely rotating single bonds in the backbone.

(d) Since mixing of L- and D-amino acids in proteins will cause loss of steric uniformity along a polypeptide chain, the encoded amino acids are not surprisingly all of the same configuration. While the peptide antibiotics gramicidin S, bacitracin, and polymyxin B contain both D- and L-amino acids [23], and fusaricidin B from *Bacillus polymyxa* contains even more D- than L-amino acids [24], these D-monomer-containing peptides are synthesized by non-ribosomal peptide synthetase. Ribosomal protein synthesis utilizes strictly L-monomers. The discovery of enantiomeric excesses of L over D in some meteoric amino acids suggests that the adoption of L-amino acids for proteins might stem from such astrochemical excesses [17].

(e) NMR evidence points to restricted motion of the γ- and δ-methyl groups of Ile and Leu relative to the ε-methyl group of norleucine. Thus, the exclusion of the straight-chain α-amino-*n*-butyric acid, norvaline, and norleucine might be due to the ability of peptides of the branched Val, Leu, and Ile to form a more highly ordered tertiary structure.

(f) Arg usefully brings a permanent positively-charged side chain to proteins. Its lower homologs α-amino-β-guanidino propionic acid and α-amino-γ-guanidino butyric acid show an undesirable tendency to cyclize.

(g) Since the lower homologs of Lys, namely α,β-diaminopropionic acid, α,γ-diaminobutyric acid, and ornithine, are prone to acyl migration or lactamization, Lys is the shortest diamino acid suitable for proteins.

These advantages of the currently encoded amino acids over potential building blocks like hydroxy-acids, alternating dicarboxylic acid–diamine units, β- or γ-amino acids, α-amino-*n*-butyric acid, norvaline, norleucine, α-amino-β-guanidino propionic acid, α-amino-γ-guanidino butyric acid, α,β-diaminopropionic acid, α,γ-diaminobutyric acid, and ornithine suggest that, with respect to the choice of its member amino acids, the genetic code is mainly a rational code. To achieve this unmistakable rationality, all primitive codes burdened with unfit building blocks must be eliminated by natural selection, i.e. through evolutionary turnovers of the code.

8.4 THE DEEP FREEZE

Despite the strong evolutionary incentive of having an increased number of encoded amino acids, the Phase 2 expansion ceased at 23 canonical amino acids plus the ter signal. A contributing factor of the cessation might be found in the two-out-of-three mode of codon reading by anticodons *in vitro* [25, 26]: owing to the stronger G–C base pairing compared with A–U base pairing, the anticodon on a primitive tRNA would tend to read all four codons in any codon box with one or two G or C in the first two codonic bases. This would limit the maximum number of encodable amino acid/ter units to two each (one to read the two NNY codons in the box and one to read the two NNR codons) for the UUN, UAN, AUN, and AAN boxes, but only one each for the remaining 12 boxes, totaling 20 encodable [amino acid + ter] units for the primitive code. While the finalized code managed to refine the reading mode and call upon a sharing of the UAG, UGA, and AUG codons to accommodate as many as 24 [amino acid + ter] units, further crowding of the code understandably would evoke rising resistance through increased noise/benefit ratios [27]. It became more advantageous instead to seek continued enrichment of the protein side-chain repertoire through PTMs rather than through packing yet more amino acids into the code.

Protein molecules contain over a hundred kinds of supernumerary amino acid residues that make their appearances not under the direction of any codon but through PTMs brought about by modifying enzymes. Examples of functionally important PTMs are plentiful, including: (i) conversion of Glu residues on prothrombin to γ-carboxyGlu to increase its affinity for chelating Ca^{2+}, thereby enhancing the anchoring of prothrombin to phospholipid platelet membranes following injury; (ii) requirement for proteolytic PTMs in blood clotting and formation of pituitary peptides from proopiomelanocortin; (iii) utilization of protein phosphorylations in a host of signal transductions; and (iv) involvement of protein glycosylations in cell–cell interactions and many other functions. The wide occurrence of PTMs attests to the magnitude of increased variety of protein side chains as an evolutionary incentive. This incentive brought about both the Phase 2 expansion and the wide ranging PTMs which constitute a Phase 3 recruitment for the side-chain repertoire of proteins [14] (Table 8.2).

All living organisms are traceable back to a last universal common ancestor (LUCA), close to the methanogenic archaeon *Methanopyrus* [28, 29]. The minimal LUCA genome contains a number of enzymes that could participate in PTMs; for example, diphthamide synthase subunit DPH2, diphthamide biosynthesis methyltransferase,

Table 8.2 Recruitment of protein sidechain repertoire

Phase	Recruitment mechanism	Recruited amino acids
1	Natural proteome-wide encoding	Gly, Ala, Ser, Asp, Glu, Val, Leu, Ile, Pro, Thr
2	Natural proteome-wide encoding	Phe, Tyr, Arg, His, Asn, Gln, Lys, Trp, Cys, Met
	Natural site-specific encoding	Sec, Pyl, fMet
3	Post-translational modification	Diphthamide, glycosyl-Asn, Cys-heme, pyro-Glu, His-flavin, desmosine, adenylyl-Tyr, phospho-Tyr, etc.
4	Synthetic proteome wide encoding	4FTrp, 5FTrp, 6FTrp, azaLeu
	Synthetic site-specific encoding	p-AminoPhe, p-azidoPhe, p-benzoylPhe, β-(2-naphthyl)Ala, α-aminocaprylic acid, dansyl side chain, etc.

deoxyhypusine synthase, signal peptidase, histone acetyltransferase, serine/threonine protein kinase, peptidyl prolyl *cis–trans* isomerase, and glycosyl transferase [30, 31]. In accord with this, diphthine, the deamidated form of diphthamide, a modified His present in eukaryotic elongation factor 2, where it can be ADP-ribosylated by diphtheria toxin [32], is found in acid protein hydrolyates from the archaeon *Halobacterium halobium* [33]. Also, deoxyhypusine synthase, which together with deoxyhypusine hydroxylase transform a conserved Lys on initiation factor IF-5A in archaea and eukaryotes to hypusine, has an archaeal origin [34]. Thus, some PTMs, including diphthamide and deoxyhypusine, are as old as LUCA, which strongly points to the cessation of natural code expansion and switch to PTMs occurring prior to the rise of LUCA around 3.6 billion years ago [3, 4]. However, the absence of many PTM enzymes from the minimal LUCA genome also suggests that the majority of current PTMs are post-LUCA in origin. Minor variations of the genetic code with respect to codon assignments are known in numerous nuclear and organelle systems, especially metazoan mitochondria [35], and likewise common post-LUCA occurrences.

In contrast to the post-LUCA PTMs and codon reassignments, the 20 proteome-wide amino acid alphabet, which does not vary among organisms or organelles, is thus estimated to be some 3.6 billion years old. The site-specific entries of Sec and Pyl into the code were also unlikely to take place much later than the pre-LUCA switch to PTMs. Even fMet is likely as old as the Bacteria domain dating from 2.5–3.2 billion years ago [36]. Accordingly, with the cessation of Phase 2 expansion and switch to PTMs, the genetic code has gone into a 3 billion year deep freeze. The restart of its expansion and turnover had to await human exploration and development of synthetic genetic codes.

8.5 SYNTHETIC CODE EXPANSION

The chemical world of L-amino acids far exceeds the "natural" amino acids found in proteins, counting both the canonical 23 and those derived from PTMs. The "unnatural" amino acids could be natural metabolites/intermediates in cells, or produced by organic synthesis. Where an unnatural amino acid resembles a canonical amino acid sufficiently to result in the charging of the latter's tRNA by the aaRS cognate to the tRNA, it will be able to participate in ribosomal protein synthesis under codon direction. Many unnatural amino acids are thus incorporated into cellular proteins [37], often finding important applications; for example, fluoro-amino acids for ^{19}F-NMR protein studies [38] and selenoMet for protein crystallography [39].

Since an unnatural amino acid analogue is structurally different from its canonical counterpart, its replacement of the latter at all positions in a proteome easily results in loss of indefinite cell propagation. For example, L-canavanine (Cav), an Arg analogue from plants, is a substrate for ArgRS and readily incorporated into proteins in place of Arg. However, the cytotoxicity of the aberrant Cav-proteins is such that Cav may be employed as a therapeutic agent against pancreatic cancer [40]. Consequently, the encoded alphabet of proteins can be expanded to admit unnatural amino acids provided that they are adequately incorporated into proteins, and the ensuing toxicities are reduced by mutating the sensitive proteins or restricting the incorporation to specific sites within the proteome. Either way, the expansion will lead to a synthetic genetic code-based Phase 4 recruitment of extra protein side chains, continuing where Phase 2 expansion stopped (Table 8.2). The scope of this Phase 4 recruitment is potentially as wide as that of unnatural amino acids activatable for bonding to tRNA by some wild type, mutated or newly designed aaRS.

8.5.1 Proteome-Wide Expansion

The 3 billion year deep freeze of the genetic code raises the question of whether it is intrinsically mutable at all. To test its mutability, experiments were conducted to change the code in *Bacillus subtilis* QB928 [41]. Although this Trp-autotrophic QB 928 strain could be propagated only on Trp and not its fluoro-analogues 4FTrp, 5FTrp, or 6FTrp, after two rounds of mutations it gave rise to the LC33 strain, which is readily propagated on both Trp and 4FTrp. The mutations have thus expanded the code to enable the proteome-wide encoding of 4FTrp as a propagation-supporting amino acid, replacing Trp at all the 12 625 Trp positions in the *B. subtilis* proteome. Recent advances furthering the establishment or facilitation of synthetic proteome-wide code expansion include:

(a) The 4FTrp-propagatable *B. subtilis* LC33 strain has been mutated to yield the LC88 strain that can be propagated on Trp, 4FTrp, 5FTrp, or 6FTrp, with Trp:4FTrp:5FTrp:6FTrp growth-rate ratios equal to 100:36.3:9.0:10.9. Thus, all three fluoro-Trp analogues, normally toxic to the cells, serve as competent protein building blocks for this strain [3, 42].

(b) *E. coli* has also been mutated to yield an *unColi* B7-3 strain that can be propagated on 4FTrp without a detectable level of Trp incorporation into proteins [43, 44].

(c) Phenotypic suppression has been introduced as an effective approach for inducing propagation dependence on an unnatural amino acid [45]: *E. coli* was rendered thymidine auxotrophic by the Arg126Leu mutation of thymidylate synthase, and without added thymidine the resultant *E. coli thyA* R126L strain could be propagated only in the presence of azaLeu, the proteome-wide incorporation of which replaces L126 with azaLeu126 sufficiently to restore an essential positive charge at that position in some of the enzyme molecules.

(d) Mutational blocking of the editing function of aaRS enhances the incorporation of unnatural amino acids into cellular proteins, as illustrated by ValRS [46, 47], LeuRS [48, 49], and IleRS [50]. In the case of IleRS, the removal of editing actually gives the mutant cells a growth yield advantage over the wild type. Adequate incorporation of an unnatural amino acid is a prerequisite to the isolation, by means of either growth rate mutation or phenotypic suppression, of an altered genetic code displaying propagation-dependence on the amino acid.

(e) Mutating the amino acid substrate specificity of aaRS to increase unnatural amino acid incorporation has been achieved with *E. coli* PheRS to enable the incorporation of *p*-chloroPhe [51].

(f) The intracellular concentration of an aaRS with low reactivity toward an unnatural amino acid can be raised by overexpression to increase incorporation of the amino acid; for example, in the successful incorporation of 2-aminohexanoic acid, selenoMet, telluroMet, ethionine [52], and *trans*-crotylglycine [53] into E. coli proteins.
(g) The adaptation of protein sequences to an unnatural amino acid has been delineated for bacteriophage Qβ in adaptation to growth on 6FTrp, providing evidence for genetic code divergence through ambiguous intermediates that can simultaneously accommodate more than one amino acid at a given codon [54].

8.5.2 Site-specific Expansion

In the Phase 2 code expansion, the then novel amino acids Phe, Tyr, Arg, His, Asn, Gln, Lys, Trp, Cys, and Met gained entry into the code by acquiring their own codons for proteome-wide encoding, whereas Sec, Pyl, and fMet each secured the use of only a shared codon for site-specific encoding. In synthetic code expansion, the exact parallel applies and, besides synthetic proteome-wide encodings, it has been possible to add novel amino acids to the code through site-specific encodings:

(a) Some aaRS is found to display strikingly low reactivity toward tRNA from other species, especially when the other species come from the other side of an aaRS–tRNA reactivity schism that exists between the Archaea–Eukarya and Bacteria blocs [55]; Table 8.3). These deficient cross-bloc reactivities between aaRS and tRNA arise from the usages of dissimilar tRNA identity elements by aaRS from the two blocs, which can be altered by even a single base change [56, 57].
(b) On the basis of such deficient cross-bloc reactivities, the strategy has been devised whereby an *orthogonal* tRNA–aaRS pair, e.g. archaeal TyrRS–tRNA^{Tyr} from *Methanococcus jannaschii*, LeuRS–tRNA^{Leu} from *Methanobacterium thermoautotrophicum*, or GluRS from *Pyrococcus horikoshii* with a consensal archaeal tRNA^{Glu} can be introduced into E. coli incurring minimal reactions with host aaRS and tRNAs [58]. By fashioning the orthogonal pair to accept an unnatural amino acid and decode a special codon such as a nonsense codon, this orthogonal aaRS–tRNA approach has brought about the site-specific encodings of over

Table 8.3 Relative aminoacylations of different sources of tRNA by *E. coli* aaRS with various amino acids (after [55])

tRNA	Phe	Leu	Asp	Lys	Arg	Tyr	Met	Val	Ser	Thr	His	Pro
Escherichia coli	100	100	100	100	100	100	100	100	100	100	100	100
Bacteria												
Agrobacterium tumefaciens	149	118	77	12	67	93	108	67	70	109	45	101
Arthrobacter luteus	47	38	2	27	58	19	61	32	27	63	15	89
Bacillus stearothermophilus	164	180	58	106	114	128	115	131	64	140	44	110
Bacillus subtilis	156	208	156	71	119	187	137	87	91	97	76	94
Micrococcus luteus	20	48	2	18	88	34	19	14	65	67	41	88
Myxococcus xanthus	176	159	114	11	78	104	99	61	72	105	29	99
Rhodopseudomonas spheroides	54	100	12	75	58	4	61	44	63	108	6	86
Thermus aquaticus	51	80	7	28	72	2	65	24	102	91	41	87
Archaea												
Halobacterium cutirubrum	2	0.4	3	2	0.4	0.5	29	30	0	69	7	1
Eukarya												
Yeast	1	2	1	14	0.5	4	43	1	4	69	10	15
Wheat germ	8	2	2	5	62	3	25	2	1	56	24	5
Rat liver	2	0.4	0.5	1	16	0.3	6	3	0	2	7	2

30 unnatural amino acids in *E. coli* or eukaryotic cells, including *p*-aminoPhe, *p*-azidoPhe, *p*-benzoylPhe, β-(2-naphthyl)Ala, α-aminocaprylic acid, dansyl sidechain, etc. [59].
(c) In the case of *p*-aminoPhe, it has also been possible to generate a biosynthetic pathway to render its incorporation into *E. coli* autonomous [60].
(d) The site-specific incorporations employing orthogonal aaRS–tRNAs can be directed by quadruplet codons in addition to nonsense codons [61].
(e) The use of a TyrRS–tRNA pair from *E. coli* engineered to favor activation of 3-iodoTyr over Tyr has facilitated the site-specific encoding of 3-iodoTyr in Chinese hamster ovarian and human embryonic kidney cells [62].
(f) A complete set of orthogonal *E. coli* GlnRS–tRNA^{Gln} suppressor pairs has been generated for site-specific encoding by the amber, ocher, and opal nonsense codons in mammalian cells [63].
(g) A protocol has been developed to reprogram aaRS to attach a desired unnatural amino acid but no natural amino acid to a UAG-decoding tRNA in eukaryotic cells [64].
(h) Orthogonal ribosome–mRNA pairs that read only messages encoding an unnatural amino acid can be employed to enhance incorporation of the amino acid via amber codon suppression [65].

8.6 SYNTHETIC CODE TURNOVER

The possible occurrence of natural code turnovers in primordial ages to rid the code of unfit amino acids finds experimental support from observations on synthetic genetic codes. The *B. subtilis* LC33 strain, which can be propagated on Trp and also more slowly on 4FTrp, gave rise to the HR7 strain, which grows faster on 4FTrp than on Trp. Another round of mutation converted HR7 to HR15, which can be propagated on 4FTrp but not on Trp [41]. For HR15 and its derivatives, Trp has in fact become an inhibitory analogue against propagation on 4FTrp [3, 13]. This displacement of Trp by 4FTrp from the alphabet of propagation-supportive amino acids in HR15 signifies a turnover of the preexistent Trp-propagatable code. If Trp can be removed from the code by mutations even after more than 3 billion years of evolutionary adaptation to Trp as a canonical building block, it would be simple matter for the code to turn over in ages past to eliminate such prebiotically available but functionally unfit building blocks as α-amino-*n*-butyric acid, α-aminoisobutyric acid, norvaline, etc. from the code. Devoid of turnovers, the primordial genetic code, once formed, would have to remain a frozen accident compelled from day one to contain all the amino acids that happened to be abundant in the prebiotic environment regardless of their functional fitness. Allowed turnovers, the code would be free to evolve toward greater competence. The functionally superb 23 amino acid alphabet of today is, therefore, the result of active selection, not blind luck.

8.7 USEFULNESS OF SYNTHETIC CODES

Synthetic genetic codes are useful for two important purposes. First, they provide unique insight into the origin and evolution of the natural code. By removing all doubts regarding the intrinsic mutability of the code, they have firmly established the feasibility of early code turnovers to eliminate unfit amino acids and code expansions to bring in new amino acids supplied by biosynthesis. It is interesting that both natural and synthetic code expansions can be implemented by either a proteome-wide or a site-specific mechanism. The natural code contains, side by side, 20 amino acid members receiving proteome-wide encodings, and three receiving site-specific encodings. Although all the synthetic codes developed so far involve either the proteome-wide or the site-specific encoding of an unnatural amino acid, a combination of both types of encoding within the same organism could be both attainable and useful [66]. Besides the study of synthetic codes, there is at present no other experimental avenue to inquire into the mechanisms of genetic code expansion and turnover in the primordial past.

Second, synthetic genetic codes widen substantially the range of protein engineering, or code engineering [67], by allowing permutations of not only the sequences in which the amino acid letters of the protein alphabet are arranged, but also the composition of the alphabet itself. This has enabled the production of unnatural amino acid-containing (Unac) proteins in the cell. For *B. subtilis* LC33, its proteins can be produced optionally containing either the fluorescent Trp or the non-fluorescent 4FTrp [68], and the 4FTrp-containing Unac proteins make possible the measurement of interactions between proteins and fluorescent ligands in the absence of any Trp fluorescence from the proteins [69]. Unac proteins from site-specific synthetic encodings are rapidly paving the way to a wide variety of potential applications; for example, preparation of photo-cross-linkable artificial extracellular matrix proteins, selective dye-labeling of newly synthesized proteins [70, 71], probes for protein conformations, redox-active proteins, and proteins with engineered metal-binding sites [59]. Therapeutic applications of Unac proteins may include the production of super-tight binding antibodies as improved vaccines against cancers, bacteria, and viruses, super-high affinity protein ligands for autoantibodies, receptors/enzymes, or tissue surfaces to block autoimmune reactions, modulate cellular responses, or to serve as surgical glues. Also, appropriate pigmented side chains might be inserted into Unac photosystem proteins to enhance the efficiency of plant and microbial photosynthesis, and so on.

8.8 DISCUSSION

The displacement of Trp by 4FTrp in the genetic code of *B. subtilis* HR15 has changed the encoded amino acid alphabet of proteins. Because the alphabet has been an invariant attribute of earthly life since time immemorial, this has led to the suggestion that *B. subtilis* HR15 represents a new type of life [72]. On account of its alternate utilization of 4FTrp in place of Trp, *E. coli* B7-3 also has been named an *unColi* [43, 44], on which basis *E. coli thyA* R126L would be an *unColi* as well, and *B. subtilis* LC33, LC88, and HR15 would be *unSubtilis*. The engineered organisms bearing genetic codes with site-specific expansion would likewise be *unColi*, *unYeast*, etc. To classify this emergent profusion of novel life forms, it is proposed that:

1. Free-living life forms that depart from earthly life with respect to one or more of the universal and, therefore, defining attributes of earthly life may be referred to as *synthetic life* to distinguish them from natural life, synthetic plasmids, synthetic viruses, synthetic parasites, synthetic organelles, and computational artificial life. The universal attributes of earthly life include:
   (a) genetic code with a canonical 23-amino acid alphabet;
   (b) TCAG deoxyribonucleotide alphabet of DNA;
   (c) UCAG ribonucleotide alphabet of RNA;
   (d) bilayer cell membranes consisting of ester- and ether-based lipids.
2. Synthetic life can exist in two different forms. In the *o-synthetic* (optional-synthetic) form, the organism can be propagated using either the standard universal set of attributes of natural life or the

SYNTHETIC GENETIC CODES AS THE BASIS OF SYNTHETIC LIFE

Table 8.4 Developed and potential genetic code-based synthetic life systems

Type of synthetic life	Mechanistic basis	System
o-Synthetic	Proteome-wide encoding	*B. subtilis* LC33, LC88, *E. coli* B7-3
m-Synthetic	Proteome-wide encoding	*B. subtilis* HR15, *E. coli thyA* R126L
o-Synthetic	Site-specific encoding	*E. coli* with optional insertion of *p*-aminoPhe or *p*-azidoPhe etc
m-Synthetic	Site-specific encoding	Potential *E. coli thyA* R126 derivative dependent on site- specific azaLeu insertion for propagation

altered o-synthetic set of attributes. In the *m-synthetic* (mandatory-synthetic) form, the use of the altered m-synthetic set of attributes is compulsory, and the organism cannot be propagated using the standard universal set of attributes of natural life.

Examples of developed and potential genetic code-based o-synthetic and m-synthetic life are shown in Table 8.4. Notably, there is hitherto a lack of m-synthetic life forms based on a site-specific insertion of unnatural amino acid, but achievement of this kind of synthetic life may not be beyond reach. For instance, the o-synthetic *E. coli thyA* R126L system [45] can be modified by converting the *thyA* R126 codon to a UAG nonsense codon: insertion of azaLeu into this position using an orthogonal aaRS–tRNA pair might then yield in the absence of thymidine an m-synthetic *unColi* displaying propagation dependence on the inserted azaLeu.

Synthetic biology spans different approaches for building new organismic constructs to advance biology and bioengineering. Among them, synthetic life is directed specifically to the development, analysis, and use of free-living life forms characterized by a departure from one or more of the universal characteristics of earthly life. The *unSubtilis* and *unColi* with proteome-wide encodings, and the array of prokaryotic and eukaryotic organisms with site-specific encodings, for an unnatural amino acid have demonstrated the attainability of free living and indefinitely propagatable synthetic life. It is expected that, in time, synthetic life based on unnatural DNA alphabets, RNA alphabets, cell membranes, or other radical innovations will also be achieved. Moreover, these different bases of synthetic life are not mutually exclusive, and synthetic life forms can come to comprise simultaneously synthetic versions of the genetic code, DNA and RNA alphabets, and cell membrane. The arrival of synthetic life, overthrowing the severe chemical constraints imposed on natural life by the invariance of its universal attributes, has begun a *sequel* to the life that has held exclusive reign over the first half of the projected lifespan of planet Earth [73].

In contrast to natural life which has gone through 3 billion years of diversification and optimization, synthetic life is just entering the nascent moments of its birth. Already it has deepened our understanding of genetic code origin and evolution. How it may unfold in the eons to come will be a fascination to behold as much as a road to chart with great care.

ACKNOWLEDGMENTS

We thank the Research Grants Council of Hong Kong (Grant No. HKUST6437/06M) for support.

REFERENCES

1. Wong, J.T. and Xue, H. (2002) Self-perfecting evolution of heteropolymer building blocks and sequences as the basis of life, in *Fundamentals of Life* (eds G. Palyi, C. Zucchi, and L. Caglioti), Elsevier, Paris, pp. 473–494.
2. Orgel, L.E. (2006) Prebiotic chemistry and the origin of the RNA world. *Critical Reviews in Biochemistry and Molecular Biology*, **39**, 99–123.
3. Wong, J.T. (2009a) Genetic code, in *Prebiotic Evolution and Astrobiology* (eds J.T. Wong and A. Lazcano), Landes Bioscience, Austin, TX, pp. 110–119.
4. Joyce, G.F. (2002) The antiquity of RNA-based evolution. *Nature*, **418**, 214–221.
5. Wong, J.T. (1991) Origin of genetically encoded protein synthesis: a model based on selection for RNA peptidation. *Origins of Life and Evolution of the Biosphere*, **21**, 165–176.
6. Weber, A.L. and Miller, S.L. (1981) Reasons for the occurrence of the twenty coded protein amino acids. *Journal of Molecular Evolution*, **17**, 273–284.
7. Wong, J.T. and Bronskill, P.M. (1979) Inadequacy of prebiotic synthesis as origin of proteinous amino acids. *Journal of Molecular Evolution*, **13**, 115–125.
8. Wong, J.T. (1984) Evolution and mutation of the amino acid code, in *Dynamics of Biochemical Systems* (eds J. Ricard and A. Cornish-Bowden), Plenum, New York, pp. 247–257.

9. Wong, J.T. (2007) Coevolution theory of the genetic code: a proven theory. *Origins of Life and Evolution of the Biosphere*, **37**, 403–408.
10. Commans, S. and Bock, A. (1999) Selenocysteine inserting tRNAs: an overview. *FEMS Microbiology Reviews*, **23**, 335–351.
11. Wong, J.T. (1988) Evolution of the genetic code. *Microbiological Sciences*, **5**, 174–181.
12. Wong, J.T. (1975) A co-evolution theory of the genetic code. *Proceedings of the National Academy of Sciences of the United States of America*, **72**, 1909–1912.
13. Wong, J.T. (2005) Coevolution theory of the genetic code at age thirty. *BioEssays*, **27**, 416–425.
14. Wong, J.T. (1981) Coevolution of the genetic code and amino acid biosynthesis. *Trends in Biochemical Sciences*, **6**, 33–35.
15. Kobayashi, K., Kaneko, T., Saito, T., and Oshima, T. (1998) Amino acid formation in gas mixtures by high energy particle irradiation. *Origins of Life and Evolution of the Biosphere*, **28**, 155–165.
16. Kobayashi, K., Tsuchiya, M., Oshima, T., and Yanagawa, H. (1990) Abiotic synthesis of amino acid and imidazole by proton irradiation of simulated primitive Earth atmospheres. *Origins of Life and Evolution of the Biosphere*, **20**, 99–109.
17. Pizzarello, S. (2009) Meteorites and the chemistry that preceded life's origin, in *Prebiotic Evolution and Astrobiology* (eds J.T. Wong and A. Lazcano), Landes Bioscience, Austin, TX, pp. 46–51.
18. Feng, L., Sheppard, K., and Namgoong, S. *et al.* (2004) Aminoacyl–tRNA synthesis by pretranslational amino acid modification. *RNA Biology*, **1**, 16–20.
19. Ambrogelly, A., Palioura, S., and Soll, D. (2007) Natural expansion of the genetic code. *Nature Chemical Biology*, **3**, 29–35.
20. Turanov, A.A., Lobanov, A.V., and Fomenko, D.E. *et al.* (2009) Genetic code supports targeted insertion of two amino acids by one codon. *Science*, **323**, 259–261.
21. Longstaff, D.G., Larue, R.C., and Faust, J.E. *et al.* (2007) A natural genetic code expansion cassette enables transmissible biosynthesis and genetic encoding of pyrrolysine. *Proceedings of the National Academy of Sciences of the United States of America*, **104**, 1021–1026.
22. Wolman, Y., Haverland, W.J., and Miller, S.L. (1972) Nonprotein amino acids from spark discharges and their comparison with the Murchsion meteorite amino acids. *Proceedings of the National Academy of Sciences of the United States of America*, **69**, 809–811.
23. Hancock, R.E.W. and Chapple, D.S. (1999) Peptide antibiotics. *Antimicrobial Agents and Chemotherapy*, **43**, 1317–1323.
24. Kajimura, Y. and Kaneda, M. (1997) Fusaricidins B, C, and D, new depsipeptide antibiotics produced by *Bacillus polymyxa* KT-8: isolation, structure elucidation and biological activity. *Journal of Antibiotics*, **50**, 230–238.

25. Lagerkvist, U. (1981) Unorthodox codon reading and the evolution of the genetic code. *Cell*, **23**, 305–306.
26. Lagerkvist, U. (1978) "Two out of three": an alternative method for codon reading. *Proceedings of the National Academy of Sciences of the United States of America*, **75**, 1759–1762.
27. Wong, J.T. (1976) The evolution of a universal genetic code. *Proceedings of the National Academy of Sciences of the United States of America*, **73**, 2336–2340.
28. Xue, H., Tong, K.L., and Marck, C. et al. (2003) Transfer RNA paralogs: evidence for genetic code–amino acid biosyntheis coevolution and an archaeal root of life. *Gene*, **310**, 59–66.
29. Wong, J.T., Chen, J., and Mat, W.K. et al. (2007) Polyphasic evidence delineating the root of life and roots of biological domains. *Gene*, **403**, 39–52.
30. Mat, W.K., Xue, H., and Wong, J.T. (2008) The genomics of LUCA. *Frontiers in Bioscience*, **13**, 5605–5613.
31. Wong, J.T. (2009b) Root of life, in *Prebiotic Evolution and Astrobiology* (eds J.T. Wong and A. Lazcano), Landes Bioscience, Austin, TX, pp. 120–144.
32. Bodley, J.W., Upham, R., and Crow, F.W. et al. (1984) Ribosyl-diphthamide: confirmation of structure by fast atom bombardment mass spectrometry. *Archives of Biochemistry and Biophysics*, **230**, 590–593.
33. Pappenheimer, A.M. Jr, Dunlop, P.C., Adolph, K.W., and Bodley, J.W. (1983) Occurrence of diphthamide in archaebacteria. *Journal of Bacteriology*, **153**, 1342–1347.
34. Brochier, C., Lopez-Garcia, P., and Moreira, D. (2004) Horizontal gene transfer and archaeal origin of deoxyhypusine synthase homologous genes in bacteria. *Gene*, **330**, 169–176.
35. Abascal, F., Zardoya, R., and Posada, D. (2006) GenDecoder: genetic code prediction for metazoan mitochondria. *Nucleic Acid Research*, **34**, w389–w393.
36. Battistuzzi, F.U., Feijao, A., and Hedges, S.B. (2004) A genomic timescale of prokaryotic evolution: insights into the origin of methanogenesis, phototrophy and the colonization of land. *BMC Evolutionary Biology*, **4**, 44–57.
37. Cowie, D.B. and Cohen, G.N. (1957) Biosynthesis by *Escherichia coli* of active altered proteins containing selenium instead of sulfur. *Biochimica et Biophysica Acta*, **26**, 252–261.
38. Sun, Z.Y., Truong, H.T., and Pratt, E.A. et al. (1993) A ^{19}F-NMR study of the membrane-binding region of D-lactate dehydrogenase of *Escherichia coli*. *Protein Science*, **2**, 1936–1947.
39. Hendrickson, W.A., Horton, J.R., and LeMaster, D.M. (1990) Selenomethionyl proteins produced for analysis by multiwavelength anomalous diffraction (MAD): a vehicle for direct determination of three-dimensional structure. *EMBO Journal*, **9**, 1665–1672.

40. Bence, A.K. and Crooks, P.A. (2003) The mechanism of L-canavanine cytotoxicity: arginyl tRNA synthetase as a novel target for anticancer drug discovery. *Journal of Enzyme Inhibition and Medicinal Chemistry*, **18**, 383–394.
41. Wong, J.T. (1983) Membership mutation of the genetic code: loss of fitness by tryptophan. *Proceedings of the National Academy of Sciences of the United States of America*, **80**, 6303–6306.
42. Mat, W.K., Xue, H., and Wong, J.T. (2004) Genetic encoding of 4-, 5-, and 6-fluorotryptophan residues: role of oligogenic barriers. American Society for Microbiology 104th Meeting, R-029.
43. Bacher, J.M. and Ellington, A.D. (2001) Selection and characterization of *Escherichia coli* variants capable of growth on an otherwise toxic tryptophan analogue. *Journal of Bacteriology*, **183**, 5414–5425.
44. Bacher, J.M. and Ellington, A.D. (2007) Global incorporation of unnatural amino acids in *Escherichia coli*. *Methods in Molecular Biology*, **352**, 23–34.
45. Lemeignan, B., Sonigo, P., and Marliere, P. (1993) Phenotypic suppression by incorporation of an alien amino acid. *Journal of Molecular Biology*, **231**, 161–166.
46. Doring, V., Mootz, H.D., and Nangle, L.A. et al. (2001) Enlarging the amino acid set of *Escherichia coli* by infiltration of the valine coding pathway. *Science*, **292**, 501–504.
47. Bock, A. (2001) Invading the genetic code. *Science*, **292**, 453–454.
48. Mursina, R.S. and Martinis, S.A. (2002) Rational design to block amino acid editing of a tRNA synthetase. *Journal of the American Chemical Society*, **124**, 7286–7287.
49. Tang, Y. and Tirrell, D.A. (2002) Attenuation of the editing activity of the *Escherichia coli* leucyl–tRNA sythetase allows incorporation of novel amino acids into proteins *in vivo*. *Biochemistry*, **41**, 10635–10645.
50. Pezo, V., Metzgar, D., and Hendrickson, T.L. et al. (2004) Artificially ambiguous genetic code confers growth yield advantage. *Proceedings of the National Academy of Sciences of the United States of America*, **101**, 8593–8597.
51. Kast, P. and Hennecke, H. (1991) Amino acid substrate specificity of *Escherichia coli* phenylalanine–tRNA synthetase altered by distinct mutations. *Journal of Molecular Biology*, **222**, 99–124.
52. Budisa, N., Steipe, B., and Demange, P. et al. (1995) High-level biosynthetic substitution of methionine in proteins by its analogues 2-aminohexanoic acid, selenomethionine, telluromethionine and ethionine in *Escherichia coli*. *European Journal of Biochemistry*, **230**, 788–796.
53. Kiick, K.L., van Hest, J.C.M., and Tirrell, D.A. (2000) Expanding the scope of protein biosynthesis by altering the methionyl–tRNA synthetase activity of a bacterial expression host. *Angewandte Chemie International Edition in English*, **39**, 2148–2152.

54. Bacher, J.M., Bull, J.J., and Ellington, A.D. (2003) Evolution of phage with chemically ambiguous proteomes. *BMC Evolutionary Biology*, **3**, 24.
55. Kwok, Y. and Wong, J.T. (1980) Evolutionary relationship between *Halobacterium cutirubrum* and eukaryotes determined by use of aminoacyl–tRNA synthetases as phylogenetic probes. *Canadian Journal of Biochemistry*, **58**, 213–218.
56. Xue, H., Shen, W., Giege, R., and Wong, J.T. (1993) Identity elements of tRNA^{Trp}: identification and evolutionary conservation. *Journal of Biological Chemistry*, **268**, 9316–9322.
57. Guo, Q., Gong, Q., and Grosjean, H. *et al.* (2002) Recognition by tryptophanyl–tRNA synthetases of discriminator base on the tRNA^{Trp} from three biological domains. *Journal of Biological Chemistry*, **277**, 14343–14349.
58. Santoro, S.W., Anderson, J.C., Lakshman, V., and Schultz, P.G. (2003) An archaealbacteria-derived glutamyl–tRNA synthetase and tRNA pair for unnatural amino acid mutagenesis of proteins in *Escherichia coli*. *Nucleic Acid Research*, **31**, 6700–6709.
59. Xie, J. and Schultz, P.G. (2006) A chemical toolkit for proteins – an expanded genetic code. *Nature Reviews Molecular Cell Biology*, **7**, 775–782.
60. Mehl, R.A., Anderson, J.C., and Santoro, S.W. *et al.* (2003) Generation of a bacterium with a 21st amino acid genetic code. *Journal of the American Chemical Society*, **125**, 935–939.
61. Anderson, J.C. and Schultz, P.G. (2003) Adaptation of an orthogonal archaeal leucyl–tRNA and synthetase pair for four-base, amber and opal suppression. *Biochemistry*, **42**, 9598–9608.
62. Sakamoto, K., Hayashi, A., and Sakamoto, A. *et al.* (2002) Site-specific incorporation of an unnatural amino acid into proteins in mammalian cells. *Nucleic Acid Research*, **30**, 4692–4699.
63. Kohrer, C., Sullivan, E.L., and RajBhandary, U.L. (2004) Complete set of orthogonal 21st aminoacyl–tRNA synthetase-amber, ochre and opal suppressor tRNA pairs: concomitant suppression of three different termination codons in an mRNA in mammalian cells. *Nucleic Acid Research*, **32**, 6200–6211.
64. Cropp, T.A., Anderson, J.C., and Chin, J.W. (2007) Reprogramming the amino acid substrate specificity of orthogonal aminoacyl–tRNA synthetases to expand the genetic code of eukaryotic cells. *Nature Protocols*, **2**, 2590–2600.
65. Wang, K., Neumann, H., Peak-Chew, S.Y., and Chin, J.W. (2007) Evolved orthogonal ribosomes enhance the efficiency of synthetic genetic code expansion. *Nature Biotechnology*, **25**, 770–777.
66. Bacher, J.M., Hughes, R.A., Wong, J.T., and Ellington, A.D. (2004) Evolving new genetic codes. *Trends in Ecology & Evolution*, **19**, 69–75.
67. Budisa, N. (2006) *Engineering the Genetic Code*, Wiley–VCH Verlag GmbH, pp. 90–183.

68. Bronskill, P.M. and Wong, J.T. (1988) Suppression of fluorescence of tryptophan residues in proteins by replacement with 4-fluorotryptophan. *Biochemical Journal*, **249**, 305–308.
69. Hogue, C.W.V., Doublie, S., and Xue, H. *et al.* (1996) A concerted tryptophanyl–adenylate-dependent conformational change in *Bacillus subtilis* tryptophanyl–tRNA synthetase revealed by the fluorescence of Trp92. *Journal of Molecular Biology*, **260**, 446–466.
70. Link, A.J., Mock, M.L., and Tirrell, D.A. (2003) Non-cannonical amino acids in protein engineering. *Current Opinion in Biotechnology*, **14**, 603–609.
71. Tirrell, D.A. (2008) Reinterpreting the genetic code: implications for macromolecular design, evolution and analysis, in *Physical Biology: From Atoms to Medicine* (ed. A.H. Zewail), Imperial College Press, pp. 165–187.
72. Hesman, T. (2000) Code breakers. Scientists are changing bacteria in a most fundamental way. *Science News*, **157**, 360–362.
73. Cohen, P. (2000) Life the sequel. *New Scientist*, **167**, 32–36.

9

Toward Safe Genetically Modified Organisms through the Chemical Diversification of Nucleic Acids

Piet Herdewijn[1] and Philippe Marliere[2]

[1] Laboratory for Medicinal Chemistry, Rega Institute for Medical Research, Minderbroedersstraat 10, Leuven-3000, Belgium
[2] Isthmus Sarl, 31 rue Saint Amand, F75015 Paris, France

9.1 Introduction 202
9.2 Specifications for an Orthogonal Episome 206
9.3 Diversification of the Backbone Motif 211
9.4 Diversification of the Leaving Group 215
9.5 Diversification of Nucleic Bases 217
9.6 Diversification of Nucleic Acid Polymerases 219
9.7 Conclusions 221
Acknowledgments 222
References 222

It is argued that nucleic proliferation should be rationally extended so as to enable the propagation *in vivo* of additional types of nucleic acids ("xeno-nucleic acids" or XNAs), whose chemical backbone motif would differ from deoxyribose and ribose and whose polymerization would not interfere with DNA and RNA biosynthesis. Because XNA building blocks do not occur in nature, they would have to be synthesized and supplied to cells which would be equipped with an appropriate enzymatic machinery for polymerizing them. The invasion of plants and animals with XNA replicons can be envisioned in the long run, but it is in microorganisms, and more specifically in bacteria, that the feasibility of such chemical systems and the establishment of genetic enclaves separated from DNA and RNA is more likely to take place. The introduction of expanded coding through additional or alternative pairing will be facilitated by the propagation of replicons based on alternative backbone motifs and leaving groups, as enabled by XNA polymerases purposefully evolved to this end.

9.1 INTRODUCTION

A seemingly inexhaustible source of innovation and evolution is provided by the chemical capability to synthesize DNA oligomers coupled with enzymatic procedures for gene assembly and amplification. The implantation of the resulting synthetic DNA constructs in chromosomes and episomes of microorganisms and plants for applied purposes in agriculture, energetics, and chemistry has triggered the proliferation of artificial genetic sequences of an ever-expanding scope. These artificial sequences are viewed by some as a threat to wild ecosystems and even suspected to lead astray the evolution of terrestrial life [1].

The first nucleic acid sequence designed and constructed to be propagated *in vivo* took the form of a synthetic gene of 77 bases encoding a suppressor tRNA that was inserted, replicated, and expressed in yeast cells [2]. This was followed by the assembly of a recombinant plasmid enabling the production of a foreign protein in *Escherichia coli* cells, dehydroquinase from *Neurospora crassa*, through the *in vitro* use of restriction endonucleases and DNA ligase [3]. The construction of artificial eukaryotic chromosomes of minimal size followed, first in yeast [4] and later in mice [5]. The advent of directed mutagenesis [6] and the elaboration of amplification techniques [7] through the judicious use of synthetic oligodeoxynucleotides provided potent and versatile protocols for genetic reprogramming of organisms [8] and directing their evolution [9, 10]. Recently, an entire bacterial chromosome of 582 970 base-pairs was synthesized through stages of hierarchical assembly in prokaryotic and then eukaryotic artificial chromosomes [11]. This synthetic chromosome is now expected to take over the genetic command of a bacterial cell *Mycoplasma genitalium*.

It is currently difficult to conceive biological experiments of scientific and industrial relevance today without genetic constructions resorting to DNA oligonucleotides. The chemical synthesis and the enzymatic amplification, assembly, and recombination of artificial genetic sequences for industrial purposes has become routine to the extent that transformation of ecosystems through the dissemination of synthetic genetic constructs (generically and misleadingly referred to as transgenes) is now a major topic of debate between the lay public and experts [12]. Indeed, the risks of genetic pollution cannot be overlooked, considering the uniformity of genetic alphabets, the universality of the genetic code, and the ubiquity of genetic interchanges throughout domesticated and wild species. Therefore, technologies for preventing or restricting genetic crosstalk between natural species and the artificial biodiversity needed for scientific and industrial progress should be designed and deployed to anticipate this challenge. Here, we argue that nucleic acid proliferation should be rationally extended so as to enable the propagation *in vivo* of additional types of nucleic acids ("xeno-nucleic acids" or XNAs), whose chemical backbone motif would differ from deoxyribose and ribose and whose polymerization would not interfere with DNA and RNA biosynthesis.

Because XNA building blocks do not occur in nature, they would have to be synthesized and supplied to cells which should be equipped with an appropriate enzymatic machinery for polymerizing them. The genetic information conveyed in XNAs would thus be expected to vanish if it could be ensured that its precursors could not be formed in existing metabolism and food chains. As a counterpart for the unavailability of such unnatural nutrients, the controlled propagation of XNA polymers would require the uptake of XNA precursors by reprogrammed host cells as well as by specific polymerization enzymes.

Despite the empirical character of biological engineering and the vast evolutionary potential of natural biodiversity, we surmise that human health and natural ecosystems will be more safely preserved by embodying genetic instructions in artificial nucleic material; that is, in replicons chemically separate from the support on which natural selection has acted so far. Natural nucleic acids originate by enzymatic polymerization of activated metabolic precursors linking a backbone motif to a nucleobase on the one hand and to a leaving group on the other hand; for example, 5'-phosphorylated D-ribose to N9 of adenine and to the oxygen atom of pyrophosphate in the RNA precursor ATP, and 5'-phosphorylated 2'-deoxy-D-ribose to the N1 of thymine and to the oxygen atom of pyrophosphate in the DNA precursor dTTP. Numerous variations have been introduced in the structure of nucleobases, culminating in the elaboration of pairing schemes that respect the

Watson–Crick base-pair geometry and can be combined with A:T and G:C pairs [13], and in the development of base-pair geometries separate from Watson and Crick's [14]. Variant backbone motifs whose polymers adopt stable double-stranded structures akin to DNA and RNA have been the subject of less systematic explorations, and activation groups deviating from pyrophosphate even less so.

Nevertheless, we believe that it is the *in vivo* creation of moieties alternative to the backbone motifs phosphoribose and phosphodeoxyribose and also to the leaving group pyrophosphate that will lead to the fastest advances, and that the introduction of expanded coding through additional or alternative pairing will be facilitated by the propagation of replicons based on alternative backbone motifs and leaving groups.

The invasion of plants and animals with XNA replicons can be envisioned in the long run, but it is in microorganisms, and more specifically in bacteria, that the feasibility of such chemical systems and the establishment of genetic enclaves separated from DNA and RNA is more likely to take place. Bacteriophages featuring modified bases could be considered as natural forays toward additional types of nucleic acid.

Known base modifications [15] have been all found to be linked to the DNA backbone and consist of close variations of the A:T and G:C pairing structures, from which they deviated, presumably to escape restriction endonucleases in host bacteria [16]. So far, only the four ribonucleoside triphosphates of A, U, C, and G, as well as the eight deoxyribonucleoside triphosphates of A, T, G, C, U, 5-hydroxymethylU,

Figure 9.1 Generalized scheme for selection of XNA biosynthesis *in vivo*. The formulas and symbols used for nucleobases (uracil and diaminopurine), leaving group (Z), backbone motif (six-membered heterocycles with X and Y variables), and import vectors (peptide) are arbitrary, and based on already existing constructions. At least two nucleobases are required for propagating *in vivo* the informational biosynthesis of XNA sequences. Propagation through replication of XNA:XNA duplexes would further imply that at least a couple of complementary nucleobases are incorporated, as depicted. The vital product could correspond to any metabolite consisting of, or that can be converted into, an indispensible building-block of the host cell and that the host cell is unable to synthesize. Alternatively, the vital product could be an essential RNA, the precursor substrates then corresponding to ribonucleoside triphosphates and the cellular co-catalyst to an RNA polymerase capable of taking XNA as template. Alternatively, the vital product could be an essential protein, the precursor substrates then corresponding to aminoacyl–tRNAs and the cellular co-catalyst to a ribosome capable of translating a messenger XNA

THE CHEMICAL DIVERSIFICATION OF NUCLEIC ACIDS

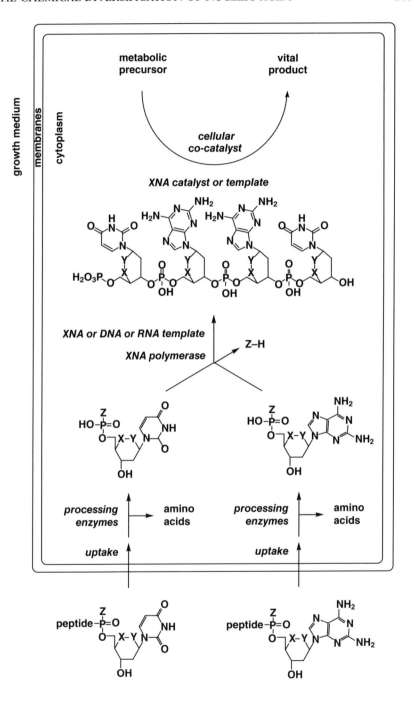

5-hydroxymethylC, and 2-aminoA, are thought to serve as substrates for nucleic acid polymerization in nature. Backbone diversification of nucleic acids thus seems entirely alien to extant living species, which suggests that the evolutionary barrier has remained too high for spontaneous innovation and natural selection to overcome.

If valid, this postulate should incite synthetic chemists and geneticists to take up the challenge, diversify nucleic acid scaffolds *in vivo* and hence access a safer level of informational transactions in engineered life forms. These artificial constructs would be known as orthogonal episomes (Figure 9.1).

9.2 SPECIFICATIONS FOR AN ORTHOGONAL EPISOME

Briefly stated, the specifications of an orthogonal episome are as follows:

1. The chemical nature of its backbone must differ from those of DNA and RNA in at least one of its two complementary strands.
2. It must carry sequences of at least two different nucleobases, either belonging to the canonical A:T and G:C pairs or to other pairs that can be combined with them.
3. It should be polymerized by a dedicated polymerase from activated precursors that do not interfere with DNA and RNA synthesis of the host cell.
4. It should convey information for at least one selectable function indispensible for growth of the host cell in specific culture media.

Several additional options can be envisaged. The orthogonal episome could be devised to act as a template that is transcribed into a functional RNA (mRNA, tRNA, or rRNA) from regular rNTPs normally present in cells by a dedicated RNA polymerase. Alternatively, it could act as a messenger that is translated by a specialized ribosome from regular aminoacyl–tRNAs into a functional protein. A third possibility is that it could directly catalyze a metabolic reaction vital to the host cell. Its activated precursors could be derived from metabolic intermediates of the host cell. Eventually, the orthogonal episome could be elaborated to encode the genes required for its replication, recombination, transcription, translation, or supply of metabolic precursors.

Seventeen different experimental scenarios combining four types of replication duplex and seven types of selection modality are listed in Table 9.1. Of the three types of replication duplex, XNA:DNA is the least attractive because the risk of illegitimate recombination of the DNA strand with the host genome, including plasmids, cannot be ruled out.

Direct translation of messenger XNA templates into proteins whose function can be selected *in vivo* by specialized versions of ribosomes is tantalizing in the long term. However, because of the complexity of ribosome assembly and function, this scheme does not seem amenable to straightforward selection protocols *in vivo* or *in vitro*. An additional difficulty would result from the opposite templating directions in translation (5' to 3') and in transcription and replication (3' to 5'), as exemplified by the intricate regulation of RNA virus cycles (Qβ).

Resorting to the smallest possible number of experimental steps of directed evolution, it would appear that the best-case scenarios in Table 9.1 would consist of:

(a) replicating XNA:XNA duplexes with one of the XNA strands acting as a metabolic enzyme; that is, a "xenozyme";
(b) replicating XNA:RNA heteroduplexes with the RNA strand acting as a metabolic ribozyme or a tRNA or rRNA essential for the host cell survival;
(c) transcribing a chromosomal or plasmidic DNA:DNA replicon into an XNA single strand which would act as a metabolic xenozyme.

In scenario (a), only one replicase enzyme would need to be evolved for reproducing the XNA:XNA duplex; that is, an XNA-dependent XNA polymerase. The same replicase could be used for PCR amplification and selection of XNA aptamers destined to function *in vivo* as xenozymes, once a reliable metabolic reaction and appropriate bacterial mutants lacking it have been validated.

In scenario (b), two polymerase enzymes would have to be evolved for replicating the XNA:RNA heteroduplex: a transcriptase and a reverse transcriptase; that is, an XNA-dependent RNA polymerase and an RNA-dependent XNA polymerase.

Both scenarios (a) and (b) correspond to proliferation modalities in which at least one XNA strand serves as template. Such is not the case in scenario (c), in which DNA serves as template for XNA, but XNA does not direct the polymerization of any nucleic acid. XNA would

Table 9.1 Divide-and-conquer plan for propagating XNA in bacterial cells

Replication duplex	XNA:XNA	XNA:RNA	XNA:DNA	DNA:DNA
XNA generation process	Replication	Transliteration of RNA template	Transliteration of DNA template	Transcription of DNA genes from promoter
Selection modality XNA strand as metabolic xenozyme	<ul><li>select *in vitro* an XNA aptamer</li><li>evolve XNA-dependent XNA polymerase</li><li>initiate and separate complementary strands</li><li>import *in vivo* XNA precursors</li></ul>	<ul><li>select *in vitro* an XNA aptamer</li><li>evolve RNA-dependent XNA polymerase</li><li>evolve XNA-dependent RNA polymerase</li><li>evolve heteroduplex specific RNase</li><li>initiate complementary strands</li><li>import *in vivo* XNA precursors</li></ul>	<ul><li>select *in vitro* an XNA aptamer</li><li>evolve DNA-dependent XNA polymerase</li><li>evolve XNA-dependent DNA polymerase</li><li>evolve heteroduplex-specific DNase</li><li>initiate complementary strands</li><li>prevent DNA recombination with host genome</li><li>import *in vivo* XNA precursors</li></ul>	<ul><li>select *in vitro* XNA aptamer</li><li>evolve DNA-dependent XNA polymerase</li><li>import *in vivo* XNA precursors</li></ul>
RNA strand as metabolic ribozyme or tRNA or rRNA	n.a.	<ul><li>select *in vitro* an RNA aptamer</li><li>evolve RNA-dependent XNA polymerase</li><li>evolve XNA-dependent RNA polymerase</li><li>initiate and separate complementary strands</li><li>import *in vivo* XNA precursors</li></ul>	n.a.	

DNA strand as metabolic enzyme	n.a.	n.a.	n.a.	• select *in vitro* a DNA aptamer • evolve DNA-dependent XNA polymerase • evolve XNA-dependent DNA polymerase • initiate and separate complementary strands • prevent DNA recombination with host genome • import *in vivo* XNA precursors
XNA strand as RNA transcription template	n.a.	• evolve XNA-dependent XNA polymerase • evolve XNA-dependent RNA polymerase • initiate and separate complementary strands • import *in vivo* XNA precursors	n.a.	• evolve DNA-dependent XNA polymerase • evolve XNA-dependent DNA polymerase • evolve XNA-dependent RNA polymerase • evolve heteroduplex-specific DNase • initiate complementary strands • prevent DNA recombination with host genome • import *in vivo* XNA precursors
DNA strand as RNA transcription template	n.a.	n.a	n.a	• evolve DNA-dependent XNA polymerase • evolve XNA-dependent DNA polymerase • initiate and separate complementary strands • prevent DNA recombination with host genome • import *in vivo* XNA precursors

(*Continued*)

Table 9.1 (Continued)

Replication duplex	XNA:XNA	XNA:RNA	XNA:DNA	DNA:DNA
XNA generation process	Replication	Transliteration of RNA template	Transliteration of DNA template	Transcription of DNA genes from promoter
XNA strand as translation template	• evolve XNA-dependent XNA polymerase • evolve specialized ribosome • initiate and separate complementary strands • prevent ribosome polymerase collision • import *in vivo* XNA precursors	• evolve RNA-dependent XNA polymerase • evolve XNA-dependent RNA polymerase • evolve specialized ribosome • initiate and separate complementary strands • prevent ribosome polymerase collision • import *in vivo* XNA precursors	• evolve DNA-dependent XNA polymerase • evolve XNA-dependent DNA polymerase • evolve specialized ribosome • evolve heteroduplex-specific DNase • initiate complementary strands • prevent DNA recombination with host genome • import XNA precursors	• evolve DNA-dependent XNA polymerase • evolve specialized ribosome • import *in vivo* XNA precursors
RNA strand as translation template	n.a.	• evolve RNA-dependent XNA polymerase • evolve XNA-dependent RNA polymerase • initiate and separate complementary strands • prevent ribosome polymerase collision • import *in vivo* XNA precursors	n.a.	n.a.

Each block lists the series of experimental tasks to be carried out in order to obtain the stages of stable selection *in vivo*. n.a., not applicable.

thus participate in the phenotype of the host cell, permitting its own synthesis, yet it would not create a genetic enclave propagating its genotype. This scenario could, nevertheless, provide a bridgehead for the launching of bona fide XNA replicons in further steps of elaboration. It would also enable the *in vivo* selection of uptake systems for XNA precursors and of diversified leaving groups.

The big advantage of scenario (c) is that it requires evolution of only a DNA-dependent XNA polymerase. It would benefit from the intrinsic ability of transcriptases to prime polymerization from a DNA signal (i.e. a promoter) to end polymerization when another DNA signal is reached (i.e. a terminator) and to disrupt the association of the polymerized XNA transcript from the DNA template.

A crucial issue in the implementation of all scenarios will be the separation of XNA strands from replication or transcription duplexes. Known possibilities elaborated by natural RNA viruses to overcome this hurdle include heteroduplex-specific RNases like RNase H, helicases, and the formation of intrastrand secondary structures of the transcript. Presumably these functional expedients could be transposed in the XNA world, by evolving XNA-heteroduplex-specific RNases, XNA helicases, and XNA stems and loops, at the cost of painstaking additional experimentation. Alternatively, backbone motifs and nucleobases could be chosen so as to destabilize XNA homoduplexes or heteroduplexes and avoid replication dead-ends.

9.3 DIVERSIFICATION OF THE BACKBONE MOTIF

The backbone motif of a third type of nucleic acid (XNA) should chemically resemble the natural polymers, but differ structurally enough from DNA and RNA so that its functions and biosynthesis could be uncoupled from those of natural polymers, which would enable the establishment of a genetic enclave.

A prerequisite might be the potential of XNA to form helical structures that are similar if not identical to that of natural nucleic acids and to hybridize with natural nucleic acids by Watson–Crick base pairing. It is now well established that several XNA backbone motifs bearing canonical bases are able to form regular homoduplexes. So far, the backbone motifs which have been validated for this purpose are anhydrohexitol nucleic acid (HNA) [17], threose nucleic acid (TNA) [18], glycerol nucleic acid (GNA) [19], and cyclohexenyl nucleic acid (CeNA) [20] (Figure 9.2).

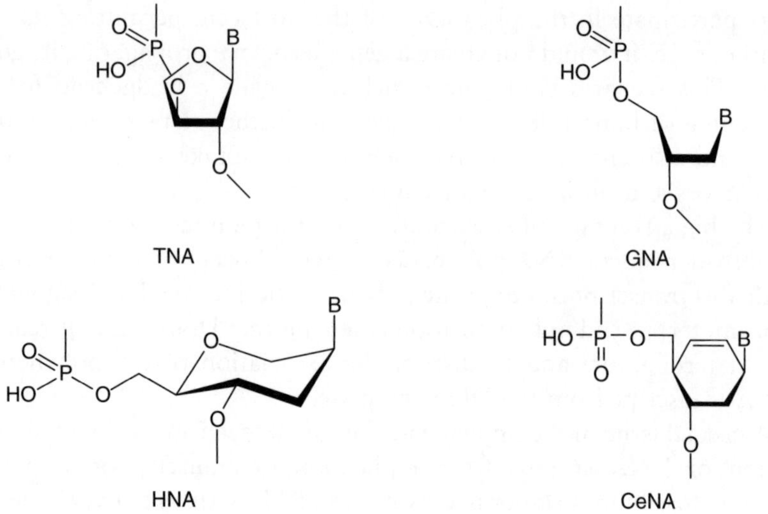

Figure 9.2 Structure of TNA, GNA, HNA, and cyclohexenyl nucleic acids (CeNA)

HNA was obtained by inserting a methylene group between the ring oxygen atom and the anomeric carbon atom of the furanose ring of a natural nucleoside. As a result, the base moiety is no longer positioned at the anomeric carbon atom. A chemically and enzymatically stable nucleic acid is obtained, which is still able to convey information to and from natural nucleic acids. HNA hybridizes with DNA and RNA and with itself [17]. An interesting observation is that the ΔT_m/modification of an HNA–RNA duplex decreases with increasing chain length, evolving from a high stabilizing effect for short duplexes to practically no stabilization/destabilization effect for long duplexes. HNA fulfills the requirement for being similar but not identical to natural nucleic acids. Its production via an appropriate metabolic pathway remains to be investigated. Likewise, it must be demonstrated that the structure of HNA is different enough from those of RNA and DNA to make it possible to construct a fully orthogonal system.

Initial DNA-dependent HNA polymerase activity, using Watson–Crick base-pair fidelity and including the potential for chain elongation, has been observed for family B DNA polymerases [21] and the M184V mutant of HIV-1 reverse transcriptase [22]. Likewise, T7 RNA polymerase and terminal transferase use hexitol nucleoside triphosphates (hNTP) as substrate for HNA synthesis, although not as efficiently as the natural nucleoside triphosphates [23]. T4 DNA ligase and T4 RNA ligase, however, do not accept the presence of HNA as donor or acceptor strand, nor as template for ligation of single-stranded DNA [23]. Marginal HNA-dependent DNA polymerase activity and HNA-dependent HNA polymerase activity was observed for DNA polymerase I from *E. coli* and for *Thermus aquaticus* polymerase [24]. The fact that hNTPs and HNA can be used by certain polymerases as both substrate and template to a limited extent shows the potential to evolve natural polymerases for affording a bona fide HNA replication system.

For performing experiments on *in vivo* replication, a selection scheme was designed to assay transmission of a genetic message from HNA to a DNA plasmid by the replication enzymes in *E. coli* [24]. *E. coli* cells bearing a chromosomal deletion of the gene for thymidylate synthase were transformed with plasmids containing HNA stretches encoding the active site region surrounding the catalytic Cys146 of the enzyme. Utilization of the HNA stretch as template was mandatory for propagation of an active gene for thymidylate synthase. Since active thymidylate synthase is absolutely required for growth of *E. coli* thyA mutants in nutrient media devoid of thymine or thymidine, this scheme can be used for the selection of strains that are able to copy HNA into DNA *in vivo*.

Prototrophic transformants of a host strain lacking an active gene for thymidylate synthase (i.e. bacteria capable of growth without added thymidine) could be obtained with the plasmid containing two HNA codons (six nucleotides). Sequencing experiments demonstrated that the message conveyed in the HNA residues was correctly copied into DNA [24].

We could also demonstrate that the replacement of two codons of a 33mer mRNA (AUG start codon and UUC second codon) by HNA (mHNA) did not influence the main steps of translation, as indicated by the same level of ribosomal binding of mRNA with hexitol residues under P-site conditions and the same yield of tRNA binding to the P- and A-sites. Both peptide formation and translocation took place on mRNA containing the two-codon-long mHNA [25]. The enzymatic stability of HNA may ensure that protein synthesis will go on for a longer time than when using mRNA.

Other successful examples of polymerization of sugar-modified nucleoside triphosphates using polymerases as catalyst on a DNA template (or polymerization of deoxynucleoside triphosphate on modified nucleic acids template) are those of GNA [26] and TNA [27, 28]. The structures of GNA and TNA, respectively featuring phosphorylated glycerol and phosphorylated tetrofuranose backbones, deviate more substantially from those of natural DNA than HNA.

The fourth case is that of cyclohexenyl nucleic acid or CeNA [20]. DNA polymerase B and HIV-1 reverse transcriptase show good DNA-dependent-CeNA polymerase and CeNA-dependent-DNA polymerase activity which designate it as a better starting point than HNA [29].

HNA featuring a cyclic ether and CeNA endowed with an unsaturated carbocyclic scaffold both appear as hardly accessible from common cell metabolites if biosynthetic pathways have to be conceived for producing them endogenously. By contrast, GNA and TNA appear more realistic in terms of biosynthetic accessibility from metabolites present in cells. Enabling the spontaneous emergence of XNA metabolic precursors will represent a feat of great scientific significance for the propagation of XNA-bearing hosts. In this case, drastic containment procedures will be mandatory for safely growing XNA prototrophs,; that is, organisms able to synthesize episomes with deviant chemistry. Therefore, an intrinsic safety device will be provided by the requirement for xenonutrients to propagate XNA in next-generation genetically modified organisms (GMOs). Metabolic inaccessibility should thus be considered as an advantage when it comes to the chemical estrangement of XNA from metabolism of extant cells and food chains of natural ecosystems. This entails that XNA precursors will have to be devised and synthesized so as to permit their uptake when they cannot be synthesized within the host cell.

A potent delivery system for nucleoside mono (di- and tri-) phosphates could make use of oligopeptide permeases, which are part of a larger group of transport systems; that is, ATP-binding cassette transporters. The periplasmic oligopeptide-binding protein is the initial receptor for the uptake of peptides by the oligopeptide permease in Gram (−) bacteria. The ligand-binding site shows a broad substrate specificity and accepts chemical groups of large diversity (it accepts di- to penta-peptides regardless of the identity of the side chains) [30]. The first conjugates for delivery of nucleotides that could function as substrate for the transporter have been synthesized [31]. They consist of a pyroglutamyl-protected tripeptide, a lateral pyridoxine moiety, and a nucleotide loaded on a serine residue (Figure 9.3). Following transport into the bacterial cytoplasm, the pyroglutamyl group could be deblocked by a specific aminopeptidase liberating a free terminal amino group that could be involved in the catalytic process to deliver the laterally attached nucleotide. Intracellular delivery of the nucleotide could indeed be accomplished by a pyridoxal-catalyzed elimination of the nucleotide, bound to the free amino group of the serine residue, via formation of a Schiff base.

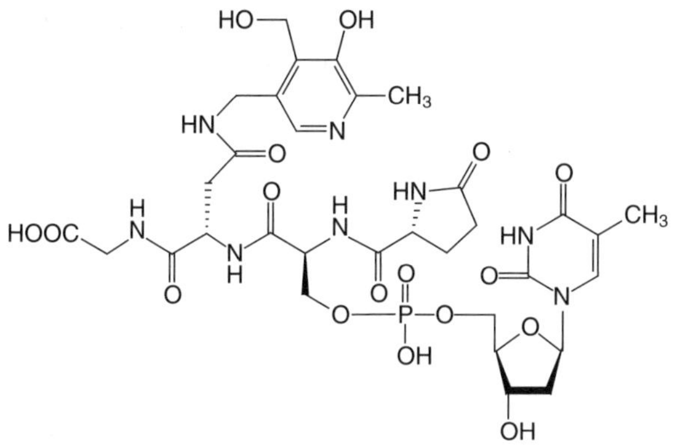

THE CHEMICAL DIVERSIFICATION OF NUCLEIC ACIDS 215

Figure 9.3 Structure of a potential delivery system for nucleotides into bacterial cells

9.4 DIVERSIFICATION OF THE LEAVING GROUP

DNA and RNA are synthesized from nucleoside triphosphates, through an iterative catalytic process where one nucleotide is attached (via its 5′-phosphate) to the 3′-hyroxyl group of the growing nucleic acid chain, releasing pyrophosphate in the process. This pyrophosphate is further hydrolyzed into two phosphate molecules, thus ensuring the irreversibility of nucleic polymerization. The nucleoside triphosphates themselves are synthesized from the monophosphates through the action of nucleotide kinases and nucleoside diphosphate kinases. The entanglement of the various functions of nucleoside triphosphates can be seen as a trap when trying to propagate an artificial biopolymer, such as novel types of nucleic acid, *in vivo*. It would be difficult indeed to install additional nucleoside triphosphates without interfering with DNA and RNA metabolism, cell energy supply via respiration, or substrate-level phosphorylation.

From a chemical standpoint, the simplest way out of this intricacy is to substitute the pyrophosphate moiety with alternative leaving groups in XNA precursors. From an informational standpoint, the use of alternative leaving groups would ipso facto result in genetic enclaves distinct from canonical nucleic acids without having to physically separate precursors of XNA from those of DNA and RNA. Activated nucleotides with alternative leaving groups should not be recognized by polymerases and other enzymes involved in cellular functions where nucleoside triphosphates play a role. If an alternative leaving group corresponded or could be converted to a common metabolite, they could be fitted into recycling pathways, and thus prevent the accumulation of by-products of XNA polymerization.

The ideal properties for XNA precursors with an alternative leaving group could be listed as follows:

- they should be very soluble and not too unstable in water;
- they should be accommodated in the active site of polymerase and react as substrate;
- they should undergo productive elongation;
- the chemical choice of the leaving group for release by polymerases should be mechanism based;
- the leaving group should be actively degraded or recycled to common metabolites so as to render polymerization irreversible;
- the nucleotide precursor equipped with the leaving group should be taken up actively in cells.

Modification of the triphosphate moiety of natural nucleosides with the aim of discovering alternative substrates for polymerases is a research area that has been little explored. Generally, such modifications have been investigated in the search for inhibitors of enzymes such as adenylate cyclase and reverse transcriptase or inhibitors of nucleotide binding receptors. The enzymatic synthesis of modified nucleic acids (phosphoramidates) using sugar-modified nucleoside triphosphates was pioneered by Letsinger *et al.* [32] using dNTTP as substrate and DNA polymerase I of *E. coli* as catalyst. The greatest amount of information about sugar- and triphosphate-modified nucleosides and their incorporation into DNA using enzymes (reverse transcriptase) can be found in the literature on potential inhibitors of HIV replication [33]. The modifications that have been investigated so far are mainly minor modifications within the triphosphate moiety itself. The most studied are methylene phosphonate [34–36], phosphoramidate [34], and thiophosphate [37–39] analogues of ATP.

The most deviant instances of a leaving group that retains the capacity to be recognized by polymerases are those of Asp-dAMP and His-dAMP in the case of reverse transcriptase [40, 41] (Figure 9.4). L-Aspartate and L-histidine were found to be efficient pyrophosphate mimics despite the structural differences. The fact that Glu-dAMP is not recognized as substrate, while L-Asp-dAMP is recognized, underscores the precise geometric and spatial arrangement of functional groups that enables the formation of a tertiary complex in the active site of the enzyme.

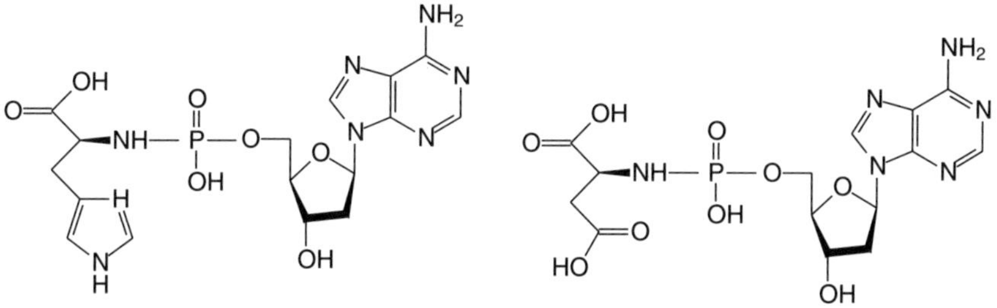

THE CHEMICAL DIVERSIFICATION OF NUCLEIC ACIDS 217

Figure 9.4 Structure of L-His-dAMP and L-Asp-dAMP

Likewise, Asp-dGMP, Asp-dCMP, and Asp-dTMP are substrates for HIV-RT, while retaining the canonical base-pair selectivity [42]. The kinetics of incorporation are, however, much slower than that of the natural substrates [40]. Encouraging results are that the primer can be extended with up to six deoxyadenines (although stalling starts after incorporation of three nucleotides), that the incorporation reaction follows Watson–Crick rules and is stereoselective (L- versus D-amino acids), and that the incorporation of the four building blocks L-Asp-dAMP, L-Asp-dCMP, L-Asp-dGMP, and L-Asp-dTMP follows Michaelis–Menten kinetics. That a moiety as simple and common as aspartate linked by a P–N bond can serve as alternative leaving group in the polymerization of DNA augurs well for the further elaboration of parallel biosynthetic and energetic pathways for XNA synthesis through metabolic engineering of host cells and directed evolution of DNA and RNA polymerases. A promising aspect of chemical diversification of the leaving group is that it could be applied first to a transcription system for generating XNA. In order to encode functional sequences, only two different monomers would have to be introduced, presumably complementary.

9.5 DIVERSIFICATION OF NUCLEIC BASES

The main synthetic effort to explore macromolecular structures of nucleic acids not known in nature has mainly focused on diversifying pairing schemes beyond those originally proposed by Watson and Crick. The genetic alphabet has been expanded from a two-base-pair system to a six-base-pair system [43, 44]. This approach is based on finding new H-bonding topologies. For example, a 6-amino-2(1H)

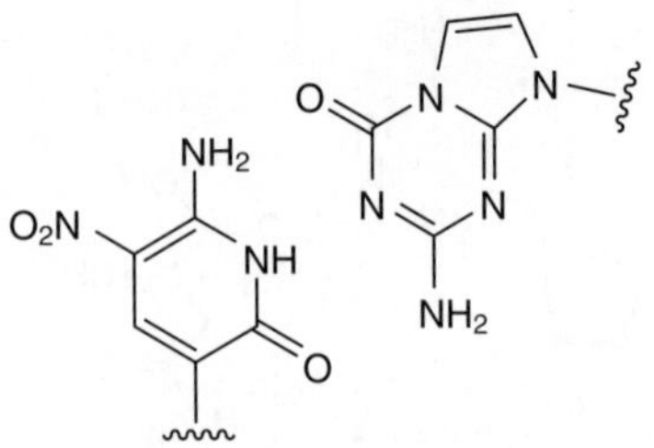

Figure 9.5 A pyDDA/puAAD system that could function as an informational system *in vivo*

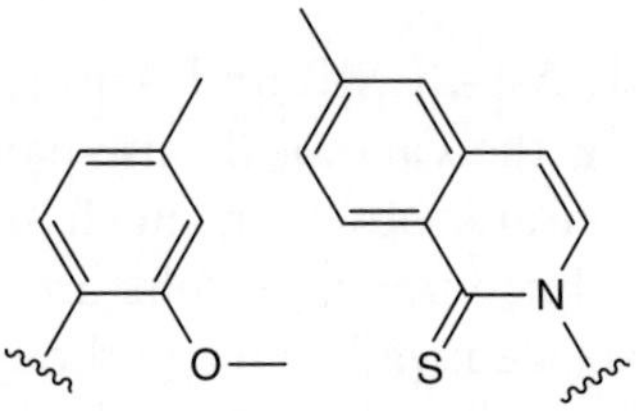

Figure 9.6 A hydrophobic base-pair system for potential use *in vivo*

pyridine:2-amino-imidazo[1,2-*a*]-triazinone base pair has been proposed for synthetic biology purposes [45] (Figure 9.5).

Others have enlarged the genetic helix with expanded-size base pairs [46, 47]. Another successful attempt involved the incorporation of hydrophobic nucleobase analogues [48–50]. Starting from a large screening effort using five different polymerases, additional hydrophobic base-pair systems were identified consisting of bases built from phenyl, isoquinoline, and the pyrrolo[2,3-*b*]pyridine scaffold [51]. The base-pair system depicted in Figure 9.6 is proposed as promising for *in vivo* use.

An additional base pair could be accommodated in XNA by providing a host cell with the corresponding activated precursors only if the replication of XNA does not involve a complementary strand made of DNA or RNA. Extensive diversification of the nucleobases might thus only fit in the XNA:XNA scenario (a) of Table 9.1.

However, base modifications compatible with natural base pairs such as diaminopurine, 5-substituted uracil, and cytosine, as well as or 8-aza-purines, 7-substituted-deaza-adenine, and guanine, would suit all scenarios. Some such modifications might be rationally introduced for modulating duplex stability, easing purification of XNA, or enhancing catalytic activities of XNA aptamers.

Backbone and leaving-group diversification thus appear more pertinent than base diversification for establishing XNA *in vivo*. Once established *in vivo*, XNA replicons will provide an opportunity to recruit alternative base pairs for protein coding or biocatalytic purposes.

9.6 DIVERSIFICATION OF NUCLEIC ACID POLYMERASES

Enzymatic activities that should be considered for propagating XNA *in vivo* following the various scenarios of Table 9.1 can be listed as follows:

- a DNA-dependent XNA polymerase, or XNA transliterase, would be needed for conveniently synthesizing XNA strands;
- an XNA-dependent DNA polymerase, or reverse XNA transliterase;
- an XNA-dependent XNA polymerase or XNA replicase;
- an XNA-dependent RNA polymerase or XNA transcriptase to copy XNA into mRNA, tRNA, or rRNA that can be used by the ribosome for protein synthesis;
- an RNA-dependent XNA polymerase to copy RNA information into XNA strands.

The XNA polymerase activities, whether templated by XNA itself, RNA, or DNA, could have the additional capability of condensing precursors with a leaving group other than pyrophosphate and to reject nucleoside triphosphates.

DNA polymerases perform a feat of molecular recognition, incorporating the correct four different nucleotide substrates as specified by the template base with minimal error rates. As a result of this high fidelity and tight geometric control within the polymerase active site, catalysis becomes very sensitive to distortions in the primer–template–dNTP tertiary complex. This precludes or diminishes the replication of modified DNA templates and restricts the enzymatic incorporation or replication of synthetic nucleic acids.

The majority of synthetic nucleic acids are poor substrates for enzymatic polymerization and studies with unnatural nucleic acids are limited to short oligomers accessible by solid-phase synthesis. Construction of artificial genetic systems, both *in vitro* and *in vivo*, implies the development of an efficient enzymatic synthesis of the nucleic acid analogues. The efficiency and selectivity of modified nucleoside triphosphate incorporation are usually low. To facilitate the development of XNA-based replicons and episomes, it will be necessary to develop mutant polymerases able to synthesize the new polymers and to use them as templates for the propagation of information. These mutants could be obtained by rational design, mutant-by-mutant construction, activity-based library screening, and enrichment [52–54]. Activity-based selection, such as compartmentalized self-replication [55, 56] and phage display [57–59], could also be exploited for this purpose. Indeed, incomplete understanding of the detailed mechanism of polymerization makes rational design approaches difficult; therefore, selection approaches and directed evolution methods have been developed [60].

Loeb and co-workers pioneered the genetic complementation approach for polymerase selection [53]. Using a similar approach, Ellington and co-workers have selected variants of T7 RNA polymerase with an increased ability to synthesize transcripts from 2′-substituted ribonucleotides [61]. Along this experimental line, one could attempt to evolve a DNA-dependent XNA polymerase responding to promoters and terminators specified by DNA, and also enzyme variants specific for leaving groups different from pyrophosphate that can be recycled in metabolism. Phage display has been used for the selection of a polymerase function by proximal display of both primer–template duplex and polymerase on the same phage particle [57] and allowed the selection of a variant of the Stoffel fragment that incorporates ribonucleoside triphosphates (rNTPs) with efficiencies approaching those of the wild-type enzyme for dNTP substrates.

A different assay for the evolution of polymerases is compartmentalized self-replication [56]. Polymerases and substrates are encapsulated in discrete noncommunicating compartments akin to artificial cells. Individual polymerase variants are isolated in separate compartments. Each polymerase replicates only its own encoding gene to the exclusion of those in other compartments. Consequently, only genes encoding active polymerases are replicated, while inactive variants disappear from the gene pool. The more active will make proportionally more copies of their own encoding gene. This method has been shown to be a powerful method for the directed evolution of polymerases. The first step towards the generation of polymerases capable of replication of unnatural nucleic acids has already been taken using this approach [62]. Unfortunately, the molecular type synthesized in this way, aS-DNA, is not sufficiently different from DNA to be treated as an orthogonal entity.

It is mandatory for an artificial genetic system to become fully and not partially disconnected from DNA replication and RNA transcription of the host cell in order to sever links of information transmission by mistemplating and misincorporation of activated monomers.

An additional feature that would need to be evolved in an XNA polymerase would be the initiation step using a tRNA primer as for reverse transcriptases, or a protein primer as for DNA polymerase from phage phi29. Another very helpful enzyme for diversifying nucleic acids *in vivo* would be an XNA ligase; that is, a ligase that accepts XNA as substrate to produce long stretches of XNA for making whole genes and episomes.

9.7 CONCLUSIONS

Propagating a third type of nucleic acid in a microbial cell will require the molecular alliance of several entities (metabolic precursors, templates, and enzymes), each of which will require systematic efforts to elaborate individually.

The successful construction of at least an uptake and processing apparatus for exogenous activated precursors of minimally two XNA precursors, an XNA polymerase and either a xenozyme (XNA-aptamer) catalyzing an essential metabolic reaction or an XNA-dependent RNA polymerase, and an XNA complement for a tRNA, an rRNA, a short mRNA, or a metabolic ribozyme (RNA-aptamer) will have to be accomplished (Table 9.1).

Research in fields as different as organic chemistry, evolutionary biotechnology, and microbiology will need to be coordinated for long periods before sustained proliferation of XNA becomes operational.

Once it is operational, an entirely new adaptive landscape will become accessible to experimental exploration through prolonged cultivation or successive selection steps for usage of deviant exogenous XNA precursors and function of XNA-encoded sequences.

The persistent requirement of vast populations of reprogrammed microorganisms for exogenous XNA precursors will establish conclusively the functional role of XNA polymers, as well as the inaccessibility of metabolic reactions for endogenously producing these XNA precursors.

Such experiments should be instrumental for estimating the risks of horizontal transfer and counteracting genetic pollution in next-generation GMOs.

In the long run, we anticipate that XNA propagation *in vivo* will lend itself to evolutionary processes of exaptation unprecedented in natural history and the experimental record, by allowing the recruitment under selection of ever-different nucleic monomers deviating in nucleobase pairing, leaving groups, and backbone motifs within the limits of organic synthesis.

Forging the limited arsenal of molecular machinery needed to launch such an open-ended and safe evolutionary regime seems feasible by relying on the theoretical understanding and experimental procedures of current nucleic acids research.

ACKNOWLEDGMENTS

We are grateful to Susan Cure for correcting and improving the text. We thank Mia Vanthienen for editorial help.

REFERENCES

1. Rifkin, J. (1998) *The Biotech Century*, Penguin Putnam, New York.
2. Khorana, H.G. (1971) Total synthesis of the gene for an alanine transfer ribonucleic acid from yeast. *Pure and Applied Chemistry*, **25**, 91.
3. Vapnek, D., Hautala, J.A., Jacobson, J.W. *et al.* (1977) *Proceedings of the National Academy of Sciences of the United States of America*, **74**, 3508.
4. Murray, A.W. and Szostak, J.W. (1983) Construction of artificial chromosomes in yeast. *Nature*, **305**, 189.
5. Giraldo, P. and Montoliu, L. (2001) Size matters: use of YACs, BACs and PACs in transgenic animals. *Transgenic Research*, **10**, 83.
6. Hutchison, C.A., III, Phillips, S., Edgell, M.H. *et al.* (1978) Mutagenesis at a specific position in a DNA sequence. *Journal of Biological Chemistry*, **253**, 6551.
7. Saiki, R.K., Scharf, S., Faloona, F. *et al.* (1985) Enzymatic amplification of beta-globin genomic sequences and restriction site analysis for diagnosis of sickle cell anemia. *Science*, **230**, 1350.
8. Stolovicki, E., Dror, T., Brenner, N., and Braun, E. (2006) Synthetic gene recruitment reveals adaptive reprogramming of gene regulation in yeast. *Genetics*, **173**, 75.
9. Posfai, G., Plunkett, G., III, Feher, T. *et al.* (2006) Emergent properties of reduced-genome *Escherichia coli*. *Science*, **312**, 1044.
10. Powell, K.A., Ramer, S.W., Cardayré, S.B.D. *et al.* (2001) Directed evolution and biocatalysis. *Angewandte Chemie, International Edition in English*, **40**, 3948.

11. Gibson, D.G., Benders, G.A., Andrews-Pfannkoch, C. et al. (2008) Complete chemical synthesis, assembly, and cloning of a Mycoplasma genitalium genome. Science, 319, 1215.
12. Delude, C.M., Mirvis, K.W., and Davidson, J. (2001) Back to school. Nature Biotechnology, 19, 911.
13. Rao, S.N. and Kollman, P.A. (1986) Hydrogen-bonding preferences in 2,6-diaminopurine: uracil (thymine) and 8-methyl adenine:uracil (thymine) complexes. Biopolymers, 25, 267.
14. Horlacher, J., Hottiger, M., Podust, V.N. et al. (1995) Recognition by viral and cellular DNA polymerases of nucleosides bearing bases with nonstandard hydrogen bonding patterns. Proceedings of the National Academy of Sciences of the United States of America, 92, 6329.
15. Warren, R.A. (1980) Modified bases in bacteriophage DNAs. Annual Review of Microbiology, 34, 137.
16. Bron, S., Luxen, E., and Venema, G. (1983) Resistance of bacteriophage H1 to restriction and modification by Bacillus subtilis R. Journal of Virology, 46, 703.
17. Hendrix, C., Verheggen, I., Rosemeyer, H. et al. (1997) 1′,5′-Anhydrohexitol oligonucleotides: synthesis, base pairing and recognition by regular oligodeoxyribonucleotides and oligoribonucleotides. Chemistry: A European Journal, 3, 110.
18. Schoning, K., Scholz, P., Guntha, S. et al. (2000) Chemical etiology of nucleic acid structure: the α-threofuranosyl-(3′→2′) oligonucleotide system. Science, 290, 1347.
19. Zhang, L., Peritz, A., and Meggers, E. (2005) A simple glycol nucleic acid. Journal of the American Chemical Society, 127, 4174.
20. Wang, J., Verbeure, B., Luyten, I. et al. (2000) Cyclohexene nucleic acids (CeNA): serum stable oligonucleotides that activate RNase H and increase duplex stability with complementary RNA. Journal of the American Chemical Society, 122, 8595.
21. Vastmans, K., Pochet, S., Peys, A. et al. (2000) Enzymatic incorporation in DNA of 1,5-anhydrohexitol nucleotides. Biochemistry, 39, 12757.
22. Vastmans, K., Froeyen, M., Kerremans, L. et al. (2001) Reverse transcriptase incorporation of 1,5-anhydrohexitol nucleotides. Nucleic Acids Research, 29, 3154.
23. Vastmans, K., Rozenski, J., Van Aerschot, A., and Herdewijn, P. (2002) Recognition of HNA and 1,5-anhydrohexitol nucleotides by DNA metabolizing enzymes. Biochimica et Biophysica Acta, 1597, 115.
24. Pochet, S., Kaminski, P.A., Van Aerschot, A. et al. (2003) Replication of hexitol oligonucleotides as a prelude to the propagation of a third type of nucleic acid in vivo. Comptes Rendus Biologies, 326, 1175.
25. Lavrik, I.N., Avdeeva, O.N., Dontsova, O.A. et al. (2001) Translational properties of mHNA, a messenger RNA containing anhydrohexitol nucleotides. Biochemistry, 40, 11777.

26. Tsai, C.H., Chen, J., and Szostak, J.W. (2007) Enzymatic synthesis of DNA on glycerol nucleic acid templates without stable duplex formation between product and template. *Proceedings of the National Academy of Sciences of the United States of America*, **104**, 14598.
27. Chaput, J.C. and Szostak, J.W. (2003) TNA synthesis by DNA polymerases. *Journal of the American Chemical Society*, **125**, 9274.
28. Kempeneers, V., Vastmans, K., Rozenski, J., and Herdewijn, P. (2003) Recognition of threosyl nucleotides by DNA and RNA polymerases. *Nucleic Acids Research*, **31**, 6221.
29. Kempeneers, V., Renders, M., Froeyen, M., and Herdewijn, P. (2005) Investigation of the DNA-dependent cyclohexenyl nucleic acid polymerization and the cyclohexenyl nucleic acid-dependent DNA polymerization. *Nucleic Acids Research*, **33**, 3828.
30. Tame, J.R., Murshudov, G.N., Dodson, E.J. et al. (1994) The structural basis of sequence-independent peptide binding by OppA protein. *Science*, **264**, 1578.
31. Marchand, A., Marchand, D., Busson, R. et al. (2007) Synthesis of a pyridoxine-peptide based delivery system for nucleotides. *Chemistry & Biodiversity*, **4**, 1450.
32. Letsinger, R.L., Wilkes, J.S., and Dumas, L.B. (1972) Enzymatic synthesis of polydeoxyribonucleotides possessing internucleotide phosphoramidate bonds. *Journal of the American Chemical Society*, **94**, 292.
33. Arzumanov, A.A., Semizarov, D.G., Victorova, L.S. et al. (1996) Gamma-phosphate-substituted 2'-deoxynucleoside 5'-triphosphates as substrates for DNA polymerases. *Journal of Biological Chemistry*, **271**, 24389.
34. Yount, R.G., Babcock, D., Ballantyne, W., and Ojala, D. (1971) Adenylyl imidiodiphosphate, an adenosine triphosphate analog containing a P–N–P linkage. *Biochemistry*, **10**, 2484.
35. Myers, T., Nakamura, K., and Flesher, J. (1963) Phosphonic acid analogs of nucleoside phosphates. I. The synthesis of 5''-adenylyl methylenediphosphonate, a phosphonic acid analog of ATP. *Journal of the American Chemical Society*, **85**, 3292.
36. Trowbridge, D.B., Yamamoto, D.M., and Kenyon, G.L. (1972) Ring openings of trimetaphosphoric acid and its bismethylene analog. Syntheses of adenosine 5'-bis(dihydroxyphosphinylmethyl) phosphinate and 5'-amino-5'-deoxyadensoine 5'-triphosphate. *Journal of the American Chemical Society*, **94**, 3816.
37. Goody, R.S. and Eckstein, F. (1971) Thiophosphate analogs of nucleoside di- and triphosphates. *Journal of the American Chemical Society*, **93**, 6252.
38. Eckstein, F. and Gindl, H. (1967) Synthesis of nucleoside 5'-polyphosphorothioates. *Biochimica et Biophysica Acta*, **149**, 35.
39. Eckstein, F. and Goody, R.S. (1976) Synthesis and properties of diastereoisomers of adenosine 5'-(O-1-thiotriphosphate) and adenosine 5'-(O-2-thiotriphosphate). *Biochemistry*, **15**, 1685.

40. Adelfinskaya, O., Terrazas, M., Froeyen, M. et al. (2007) Polymerase-catalyzed synthesis of DNA from phosphoramidate conjugates of deoxynucleotides and amino acids. *Nucleic Acids Research*, **35**, 5060.
41. Adelfinskaya, O. and Herdewijn, P. (2007) Amino acid phosphoramidate nucleotides as alternative substrates for HIV-1 reverse transcriptase. *Angewandte Chemie, International Edition in English*, **46**, 4356.
42. Terrazas, M., Marliere, P., and Herdewijn, P. (2008) Enzymatically catalyzed DNA synthesis using L-Asp-dGMP, L-Asp-dCMP, and L-Asp-dTMP. *Chemistry & Biodiversity*, **5**, 31.
43. Benner, S.A. (2004) Understanding nucleic acids using synthetic chemistry. *Accounts of Chemical Research*, **37**, 784.
44. Geyer, C.R., Battersby, T.R., and Benner, S.A. (2003) Nucleobase pairing in expanded Watson–Crick-like genetic information systems. *Structure*, **11**, 1485.
45. Yang, Z., Sismour, A.M., Sheng, P. et al. (2007) Enzymatic incorporation of a third nucleobase pair. *Nucleic Acids Research*, **35**, 4238.
46. Lynch, S.R., Liu, H., Gao, J., and Kool, E.T. (2006) Toward a designed, functioning genetic system with expanded-size base pairs: solution structure of the eight-base xDNA double helix. *Journal of the American Chemical Society*, **128**, 14704.
47. Liu, H., Gao, J., Lynch, S.R. et al. (2003) A four-base paired genetic helix with expanded size. *Science*, **302**, 868.
48. Matsuda, S., Fillo, J.D., Henry, A.A. et al. (2007) unnatural base pair structure–activity relationships. *Journal of the American Chemical Society*, **129**, 10466.
49. Henry, A.A., Olsen, A.G., Matsuda, S. et al. (2004) Efforts To expand the genetic alphabet: identification of a replicable unnatural DNA self-pair. *Journal of the American Chemical Society*, **126**, 6923.
50. Leconte, A.M., Hwang, G.T., Matsuda, S. et al. (2008) Discovery, characterization, and optimization of an unnatural base pair for expansion of the genetic alphabet. *Journal of the American Chemical Society*, **130**, 2336.
51. Hwang, G.T. and Romesberg, F.E. (2008) Unnatural substrate repertoire of A, B, and X family DNA polymerases. *Journal of the American Chemical Society*, **130**, 14872.
52. Sweasy, J.B. and Loeb, L.A. (1993) Detection and characterization of mammalian DNA polymerase beta mutants by functional complementation in *Escherichia coli*. *Proceedings of the National Academy of Sciences of the United States of America*, **90**, 4626.
53. Suzuki, M., Baskin, D., Hood, L., and Loeb, L.A. (1996) Random mutagenesis of *Thermus aquaticus* DNA polymerase I: concordance of immutable sites *in vivo* with the crystal structure. *Proceedings of the National Academy of Sciences of the United States of America*, **93**, 9670.

54. Shinkai, A., Patel, P.H., and Loeb, L.A. (2001) The conserved active site motif A of *Escherichia coli* DNA polymerase I is highly mutable. *Journal of Biological Chemistry*, **276**, 18836.
55. Tawfik, D.S. and Griffiths, A.D. (1998) Man-made cell-like compartments for molecular evolution. *Nature Biotechnology*, **16**, 652.
56. Ghadessy, F.J., Ong, J.L., and Holliger, P. (2001) Directed evolution of polymerase function by compartmentalized self-replication. *Proceedings of the National Academy of Sciences of the United States of America*, **98**, 4552.
57. Jestin, J., Kristensen, P., and Winter, G. (1999) A method for the selection of catalytic activity using phage display and proximity coupling. *Angewandte Chemie, International Edition in English*, **38**, 1124.
58. Fa, M., Radeghieri, A., Henry, A.A., and Romesberg, F.E. (2004) Expanding the substrate repertoire of a DNA polymerase by directed evolution. *Journal of the American Chemical Society*, **126**, 1748.
59. Xia, G., Chen, L., Sera, T. *et al.* (2002) Directed evolution of novel polymerase activities: mutation of a DNA polymerase into an efficient RNA polymerase. *Proceedings of the National Academy of Sciences of the United States of America*, **99**, 6597.
60. Holmberg, R.C., Henry, A.A., and Romesberg, F.E. (2005) Directed evolution of novel polymerases. *Biomolecular Engineering*, **22**, 39.
61. Chelliserrykattil, J. and Ellington, A.D. (2004) Evolution of a T7 RNA polymerase variant that transcribes 2′-O-methyl RNA. *Nature Biotechnology*, **22**, 1155.
62. Ghadessy, F.J., Ramsay, N., Boudsocq, F. *et al.* (2004) Generic expansion of the substrate spectrum of a DNA polymerase by directed evolution. *Nature Biotechnology*, **22**, 755.

10

The Minimal Ribosome

Hiroshi Yamamoto[1], Markus Pech[1], Daniela Wittek[1], Isabella Moll[2], and Knud H. Nierhaus[1]

[1] Max-Planck-Institut für Molekulare Genetik, AG Ribosomen, Ihnestr. 73, D-14195 Berlin, Germany
[2] Max F. Perutz Laboratories, Department of Microbiology, Immunobiology, and Genetics, Center for Molecular Biology, Dr. Bohrgasse 9/4, 1030 Vienna, Austria

10.1 Introduction 228
10.2 A Brief Description of the Ribosomal Structure and Function 228
10.3 Non-Essential Ribosomal Proteins: Mutational Approach 231
10.4 A Surprising Observation: The Kasugamycin Particle 233
10.5 A Minimal Core of the Large Ribosomal Subunit still Active in Peptide-Bond Formation 240
10.6 Cutting Down the rRNAs 241
10.7 Conclusions 242
Reference 243

The chapter starts with a short description of the ribosomal structure and functions. This is related to the origin of cells to evolution, since at the time of the first cells about 3.5 billion years ago the ribosome complexity could not have been as high as we see in recent organisms. It follows a consideration of the question of nonessential ribosome proteins, with a mutational approach. We then discuss these evolutionary aspects at the level of the last universal common ancestor, arriving at the surprising observation of the kasugamycin 61S particle.

Chemical Synthetic Biology, First Edition. Edited by Pier Luigi Luisi and Cristiano Chiarabelli.
© 2011 John Wiley & Sons, Ltd. Published 2011 by John Wiley & Sons, Ltd.

This particle might mirror an ancient form of the small subunit present in proto-ribosomes before domain separation about 3 billion years ago. An important observation is that a minimal core of the large ribosomal subunit is still active in peptide-bond formation. A subsequent section concerns the mitochondrial rRNAS. We conclude by discussing the difficulties of clarifying and possibly constructing a minimal ribosome form.

10.1 INTRODUCTION

Genetic information can be divided into three classes. Above all is the information about the protein structure, followed by that of the rRNAs and tRNAs of the translational apparatus, which translates the genetic information into protein structure, and finally the information of regulatory elements, usually small RNAs that regulate transcription, splicing, and translation. The ribosome is the center of the translational apparatus surrounded by hundreds of factors, like tRNAs and proteins, which are involved in the assembly of ribosomes, charging of tRNAs, and regulation of the ribosomal functions such as factors. Here, we will start with a short description of the ribosomal functions before we analyze the questions about a core ribosome, which should be still functional concerning the basic ribosomal activities for protein synthesis. This question is related to evolution, since at the time of the first cells about 3.5 billion years ago and already containing a genetic code the ribosome could not be of the complexity that we see in recent organisms. Therefore, we will also consider aspects of evolutionary relics in the present-day translational apparatus.

10.2 A BRIEF DESCRIPTION OF THE RIBOSOMAL STRUCTURE AND FUNCTION

The ribosome is one of the most complex structures known in biology. For example, the 70S ribosome of the bacterial model organism *Escherichia coli* consists of 54 proteins and three rRNAs distributed over the two ribosomal subunits 50S and 30S (Figure 10.1). In spite of the small number of rRNA molecules the rRNA makes up two-thirds of the ribosome mass. All proteins are present in one copy per 70S with the exception of L7/L12 derived from the same gene (L7 is the acetylated form of L12), which is present in four copies and in some organisms even in six copies. Most proteins are located at or near the ribosomal

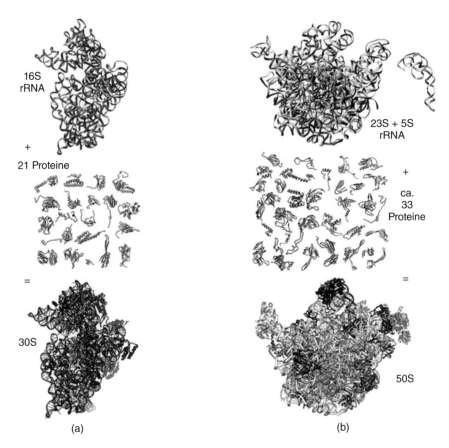

Figure 10.1 Components of the bacterial ribosome (*Escherichia coli*). The 21 proteins of the small subunit carry the prefix "S" are numbered from S1 to S21 roughly according to their molecular mass beginning with the largest and ending with the smallest protein. Accordingly, the 33 proteins of the large subunit carry the prefix "L". See plate section for a colour version of this figure

surface, with their globular domain(s) sending long extensions towards the center of the ribosomal subunit.

This structural complexity reflects a complicated net of functions, which fall into four phases, each of which has its own set of factors: (i) the start of protein synthesis (initiation, Figure 10.2a and b), (ii) the prolongation of the nascent peptide chain (elongation cycle), which represents a cycle of reactions prolonging the nascent peptide chain by one amino acid (Figure 10.2c–g), (iii) the termination, where the protein is released from the ribosome, and (iv) the recycling phase, where the

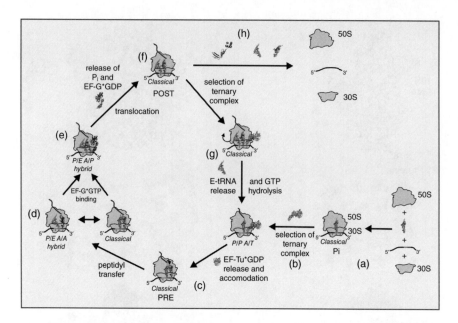

Figure 10.2 Sketch of the functional phases of the ribosome. The termination and recycling phases are combined. See text for explanations. See plate section for a colour version of this figure

ribosome leaves the mRNA to start the next initiation event (Figure 10.2h).

Recognition and deciphering of the codons of the mRNA by the complementary anticodon of the tRNA occurs at the "decoding site" located on the small subunit. tRNAs can occupy three sites on the ribosome, the A, P, and E sites. According to the codon displayed at the A site, the corresponding aminoacyl-tRNA (aa-tRNA) binds to this site and brings the new amino acid to extend the growing polypeptide chain. The peptidyl-tRNA, which carries the nascent polypeptide chain, is bound at the P site before peptide-bond formation. The E site is the exit site for the deacylated or uncharged tRNA. During translation, tRNAs move through each of the sites sequentially, starting at the A site and passing through the P site to the E site, before leaving the ribosome. The exception is the binding of the very first tRNA, the initiator tRNA, which binds directly to the P site (Figure 10.2a).

Initiator tRNAs decode the start codon, usually AUG, and carry the amino acid formyl-methionine in bacteria or methionine in eukaryotes (including Archaea). In the next step, aa-tRNAs are delivered to the A site according to the displayed codon (Figure 10.2b). They enter the

A site in the form of a ternary complex consisting of an elongation factor (EF-Tu in bacteria and EF1a in eukaryotes), GTP and the aa-tRNA. After GTP hydrolysis EF-Tu·GDP is released from the ribosome and the aa-tRNA docks into the A site (Figure 10.2c). Peptide-bond formation occurs on the large subunit at the peptidyl-transferase (PTF) center and involves the transfer of the peptidyl moiety of the P-tRNA to the aminoacyl moiety of the A-tRNA; note that the whole polypeptide chain is added to the new amino acid rather than the addition of the new amino acid to the chain. Following peptide-bond formation, the tRNAs are in an equilibrium between the classical state with tRNAs at the A and P sites and the hybrid state with tRNAs at the A/P and P/E sites respectively (P/E, for example, means that the deacylated tRNA is at the P site on the small subunit and at the E site on the large subunit; Figure 10.2d). The A/P state is almost identical with the peptidyl-tRNA in the classical A site with the exception of the CCA3′-end, which extends to the P-site region of the PTF center, whereas the position of the deacylated tRNA at the P/E state is dramatically different form the classical P site with a movement of the CCA3′-end of about 50Å [1].

Next, translocation of the A- and P-tRNAs to the P and E sites respectively is mediated by a second elongation factor (EF-G in bacteria and EF2 in eukaryotes; Figure 10.2e). The elongation factors accelerate the elongation cycle to achieve the rate of up to 50 ms per elongation cycle. Without elongation factors the rate is about four orders of magnitude slower [2] due to the high energy barrier (120 kJ/mol) that separates the pre- and post-translocational states (Figure 10.2c and f respectively) in *E. coli* ribosomes [3]. Translocation means a movement of the tRNA2·mRNA; therefore, the A site is now free for binding the next aa-tRNA (Figure 10.2f), which leads to the release of the E-tRNA (Figure 10.2g). This the cycle repeats (i.e. back to Figure 10.2c) until a stop codon enters the A site. At this point protein termination factors release the completed polypeptide and dissociate the ribosome into subunits in preparation for the next round of translation (Figure 10.2h).

10.3 NON-ESSENTIAL RIBOSOMAL PROTEINS: MUTATIONAL APPROACH

Not all the ribosomal components are essential, about one-third of the ribosomal proteins can be deleted without affecting viability. In the late 1970s Eric Dabbs isolated *E. coli* mutants that lacked one or two

Table 10.1 Mutants of *E. coli* lacking ribosomal proteins

Protein	Mutant designation	Phenotype
L1	RD19, MV17-10	–
L11	AM68, AM76, AM77	–
L15	AM16–98	cs
L19	AM149	–
L24	AM290	ts
L27	AM125	cs
L28	AM81, AM 108	cs
L29	AM111	–
L30	AM10	–
L33	AM90, AM 108	cs
S1	VTS03	–
S6	AM80	– unpublished
S9	AM83	cs
S13	AM109–113	– unpublished
S17	AM 111	ts
S20	VT514	ts

cs, cold sensitive; ts, temperature sensitive. Data taken from Ref. [5].

ribosomal proteins [4] (Table 10.1). Such a mutation is deleterious for the cell, since the growth rate is reduced seven to ten times compared with the wild type, and in some cases the mutation is even conditionally lethal; that is, the mutant is temperature sensitive and the non-permissive conditions could be higher or lower temperature (ts and cs respectively in Table 10.1). Consideration of a minimal ribosome should target these proteins first, although it is questionable whether all these proteins can be omitted simultaneously without threatening viability. Interestingly, the shortest rRNAs known are those of mitochondrial ribosomes from higher eukaryotes (not from yeast, which are similar in size to *E. coli*), which are ~30% shorter than the *E. coli* rRNAs. These mitochondrial ribosomes compensate for the lack of rRNA sequences by the presence of additional proteins [6], so that the overall size of a mitochondrial ribosome is even larger than the corresponding one from *E. coli*. Therefore, it does not seem to be very likely that a ribosome could be derived from recent ribosomal structures, where both the number of ribosomal proteins is reduced according to the Dabbs study and the rRNA shortened according to the mitochondrial rRNAs. However, a recent surprising observation has changed our view significantly, as described in the next section.

10.4 A SURPRISING OBSERVATION: THE KASUGAMYCIN PARTICLE

In this section we will discuss some features of a translational apparatus at early stages of life development. Chemical evolution leading to the first cells cannot have started before 4×10^9 years ago, since at this time stage the Earth's surface temperature fell below 1000 °C, allowing the first rocks to form; for example, the gneiss in North Canada [7]. At this or higher temperatures life cannot exist, since covalent energy-rich bonds start to break at 130 °C, and ordinary covalent bonds of cellular molecules at ~180 °C. The next important time point is the appearance of the first cells, which happened not before 3.75×10^9 years ago, which was reported as the oldest mats of ancient cells, so-called "stromatolites" from rocks of the Isua Supracrustal Belt, Greenland. More solid data exist about formations that are 3.5×10^9 years old containing remnants of ancient cells. (i) One is the Pilbara region of Western Australia with an age of 3.43×10^9 years. A recent report supports the suggestion that these Pilbara-Craton structures might be of biotic origin [8]. (ii) The other one, with about the same age, is seen in microbial biomarkers in pillow lavas from the Baberton Greenstone Belt in South Africa [9]. These cells must have had a translational apparatus governed by a genetic code, which might have carried the genetic information in the form of RNA.

Today it is fashionable to speak about an "RNA world," which means that information storage and the executing molecules all were RNA [10]. Support of this view was gained by the detection of enzymatic activities detected in RNA molecules (for a review, see Ref. [11]) and, in particular, by the finding that two important centers of the ribosome, the decoding center on the small ribosomal subunit and the peptidyl transferase center on the large one, are dominated by ribosomal RNA [12, 13]. But from the data mentioned in the preceding paragraph it is clear that an RNA world, if it existed at all, plus a chemical evolution, acquisition of a membrane, and development of the genetic code, all of which are important features, must have happened in the relative short time window of $(250-500) \times 10^6$ years, namely between 4×10^9 and 3.5×10^9 years ago.

Cells might have achieved a remarkable perfection before the cell lines separated into the three fundamental domains Bacteria, Archaea, and Eukarya accommodating all organisms in past and present [14]. This separation happened about 3×10^9 years ago [15, 16], and we will consider now the last universal common ancestor (LUCA) before domain separation, since recently a new kind of ribosome was described, the 61S ribosome, which might reflect the translational apparatus of LUCA.

There are good reasons to assume that tRNAs, synthetases, and ribosomes co-evolved. Two domains of recent tRNAs can be distinguished, the amino acid acceptor arm and the anticodon arm [17] (Figure 10.3a).

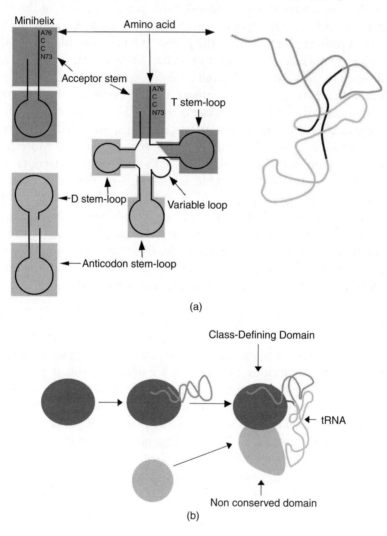

Figure 10.3 Co-evolution of tRNAs, synthetases, and ribosomes. (a) The two domains of tRNAs. The older one is the amino acid arm (minihelix), which can be charged separately in the case tRNAAla, according to Ref. [17]; modified. (Reprinted from [17] with permission from Elsevier) (b) The two functional domains of a synthetase, the evolutionary old catalytic one and the younger recognition domain. (c) Different architecture of the small and the large ribosomal subunits according to Ref. [18]. See plate section for a colour version of this figure

THE MINIMAL RIBOSOME

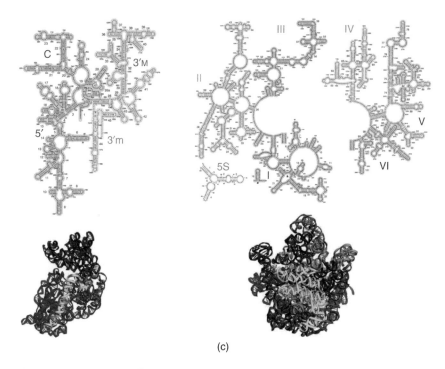

Figure 10.3 (*Continued*)

Evolution might have started with the first one, and in the case of tRNAAla the corresponding so-called mini-helix can be charged specifically with Ala by the alanine synthetases (AlaRSs). A similar and corresponding two-step evolution can also be stated with both the synthetases and the ribosome. The synthetases consist of two functional subunits/domains, the catalytic and the recognizing subunit, and they are ordered in two classes, each of which has 10 members that are paralogs, namely the catalytic domains of either class have one common ancestor (Figure 10.3b). Members of class I attach the amino acid to the 2′-OH and those of class II to the 3′-OH of the terminal ribose (Table 10.2). The catalytic domains are old; the recognition domains are evolutionarily younger. In contrast to the catalytic domain, the recognition domain does not have a common origin; each of them is different and has its own evolutionary history. Sixteen out of 20 synthetases contact the anticodon via the recognition domain and exploit it as a recognition feature (identity elements). This evolutionary dichotomy extends to the ribosome, where the large subunit contacts the amino acid arm and the small subunit the anticodon arm. Accordingly, the

Table 10.2 Classification of synthetases

Class I	Class II
1a	**2a**
Leucin	Serine
Isoleucine	Threonine
Valine	Alanine
Arginine	Glycine
Cysteine	Proline
Methionine	Histidine
1b	**2b**
Glutamate	Aspartate
Glutamine	Asparagine
Lysine-I	Lysine-II
1c	**2c**
Tyrosine	Phenylalanine
Tryptophane	
2′ Aa-attachment	**3′ Aa-attachment**

large subunit is considered to be older and the small subunit younger. This view is also supported by the different architecture of the two subunits: the four domains of the secondary 16S-type structure also represent easily recognizable domains of the folded 30S subunit, an indication of a relative young age, whereas the domains of the 23S-type secondary structure are intermingled in a complicated fashion in the mature 50S subunit (Figure 10.3c). Collectively, the evidence hints to a joint and co-evolving process beginning with the tRNA amino acid arm, the catalytic subunit of the synthetases, and the large ribosomal subunit (i.e. the ancient ribosome might have started with one subunit only), whereas the anticodon arm, the recognition subunit, and the small ribosomal subunit were added later.

The ribosome of LUCA was translating a simple mRNA, much simpler than the recent mRNAs with 5′-UTR-containing recognition and regulation elements, start signals, and termination signals of the decoding region and the decoding mRNA itself. In contrast, the mRNA for a LUCA ribosome was leaderless (Figure 10.4), lacking signals for initiation and termination and started with just the 5′-codon. This is indicated by the fact that the initiation and termination modes of Bacteria versus those of Archaea and Eukarya are quite different, requiring different factors that are usually evolutionarily nonrelated. For example, the two

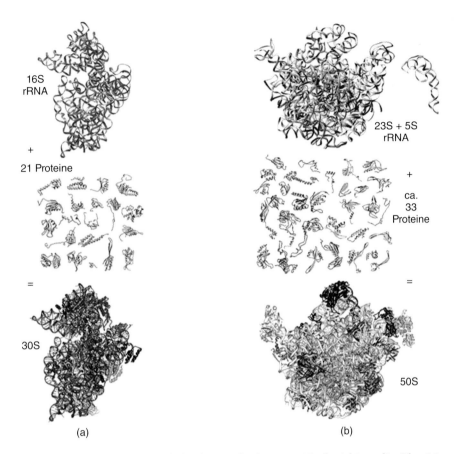

Figure 10.1 Components of the bacterial ribosome (*Escherichia coli*). The 21 proteins of the small subunit carry the prefix "S" are numbered from S1 to S21 roughly according to their molecular mass beginning with the largest and ending with the smallest protein. Accordingly, the 33 proteins of the large subunit carry the prefix "L"

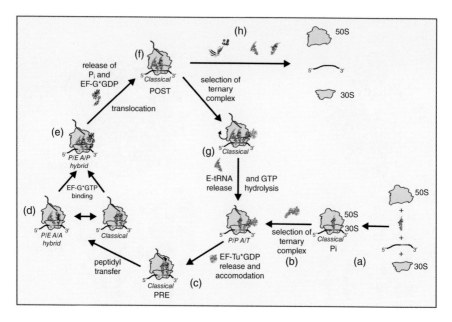

Figure 10.2 Sketch of the functional phases of the ribosome. The termination and recycling phases are combined. See text for explanations

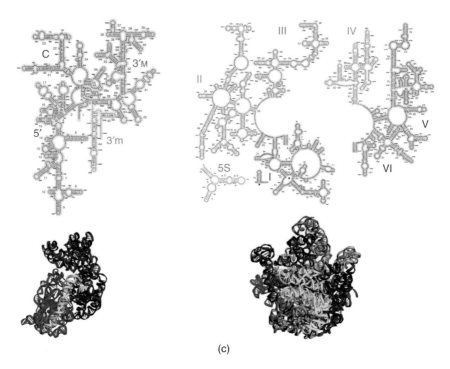

Figure 10.3 (c) Different architecture of the small and the large ribosomal subunits according to Ref. [18]

usual mRNA in eukaryotes

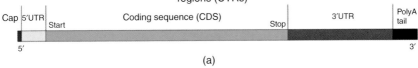

(a)

leaderless mRNA present in organisms of all three domains

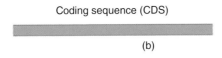

(b)

Figure 10.4 Structures of (a) recent and (b) leaderless mRNA

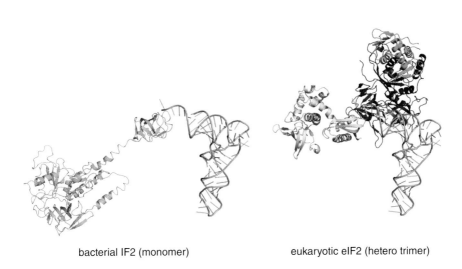

bacterial IF2 (monomer)   eukaryotic eIF2 (hetero trimer)

Figure 10.5 Structures of (a) the bacterial fMet-tRNA·IF2·GTP complex and (b) the eukaryotic Met-tRNAi·eIF2·GTP

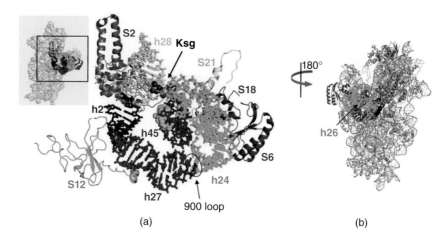

Figure 10.6 The platform of the 30S subunit is made of a ring of four helices; Ksg binds at the junction of h24 and h28. (See text for full caption.)

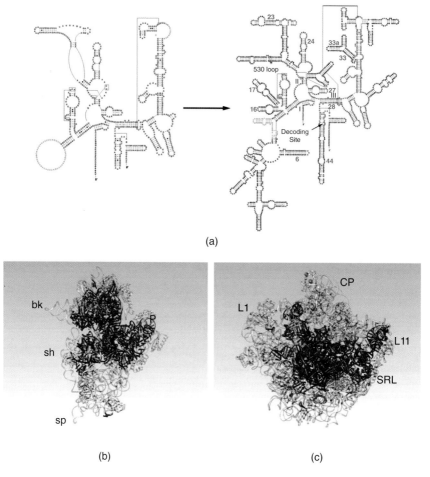

Figure 10.7 Comparison of rRNAs from mitochondria (*C. elegans*) and bacteria. (See text for full caption.)

usual mRNA in eukaryotes

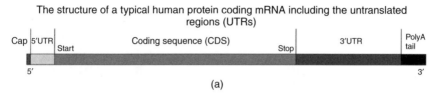

(a)

leaderless mRNA present in organisms of all three domains

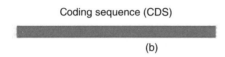

(b)

Figure 10.4 Structures of (a) recent and (b) leaderless mRNA. See plate section for a colour version of this figure

similar bacterial release factors RF1 and RF2 have nothing in common with the eukaryotic counterpart eRF1, except at the amino acid triplet GGQ interacting with the peptidyl transferase center, an example of convergent evolution. Likewise, bacterial IF2 and eukaryotic eIF2 have similar functions, in that they mediate the binding of the initiating (f)Met-tRNA but have a completely different structure: The bacterial is a monomer and the eukaryotic eIF2 is a hetero trimer (Figure 10.5).

Leaderless mRNAs (lmRNAs) are found in all organisms of the three domains and can be considered as molecular fossils. In the early 1990 Gottesman and coworkers reported a strange observation, namely that cI lmRNA from the lambda phage was translated in the presence of initiation inhibitors including the antibiotic kasugamycin (Ksg) [19]. Translation of lmRNA does not start with the canonical 30S *de novo* initiation, but by 70S ribosomes, indicated by the observation that overexpression of the anti-association factor IF3 inhibits lmRNA initiation [20] and most convincingly shown by the fact that cross-linked 70S ribosomes unable to dissociate could initiate lmRNA [21].

Administering Ksg to growing *E. coli* cells shortly after strong expression of lmRNA led to an intriguing observation: a new type of ribosome was formed sedimenting with 61S [22]. The formation was significant and allowed isolation of the 61S particles, the analysis of which revealed a number of surprising results:

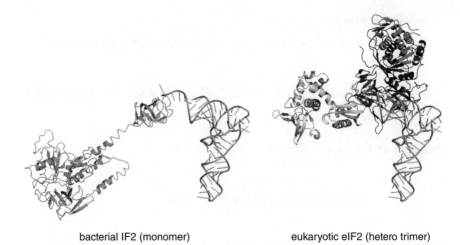

bacterial IF2 (monomer)　　　　　　　eukaryotic eIF2 (hetero trimer)

Figure 10.5 Structures of (a) the bacterial fMet-tRNA·IF2·GTP complex and (b) the eukaryotic Met-tRNAi··eIF2··GTP. See plate section for a colour version of this figure

1. The small subunit completely lacked six and partially lacked five ribosomal proteins (Table 10.3). Among these are important ribosomal proteins such as S1 (important for mRNA catching), S2 (which mediates S1 binding; S2 mutations can confer resistance against Ksg [23]), and S12 (which plays an important role in the decoding process). Mutations of S12 confer resistance against streptomycin (for a review, see Ref. [24]). The 50S subunit was unaltered.
2. Undermethylation of the universally conserved residues A1518 and A1519 of the 16S rRNA mediates Ksg resistance. However, these residues in the 16S rRNA of the 61S ribosome were fully methylated, so that the Ksg resistance of the 61S particle is probably due to the absence of S2.
3. The absence of S12 made the translating 61S particle also resistant against streptomycin *in vivo*.
4. The 61S ribosomes hardly dissociated into subunits under conditions where 70S are fully dissociated. This is reminiscent of the one-subunit ancient ribosome.
5. Ksg binds at the junction of a ring of helices consisting of h28, h2, h27, and h24 (Figure 10.6). Binding of the drug changes the conformation of the platform ring. This might trigger the release

Table 10.3 Protein content of the small ribosomal subunit of 61S ribosomes

Protein	QI	MS
S1	–	–
S2	–	–
S3	↓	↓
S4	+	+
S5	↓	↓
S6	–	–
S7	+	+
S8	+	+
S9	+	+
S10	+	+
S11	↓	↓
S12	–	–
S13	+	+
S14	+	↓
S15	+	n.d.
S16	↓	↓
S17	↓	n.d.
S18	–	n.d.
S19	n.d.	+
S20	+	↓
S21	–	–

QI, quantitative immuno detection; MS, mass spectroscopy; ↓, present in reduced amounts; n.d., not determined.

of some S-proteins, since all lacking proteins are bound directly or indirectly to this ring structure.

In summary, translation of lmRNA requires ribosomes that master the elongation phase of protein synthesis rather than initiation and termination, as seen still today in efficient translation of poly(U) mRNA in the absence of initiation and termination factors. As mentioned above, essential features of the initiation and termination phases developed after domain separation [27]. One can thus speculate that the 61S-particles mirror an ancient form of the small subunit present in proto-ribosomes before domain separation about 3 billion years ago.

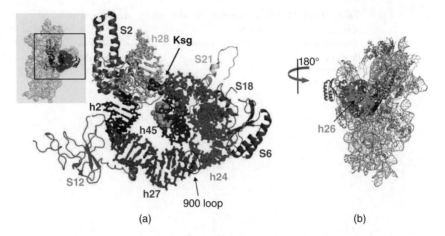

Figure 10.6 The platform of the 30S subunit is made of a ring of four helices; Ksg binds at the junction of h24 and h28. (a) The 30S subunit as seen from the subunit interface and enlargement of the region affected by Ksg *in vivo*. The helices which are affected by Ksg are highlighted. h26 has been omitted for clarity. Binding of Ksg (sphere model) between the G926 of h28 and the 790 loop of h24 is shown [25] [26]. The r-proteins S2 (blue), S6 (purple), S12 (green), S18 (pink), and S21 (yellow) are absent in 61S particles. Nucleotides A1518/A1519 in h45 are shown as light-blue spheres. (b) The 30S subunit shown from the solvent side. h26 (cyan) is located on the platform interacting with proteins S18 (pink) and S1 (not shown). Taken from Ref. [22]. See plate section for a colour version of this figure (Reptinted from [22] with permission from Elsevier)

The severely reduced small subunit in the 61S particle still allowing efficient protein synthesis is, therefore, a welcome starting point for the construction of the minimal ribosome. But what can be said about reducing the components of the large ribosomal subunit?

10.5 A MINIMAL CORE OF THE LARGE RIBOSOMAL SUBUNIT STILL ACTIVE IN PEPTIDE-BOND FORMATION

Peptide-bond formation is the essential catalytic activity of the large subunit of all ribosomes. A combined approach of single omission tests and mass reductions of groups of L-proteins has been used applying total reconstitution of the large ribosomal subunit [28, 29]. The result was that 16 out of 34 distinct ribosomal proteins could be omitted, including the 5S rRNA (Table 10.4). To reconstitute this core, a more subtle reconstitution technique had to be applied. Instead of the stand-

THE MINIMAL RIBOSOME

Table 10.4 A minimal set of ribosomal L-proteins and the 23S rRNA forming a core of the large ribosomal subunit still active in peptide-bond formation

L2	L10	L16	L21
L3	L11	L17	L22
L4	L13	L18	L25
L6	L15	L20	L27

5S rRNA is not required.

ard procedure with two incubations at 42 °C and 50 °C for 20 min and 50 min respectively [30], the incubation temperatures had to be lowered to 36 °C and 40 °C at the first and second incubations, whereas there was no need to change the ionic conditions.

10.6 CUTTING DOWN THE RRNAS

The shortest rRNAs can be found in mitochondria of animals, in particular of worms such as the nematode *Caenorhabditis elegans*. Therefore, a comparison of mitochondrial rRNAs with those of bacteria gives a first hint of rRNA regions which are not essential and, thus, are candidates for removal. Figure 10.7a shows, for example, the secondary structure of the rRNA of the small ribosomal subunit (16S rRNA type). On the right side the mitochondrial nucleotides in red are aligned with those from the bacteria *Thermus thermophilus*; those nucleotides, which could not be aligned are shown in yellow. Similarly, the mitochondrial rRNA is projected onto the crystal structures of small and large ribosomal subunits from *T. thermophilus* and the archaeon *Haloarcula marismortui*.

Although the rRNA of mitochondrial ribosomes is the shortest known, the ribosome itself is not smaller than that of bacterial origin but rather is larger. The reason is that the lack of rRNA regions is even overcompensated by ribosomal proteins: the individual protein is larger than the corresponding one from bacteria, and the total number of proteins is also larger due to proteins specific for this organelle. For example, an *E. coli* ribosome has 54 ribosomal proteins; the respective number for mitochondrial ribosomes is greater than 70. Therefore, it cannot be predicted to what extent the rRNA can be shortened after the number of ribosomal proteins has been minimized.

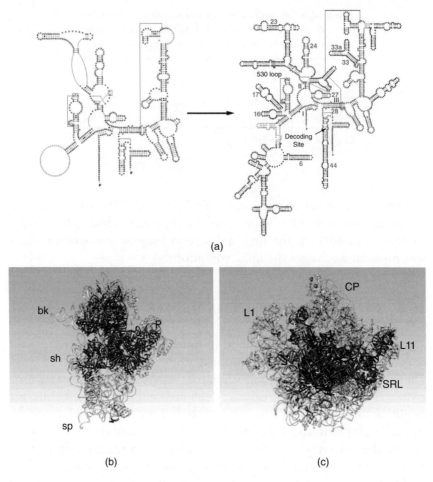

Figure 10.7 Comparison of rRNAs from mitochondria (*C. elegans*) and bacteria. (a) Secondary structure of the 16S rRNA-type (small ribosomal subunit) from mitochondria of *C. elegans* (left) and *T. thermophilus* (right). (b and c) Alignment of the *C. elegans* rRNA on the crystal structure of the small subunit from *T. thermophilus* and the large one from *H. marismortui* respectively. Red: *C. elegans* nucleotides that align with the bacterial rRNAs; yellow: *C. elegans* nucleotides that do not align. Taken from Ref. [31]. See plate section for a colour version of this figure (Reprinted from [30] 2002 with permission from Elsevier)

10.7 CONCLUSIONS

The construction of a minimal ribosome, which still should be functional in the translation of mRNAs or lmRNAs, is a challenging and ambitious enterprise. One has to consider and explore all different kinds of information. We mention here mutants lacking a single protein, which altogether comprise one-third of the ribosomal proteins of a bacterial ribosome (Table 10.1). The problem here is that practically all the mutations severely affect cell growth, which makes it prohibitively difficult to omit all in one step. A better situation is seen with the reconstitution of a minimal core of the large ribosomal subunit, where 16 L-proteins and 5S rRNA could successfully be omitted and still yield a particle with significant activity for peptide-bond formation (Table 10.4). A very elegant way is seen with the recently detected and special 61S ribosome, which might have opened a window to a protein synthesis machinery of ancient times. The 61S ribosome admittedly contains a normal large ribosomal subunit; however, the small subunit lacks six proteins completely and a further five proteins partially and still is active in translating lmRNA efficiently. If it turns out that the partially lacking five S-proteins also can still be removed and the activity preserved, a particle will be obtained containing only 10 proteins and 16S rRNA, a surprising situation. After the number of the ribosomal proteins has been reduced, the next step would be cutting down the ribosomal rRNAs. Here, a comparison of the *E. coli* rRNAs with those of the mitochondrial rRNA could work as a guideline: mitochondrial ribosomes of higher animals contain rRNAs that are shorter than the bacterial ones by one-third.

REFERENCES

1. Agirrezabala, X., Lei, J., Brunelle, J.L. *et al.* (2008) Visualization of the hybrid state of tRNA binding promoted by spontaneous ratcheting of the ribosome. *Molecular Cell*, **32**, 190–197.
2. Gavrilova, L.P., Kostiashkina, O.E., Koteliansky, V.E. *et al.* (1976) Factor-free ("non-enzymic") and factor-dependent systems of translation of polyuridylic acid by *Escherichia coli* ribosomes. *Journal of Molecular Biology*, **101**, 537–552.
3. Schilling-Bartetzko, S., Bartetzko, A., and Nierhaus, K.H. (1992) Kinetic and thermodynamic parameters for transfer RNA binding to the ribosome and for the translocation reaction. *Journal of Biological Chemistry*, **267**, 4703–4712.
4. Dabbs, E.R. (1978) Mutational alterations in 50 proteins of the *Escherichia coli* ribosomes. *Molecular and General Genetics*, **165**, 73–78.
5. Dabbs, E.R. (1991) Mutants lacking individual ribosomal proteins as a tool to investigate ribosomal properties. *Biochimie*, **73**, 639–645.
6. O'Brien, T.W., Denslow, N.D., Anders, J.C., and Courtney, B.C. (1990) The translation system of mammalian mitochondria. *Biochimica et Biophysica Acta*, **1050**, 174–178.

7. Windley, B.F. (2007) Geological history of Earth. In *Encyclopaedia Britannica*, Encyclopædia Britannica, Inc.
8. Allwood, A.C., Walter, M.R., Kamber, B.S. *et al.* (2006) Stromatolite reef from the Early Archaean era of Australia. *Nature*, **441**, 714–718.
9. Banerjee, N.R., Furnes, H., Muehlenbachs, K. *et al.* (2006) Preservation of ~3.4–3.5 Ga microbial biomarkers in pillow lavas and hyaloclastites from the Barberton Greenstone Belt, South Africa. *Earth and Planetary Science Letters*, **241**, 707–722.
10. Gilbert, W. (1986) Origin of life: the RNA world. *Nature*, **319**, 618.
11. Joyce, G.F. (1991) The rise and fall of the RNA world. *New Biology*, **3**, 399–407.
12. Ogle, J.M., Murphy, F.V., Tarry, M.J., and Ramakrishnan, V. (2002) Selection of tRNA by the ribosome requires a transition from an open to a closed form. *Cell*, **111**, 721–732.
13. Moore, P.B. and Steitz, T.A. (2002) The involvement of RNA in ribosome function. *Nature*, **418**, 229–235.
14. Woese, C.R., Kandler, O., and Wheelis, M.L. (1990) Towards a natural system of organisms: proposal for the domains Archaea, Bacteria, and Eucarya. *Proceedings of the National Academy of Sciences of the United States of America*, **87**, 4576–4579.
15. Brocks, J.J., Logan, G.A., Buick, R., and Summons, R.E. (1999) Archean molecular fossils and the early rise of eukaryotes. *Science*, **285**, 1033–1036.
16. Knoll, A.H. (1999) A new molecular window on early life. *Science*, **285**, 1025–1026.
17. Schimmel, P. and Depouplana, L.R. (1995) Transfer RNA: from minihelix to genetic code. *Cell*, **81**, 983–986.
18. Yusupov, M.M., Yusupova, G.Z., Baucom, A. *et al.* (2001) Crystal structure of the ribosome at 5.5 Å resolution. *Science*, **292**, 883–896.
19. Chin, K., Shean, C.S., and Gottesman, M.E. (1993) Resistance of lambda cI translation to antibiotics that inhibit translation initiation. *Journal of Bacteriology*, **175**, 7471–7473.
20. Grill, S., Moll, I., Hasenohrl, D. *et al.* (2001) Modulation of ribosomal recruitment to 5′-terminal start codons by translation initiation factors IF2 and IF3. *FEBS Letters*, **495**, 167–171.
21. Moll, I., Hirokawa, G., Kiel, M.C. *et al.* (2004) Translation initiation with 70S ribosomes: an alternative pathway for leaderless mRNAs. *Nucleic Acids Research*, **32**, 3354–3363.
22. Kaberdina, A.C., Szaflarski, W., Nierhaus, K.H., and Moll, I. (2009) An unexpected type of ribosomes induced by kasugamycin: a look into ancestral times of protein synthesis? *Molecular Cell*, **33**, 227–236.
23. Yoshikawa, M., Okuyama, A., and Tanaka, N. (1975) A third kasugamycin resistance locus, ksgC, affecting ribosomal protein S2 in *Escherichia coli* K-12. *Journal of Bacteriology*, **122**, 796–797.

24. Wilson, D.N. and Nierhaus, K.H. (2005) Ribosomal proteins in the spotlight. *Critical Reviews in Biochemistry and Molecular Biology*, **40**, 243–267.
25. Schluenzen, F., Takemoto, C., Wilson, D.N. *et al.* (2006) The antibiotic kasugamycin mimics mRNA nucleotides to destabilize tRNA binding and inhibit canonical translation initiation. *Nature Structural & Molecular Biology*, **13**, 871–878.
26. Schuwirth, B.S. Borovinskaya, M.A., Hau, C.W. *et al.* (2005) Structures of the bacterial ribosome at 3.5 A resolution. *Science*, **310**, 827–834.
27. Londei, P. (2005) Evolution of translational initiation: new insights from the Archaea. *FEMS Microbiology Reviews*, **29**, 185–200.
28. Hampl, H., Schulze, H., and Nierhaus, K.H. (1981) Ribosomal components from E. coli 50S subunits involved in the reconstitution of peptidyltransferase activity. *Journal of Biological Chemistry*, **256**, 2284–2288.
29. Schulze, H. and Nierhaus, K.H. (1982) Minimal set of ribosomal components for the reconstitution of the peptidyltransferase activity. *EMBO JOURNAL*, **1**, 609–613.
30. Dohme, F. and Nierhaus, K.H. (1976) Total reconstitution and assembly of 50S subunits from E. coli ribosomes *in vitro*. *Journal of Molecular Biology*, **107**, 585–599.
31. Mears, J.A., Cannone, J.J., Stagg, S.M. *et al.* (2002) Modeling a minimal ribosome based on comparative sequence analysis. *Journal of Molecular Biology*, **321**, 215–234.

11

Semi-Synthetic Minimal Living Cells

Pasquale Stano[1], Francesca Ferri[2], and Pier Luigi Luisi[1]

[1]University of Roma 3, Biology Department, Viale G. Marconi 446, 00146 Roma, Italy
[2]University of Parma, Neuroscience Department, Via Volturno 39, 43100 Parma, Italy

11.1 The Notion of Minimal Cell 248
11.2 The Minimal Genome 250
    11.2.1 Further Reductions 256
11.3 The Minimal RNA Cell 260
11.4 The Minimal Size of Cells 262
11.5 Current Experimental Approaches for the Construction of Minimal Cells 263
    11.5.1 Reactivity in Vesicles 265
    11.5.2 Protein Expression in Liposomes 269
    11.5.3 Future Developments towards Minimal Cells 275
11.6 Concluding Remarks 278
Acknowledgments 280
Reference 280

In this chapter we aim to introduce the reader to the issue of minimal living cells, defined as semi-synthetic cells having the minimal and sufficient number of components/functions to be considered alive. Semi-synthetic cells are built by a synthetic procedure that functionally combines nonliving components (DNAs, RNAs, enzymes) into a multimolecular confined system, enclosed within a semipermeable membrane (the lipid bilayer). Describing the concepts and experiments at the basis of these constructions, we will point out that an operational definition of minimal cell does not define a single species, but rather a broad family of interrelated cell-like structures. The relevance of this research is explained and discussed, also taking into account that minimal cells may functionally resemble early simple cells (i.e. protocells); therefore, emphasis will be given to the significance of these studies in the origins of life and evolution contexts. From this point of view, the bioengineering of minimal living cells, typical of a synthetic biology approach, complements the more fundamental aspects of the origins of early cells, driven by self-organization and selection.

In addition, we discuss the minimal genome and the minimal size of living cells. Although several authors agree on setting the theoretical, full-fledged minimal genome to a figure between 200 and 300 genes, further assumptions may significantly reduce this number; that is, by eliminating ribosomal proteins and limiting the DNA and RNA polymerases to only a few, less-specific molecular species. These considerations can be pertinent from the evolutive viewpoint, by considering the increase of complexity as an outcome of function refinements instead of the introduction of an essentially new pattern. Although hypothetic, such a "reduced" minimal genome represents a theoretical basis for understanding minimal and perhaps "limping" life. Together with the genome reduction, minimal life also implies minimal size, since it is conceivable that smaller genomes and minimal metabolism do not require large physical cell extension.

To date, the experimental approach to minimal cells consists in designing liposome-based bioreactors composed of genes/enzymes corresponding to the minimal cellular functions. Recent developments indicate that functional proteins can be successfully synthesized inside liposomes, paving the way to more complex minimal cells, which will ultimately lead to minimal living cells. The present constructs are still rather far from a minimal cell, and experimental as well as theoretical difficulties opposing further reduction of complexity are discussed. While most of these minimal cell constructions may represent relatively poor imitations of a modern full-fledged cell, further studies will begin precisely from these constructs. In conclusion, we give a brief outline of the next possible steps on the road map to a minimal cell.

11.1 THE NOTION OF MINIMAL CELL

The simplest living cells existing on our Earth have several hundred genes, with hundreds of expressed proteins, which, more or less simultaneously, catalyze hundreds of reactions within the same tiny compartment – a maze of enormous complexity.

This picture elicits the question of whether or not such complexity is really essential for life – or if cellular life might be possible with a much smaller number of components and functions. In fact, the enormous complexity of modern cells is probably the result of billions of years of evolution in which a series of feedback mechanisms, redundancies, and metabolic loops were developed. Clearly, early cells could not have been as complex as modern ones. In addition to this consideration, we must point out that under highly permissive conditions many of such mechanisms might not be necessary. These considerations led to the notion of a minimal cell, here broadly defined here as a cell having the minimal and sufficient number of components to be considered alive. This automatically precedes the next fundamental question: what does "alive" mean? A complex question. One may choose quite a general definition, defining life at a cellular level as the concomitance of three basic properties: self-maintenance (metabolism), self-reproduction, and evolvability.

Notice, however, that self-maintenance and self-reproduction are properties of individual cells, whereas evolvability is a Darwinian notion. As such, it refers to populations rather than individual cells. Consequently, one should take into consideration an entire family of minimal cells in the stream of environmental pressure and corresponding genetic evolution.

The trilogy defining cellular life may not be perfectly implemented, particularly in synthetic constructs, and several approximations to cellular life can be envisaged. For example, we may have protocells capable of self-maintenance but not self-reproduction, or vice versa. Or we might have protocells in which self-reproduction is active for only a few generations, or systems that are not capable of evolvability. In any given type of minimal cell – for example, one with all three attributes – there may be quite different ways of implementation and sophistication. So, clearly the term "minimal cell" depicts large families of possibilities, not simply one particular construct. The idea that the minimal forms of life are not univocally defined, and correspond rather to a large family, is not new in the field of the origin of life and early evolution. However, it is important to keep in mind that we are not simply considering theoretical possibilities, but something new: a synthetic biology approach and the particular methodology of experimental implementation.

Harold J. Morowitz [1], in his book *Beginning of Cellular Life*, presents a systematic discussion on minimal cells and on the origin of cellular life, "from the above" and "from the bottom." In the approach "from the above," the search for primitiveness stems from the analysis of modern living systems and goes backward in time (and complexity) to define or to clarify the nature of early cells. In the approach "from the bottom," the fundamental laws of physics and chemistry are used to depict a scenario for the origins of cellular life. Additional significant insights in the field are those by Jay and Gilbert [2], as well as by Woese [3] and Dyson [4]. More recently, the reviews by Pohorille and Deamer [5], Szathmáry et al. [6], Forster and Church [7], and LeDuc amd coworkers [8] have discussed the issue of minimal cells, also revealing the possible role of minimal synthetic cells in biotechnology. Our group has been interested in minimal cells from the early 1990s, by carrying out experimental research focused on reaction in compartments (reverse micelles, water-in-oil emulsion, and liposomes). Reviews on this work and on our approach were already published before the advent of synthetic biology [9, 10]. In recent years, however, there has been a significant revival of interest in the field of the minimal cell; it probably witnesses a sort of paradigm change in experimental biology, due to the affirmation of systems biology and synthetic biology.

Great interest has been focused on the semi-synthetic minimal cells, a concept that was present in our seminal work at the ETH in Zurich [9, 10], and further developed in recent studies [11–16]. In this chapter, which is largely based on the discussion published in *Naturwissenschaften* [12], we wish to review the recent theoretical and experimental approaches for the construction of such semi-synthetic minimal cells. The issue of "minimal genome" will be introduced first, followed by a short discussion on the issue of minimal cellular size. Greater emphasis will be given to experimental approaches for synthesizing cells, summarizing successful achievements, and envisaging future developments, also discussing the complications that must be faced before attaining the first man-made living cell. Our approach is now considered within the mainstream of synthetic biology as one at the side of "total reconstruction," namely not based on manipulation of already-existing organisms. Its dual relevance as a conceptual tool and as a generator of potential biotechnological devices is well recognized. The minimal cell of Craig Venter and collaborators, in contrast, stems from the combination of a minimal genome, realized by chemical synthesis, with an existing cell deprived of its own "natural" genome. We will not discuss this approach here; the interested reader should refer to original publications [17–19] [100].

11.2 THE MINIMAL GENOME

The issue of the minimal genome is the first aspect we would like to discuss within the framework of minimal cells. In order to put this discus-

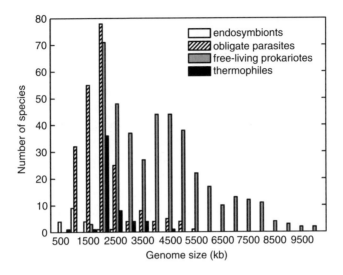

Figure 11.1 Prokaryotic genome size distribution ($n = 641$). Genome sizes, complete proteomes, and the number of open reading frames were all retrieved from the National Center for Biotechnology Information (http://www.ncbi.nlm.nih.gov) [20] (Reprinted from [12] with kind permission from Springer Science+Business Media)

sion into a more concrete perspective, we will follow the classical view, which originates from a comparative genomics approach focused at the smallest unicellular organisms. Then, we will evaluate if further reductions are possible – at least in principle – introducing hypotheses that are more oriented towards the origin and developments of primitive cells.

Figure 11.1 compares the genome size distribution calculated under a series of assumptions [20], of free-living prokaryotes, obligate parasites, thermophiles, and endosymbionts. The values of DNA content of free-living prokaryotes can vary over a tenfold range, from the 1450 kb of *Halomonas halmophila* to the 9700 kb genome of *Azospirillum lipoferum* Sp59b. In comparison, consider that *Escherichia coli* K-12 has a genome size of approximately 4640 kb and *Bacillus subtilis* is 4200 kb.

Classification of endosymbionts as a separate group shows that their DNA content may be significantly smaller; the smallest sizes are then those of *Mycoplasma genitalium* (580 kb) and *Buchnera aphidicola* BCc (420 kb), with values that confirm the predictions of Shimkets [21], according to whom the minimum genome size for a living organism should be approximately 600 kb. It is argued that these two organisms have undergone massive gene losses and that their limited encoding capacities are due to their adaptation to the highly permissive intracellular environments provided by the hosts [20].

What do these figures mean in terms of minimal gene numbers? Table 11.1, also taken from Islas *et al.* [20], reports the number of coding regions in some small genomes. The table also gives an account of the redundant genes, amounting to an average of 6–20% of the whole genome. How is it then possible to define a "universal" minimal cellular genome from current genomic knowledge? We will follow here a recent review of Moya *et al.* [22] that discusses in detail the theoretical aspects of defining a minimal genome. The first attempt was made by comparative genomics [23], soon after the first two bacterial genomes, from *Haemophilus influenzae* and *M. genitalium*, were sequenced. According to their analysis, a minimal genome of 256 genes was considered as "essential" to cellular life. However, as the authors pointed out, since

Table 11.1 Genetic redundancies in small genomes of endosymbionts and obligate parasites[a]

Proteome	Genome size (kb)	No. of ORFs	No. of redundant sequences	Redundancy (%)
Mycoplasma genitalium	580	480	52	10.83
Mycoplasma pneumoniae	816	688	134	19.48
Buchnera sp. APS	640	574	67	11.67
Ureaplasma urealyticum	751	611	105	17.18
Chlamydia trachomatis	1000	895	60	6.70
Chlamydia muridarum	1000	920	60	6.52
Chlamydophila pneumoniae J138	1200	1070	148	13.83
Rickettsia prowazekii	1100	834	49	5.88
Rickettsia conorii	1200	1366	189	13.84
Treponema pallidum	1100	1031	78	7.57

[a]Genome sizes, complete proteomes, and the number of ORFs (open reading frames) were all retrieved from NCBI (http://www.ncbi.nlm.nih.gov). Reprinted from [12] with kind permission from Springer Science+Business Media.

many essential functions may be carried out by nonorthologous proteins, the comparative approach may not be sufficient to determine the minimal genome correctly. Three different experimental approaches have been used to identify genes that are essential under particular growth conditions: massive transposon mutagenesis strategies (the most widely used approach), the use of antisense RNA to inhibit gene expression, and the systematic inactivation of each individual gene present in a genome (a summary of such work, which does not pretend to be complete, is shown in Table 11.2. Further reference to original work can be found in [33]).

By the combination of all published results using computational and experimental methods, including reduced genomes from insect endosymbionts, Moya and coworkers defined the minimal gene set of 206 genes for a free-living bacterium thriving in a chemically rich environment [33, 34]; see Table 11.3 and Figure 11.2.

Table 11.2 Works on the minimal genome

Description of the system	Main goal and results	Ref.
The complete nucleotide sequence (580 070 base pairs) of the *Mycoplasma genitalium* genome has been determined by whole-genome random sequencing and assembly.	Only 470 predicted coding regions were identified (genes required for DNA replication, transcription and translation, DNA repair, cellular transport, and energy metabolism).	[24]
Site-directed gene disruption in *Bacillus subtilis*	The values of viable minimal genome size were inferred	[25]
The 468 predicted *M. genitalium* protein sequences were compared with the 1703 protein sequences encoded by the other completely sequenced small bacterial genome, that of *Haemophilus influenzae*	A minimal self-sufficient gene set: the 256 genes that are conserved in these two bacteria, Gram-positive and Gram-negative respectively, are almost certainly essential for cellular function	[23]
Computational analysis (quantification of gene content, of gene family expansion, of orthologous gene conservation, as well as their displacement)	A set close to 300 genes was estimated as the minimal set sufficient for cellular life	[26]

(*Continued*)

Table 11.2 (*Continued*)

Description of the system	Main goal and results	Ref.
Global transposon mutagenesis was used to identify nonessential genes in *Mycoplasma* genome	265 to 350 of the 480 protein-coding genes of *M. genitalium* are essential under laboratory growth conditions, including about 100 genes of unknown function	[27]
Several theoretical and experimental studies are reviewed	The minimal-gene-set concept	[28]
The article focuses on the notion of a DNA minimal cell	The conceptual background of the minimal genome is discussed	[29]
A technique for precise genomic surgery was developed and applied them to deleting the largest K-islands of *Escherichia coli*, identified by comparative genomics as recent horizontal acquisitions to the genome	12 K-islands were successfully deleted, resulting in an 8.1% reduced genome size, a 9.3% reduction of gene count, and elimination of 24 of the 44 transposable elements of *E. coli*. The goal was to construct a maximally reduced *E. coli* strain to serve as a better model organism	[30]
Physical mapping of *Buchnera* genomes obtained from five aphid lineages	They suggest that the *Buchnera* genome is still experiencing a reductive process toward a minimum set of genes necessary for its symbiotic lifestyle	[31]
Computational and experimental methods on comparative genomics	60 proteins are common to all cellular life. A core of 500–600 genes should represent the gene set of the last universal common ancestor	[32]
Buchnera and other organism genomes were compared	206 genes were identified as the core of a minimal bacterial gene set	[33]
Comparative genomics	Estimates of the size of minimal gene complement were done to infer the primary biological functions required for a sustainable, reproducing cell nowadays and throughout evolutionary times	[20]

Reproduced, with permission, from Luisi *et al.* [12]. Reprinted from [12] with kind permission from Springer Science+Business Media.

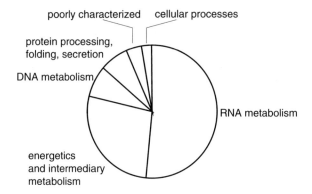

Figure 11.2 Pie chart representing the functions of the 206 genes composing the minimal genome as defined by Gil et al. [33]

Table 11.3 Core of a minimal bacterial gene set[a]

DNA metabolism		16
Basic replication machinery	13	
DNA repair, restriction, and modification	3	
RNA metabolism		106
Basic transcription machinery	8	
Translation: aminoacyl-tRNA synthesis	21	
Translation: tRNA maturation and modification	6	
Translation: ribosomal proteins	50	
Translation: ribosome function, maturation, and modification	7	
Translation factors	12	
RNA degradation	2	
Protein processing, folding, and secretion		15
Protein post-translational modification	2	
Protein folding	5	
Protein translocation and secretion	5	
Protein turnover	3	
Cellular processes		5
Energetic and intermediary metabolism		56
Poorly characterized		8
Total		206

[a]From Gil et al. [33]. Reprinted from [12] with kind permission from Springer Science+Business Media.

This rather low number of genes is already a considerable simplification of the gene number in existing cells. How would a cell behave with such a minimal genome? By the analysis of the genes present in the minimal set [33], we can derive a set of minimal functions implemented by the corresponding (hypothetical) minimal cell. It is clear from the beginning that such minimal metabolic complexity is not "absolute," but is a function of the chemical compounds and the primary energy source(s) available to the living system in a given environment [1]. According to these hypotheses [33], then, the minimal genome includes: (i) a virtually complete DNA replication machinery (13 genes); (ii) a rudimentary DNA repair system (three genes); (iii) a virtually complete transcription machinery (eight genes), not regulated; (iv) a nearly complete translational system (98 genes, 50 of which are ribosomal proteins); (v) genes for protein processing, folding, secretion, and degradation (15 genes); (vi) one gene for cell division; (vii) a basic (although incomplete) substrate transport machinery (four genes); (viii) genes for energetic metabolism – that is, glycolysis and ATP synthesis (19 genes); (ix) the nonoxidative branch of the pentose pathway (three genes); (x) lipid, nucleotides, and cofactors biosynthesis (7, 15, and 12 genes respectively), whereas amino acids, fatty acids, adenine, guanine, uracil, and cofactors precursors are supposed as provided by the environment; (xi) eight genes of unknown function.

Clearly, the theoretical discussion of minimal self-maintenance does not furnish details on the number of elements required to accomplish such a function. It very much depends on the nature of biopolymers used by the cell and by the "richness" of the environment in terms of biochemicals. In other words, the minimal genome (and, therefore, the definition of the components/functions required to implement cellular life) depends strongly on the definition of "available" substances and on the material implementation of a minimal cell. The reader will see some examples that clarify this point. In the next section we discuss a possible further reduction to the already low number of 206 genes, from the point of view of primitiveness, an aspect that cannot be derived from comparative genomics. A radically different case will be presented later, the RNA (ribozyme)-based minimal cell, where the properties of RNA (in terms of a combination of replicator and enzyme-like catalyst) are exploited to design a minimal chemical machine that obeys autopoietic logic and can be considered alive according to the three criteria mentioned above.

11.2.1 Further Reductions

In this section we would like to speculate on further reductions to the minimal gene set (206 genes) defined by Moya and coworkers [33]. Imagine a kind of theoretical knock down of the genome, reducing, at the same time, cellular complexity and part of the nonessential functions [29].

The first pit stop of this intellectual game is to imagine that low molecular weight compounds, including nucleotides, amino acids, fatty acids, cofactors, etc., were available in the surrounding medium and able to permeate into the cell membrane, even if at a low rate. This would be a semipermeable minimal cell, where low molecular weight compounds are able to enter/exit the cell according to chemical gradients, whereas (and this is important) high molecular weight compounds (which carry function and information) are not. This finally brings us to a cell able to perform protein and lipid biosynthesis by a modern ribosomal system, but limited to a rather restricted number of enzymes – see Table 11.4 [29]. This cell would have approximately 25 genes for the entire DNA/RNA synthetic machinery, approximately 120 genes for the entire protein synthesis (including RNA synthesis and the 55 ribosomal proteins), and four genes for the synthesis of the membrane. Which brings us to a total of about 150 genes.

Thanks to the outside supply of substrates, such a cell should be capable of self-maintenance and of self-reproduction, including replication of the membrane's components. However, it would neither synthesize low molecular weight compounds nor have redundancies for its own defense and security (in fact, all self-repair mechanisms are missing). Also, cell division would simply be due to a physically based statistical process.

There is, of course, no proof that this theoretical construct would be viable – but this also goes for Moya's 206 genes. It is, nevertheless, instructive to take these theoretical knock-down experiments further, the next victims being ribosomal proteins. Can we take them out? Some indications suggest that ribosomal proteins may not be essential for protein synthesis [35], and there are other suggestions about an ancient and simpler translation system [36, 37].

Of course, this sort of discussion takes us directly into the scenario of early cells at the origin of life, and, in fact, there are some claims that the first ribosomes consisted of rRNAs associated simply with basic peptides [38]. If we accept this, and take out the 55 genes for ribosomal proteins and some other enzymes, we would then land at a number of around 110 genes.

A large portion of the foreseen genes correspond to RNA and DNA polymerases. A number of data [39–42] suggest that a simplified replicating enzymatic repertoire – as well as a simplified version of protein synthesis – might be possible. In particular, there is the idea that it is

Table 11.4 A hypothetical list of gene products that defines the minimal cells according to the definitions used in this paper, sorted by functional category

Gene product	Number of genes		
	Minimal DNA cell[a]	"Simple-ribosome" cell	Extremely reduced cell
DNA/RNA metabolism			
DNA polymerase III	4[b]	4[b]	1
DNA-dependent RNA polymerase	3[c]	3[c]	1
DNA primase	1	1	
DNA ligase	1	1	1
Helicases	2–3	2–3	1
DNA gyrase	2[d]	2[d]	1
ssDNA-binding proteins	1	1	1
Chromosomal replication initiator	1	1	
DNA topoisomerase I and IV	1 + 2[d]	1 + 2[d]	1
ATP-dependent RNA helicase	1	1	
Transcription elongation factor	1	1	
RNases (III, P)	2	2	
DNases (endo/exo)	1	1	
Ribonucleotide reductase	1	1	1
Protein biosynthesis/translational apparatus			
Ribosomal proteins	51	0	0
Ribosomal RNAs	1[e]	1[e]	1[e], self-splicing
aa-tRNA synthetases	24	24	14[f]
Protein factors required for biosynthesis and synthesis of membrane proteins	9–12[g]	9–12[g]	3

tRNAs	33	33	16[b]
Lipid metabolism			
Acyltransferase "plsX"	1	1	1
Acyltransferase "plsC"	1	1	1
PG synthase	1	1	1
Acyl carrier protein	1	1	1
Total	**146–150**	**105–107**	**46**

[a]Based on *M. genitalium*.
[b]Subunits a, b, y, tau.
[c]Subunits a, b, b'.
[d]Subunits a, b.
[e]One operon with three functions (rRNAs).
[f]Assuming a reduced code.
[g]Including the possible limited potential to synthesize membrane proteins.
[h]Assuming the third base to be irrelevant.
Reprinted from [12] with kind permission from Springer Science+Business Media.

conceivable that a single polymerase could play multiple roles as a DNA polymerase, transcriptase, and primase in very early cells [29].

The game could go on by assuming that, at the time of the early cells, not all "our" 20 amino acids were involved – and a lower number of amino acids would reduce the number of amino acyl-tRNA-synthetases and the number of t-RNA genes.

All these considerations may help to decrease the number of genes down to a happy number of, say, 45–50 genes (see Table 11.4) for a living, although certainly limping, minimal cell [29].

This number is significantly lower than the one proposed by Moya in Table 11.3, but is, of course, based on a higher degree of speculation. Many authors would doubt that a cell with only 45–50 genes would be able to work. But again, the consideration goes on to the early cells, and to the consideration that the first cells could have not started with dozens of genes from the very beginning in the same compartment.

This last consideration permits a logical link with the notion of compartments, and on their role in the origins of functional cells.

Suppose that these 45–50 macromolecules (or their precursors) developed first in solution. Then, in order to start cellular life, compartmentation would have come later on, and one would then have to assume the simultaneous entrapment of all these different genes in the same vesicle. This could indeed be regarded as highly improbable; in fact, a scenario in which the complexity of cellular life evolved from within the compartment is more reasonable – a situation namely where the 45 (or 206) macromolecules were produced and evolved from a much smaller group of components from the inside the protocell.

Until now we have speculated on "normal" protein/DNA/RNA cells, the ones we know in nature. In a further speculative leap we could ask the question: what about a theoretical RNA cell? Let us briefly consider this question before proceeding further with the discussion on the minimal size of cells.

11.3 THE MINIMAL RNA CELL

One of the simplest constructs that responds to the criteria of evolvability, self-maintenance, and reproduction is the so-called "RNA cell" (Figure 11.3). This object, not yet observed or realized in the laboratory, was developed in a landmark paper by Szostak *et al.* [43]. It represents an attempt to furnish a synthesis of the RNA-world and the compartment-world models.

SEMI-SYNTHETIC MINIMAL LIVING CELLS 261

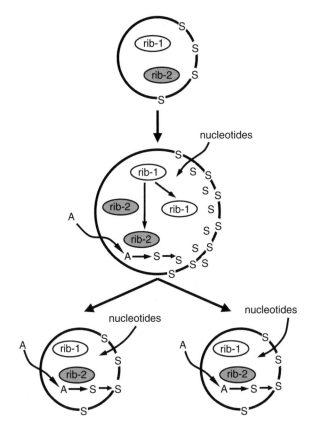

Figure 11.3 The RNA cell, containing two ribozymes. Rib-1 is an RNA replicase capable of reproducing itself and making copies of Rib-2. Rib-2 is capable of synthesizing the cell membrane by converting precursor A to membrane-forming S. All necessary low molecular weight components required for macromolecular synthesis are provided from the surrounding medium and are capable of permeating the membrane. For the sake of simplicity, an ideal cell division is represented in this figure, where all core components are equally shared between new vesicles (Reprinted from [12] with kind permission from Springer Science+Business Media)

In this case, the combined "genetic" and catalytic properties of ribozymes play a central role. The RNA cell consists of a vesicle containing two idealized ribozymes: a replicase and a lipid synthase. The first ribozyme is capable of replicating a copy of itself, producing a complementary RNA strand that is replicated again (as in the mechanism of RNA replication catalyzed by Qβ replicase, an RNA-dependent RNA polymerase). The second ribozyme is replicated by the first one by a similar mechanism. The second ribozyme, however, being a lipid synthase, is able to transform lipid precursors into the membrane-forming compound, allowing the growth and the subsequent division of the parent vesicle. In this way, a concerted core-and-shell reproduction of the entire construct could be obtained.

As mentioned previously, this is a hypothetical scheme, based on not-yet-existing ribozymes and a series of additional assumptions; for example, the full permeability of the membrane to the precursor A and nucleotides, both present in large excesses in the environment, or the assumption that the cell divides distributing both kinds of ribozyme in the daughters (so that in each cell there are always first and second ribozyme types). Moreover, it is not clear how the cell may manage the increase of concentration of complementary RNA strands, which accumulate at each cycle.

The construct of Figure 11.3, though quite exciting for its simplicity (only two genes!), remains a theoretical model; also, bearing in mind that the two ribozymes required for this construction are still nonexistent. Clearly, the RNA cell – in a historical scenario – must eventually evolve into the DNA/protein cell. Nevertheless, if we focus on the possibility of realizing a synthetic cell, RNA-based cells are simpler than the corresponding DNA/proteins-based ones, and it will be certainly possible (in the near future) to design and implement simple ribozyme-based compartmentalized systems capable of reproducing interesting cellular patterns.

11.4 THE MINIMAL SIZE OF CELLS

The other side of cell minimization concerns the minimal physical dimension of the cell – namely, the dimension that still permits cellular life. These two aspects, minimal genome and minimal size, are obviously connected to one another, being also related to evolutionary paths and to the environment conditions.

The issue of size limits of very small microorganisms has recently been discussed by focusing on the complexity of modern and early cells, its relation to viability, biochemical requirements, as well as physical and evolutionary constraints [44]. As we have seen, microorganisms that receive many basic nutrients and metabolites from their "environment" need fewer genes, as observed in intracellular mutualists, or host-associated parasites. The sizes of such microorganisms range from about 0.3 to 0.5 µm.

Together with a reduction of number of molecular species, a reduction in size is possible only if the number of copies of each species is reduced.

It follows that the efficiency and the reproduction rate of hypothetical small cells should decrease. According to stringent assumptions (e.g. one ribosome, one t-RNA set, one m-RNA for each of 100 non-ribosomal proteins, each present in 10 copies), the diameter of a spherical cell compatible with a modern system of genome expression would be between 200 and 300 nm [44]. Small cells may be favored in diffusion-limited growth conditions, thanks to their high surface-to-volume ratio, and also by the fact that (at a given membrane composition) small cells can sustain typical values of bacterial osmotic pressure and, therefore, avoid the construction of cell walls (this, in turn, would reduce the amount of DNA, i.e., the sequences that codify for the construction of the cell walls) [45].

In addition to these theoretical considerations, we should mention the controversial reports on "nanobacteria" [46], which are particles found in human and cow blood, with dimensions around 200 nm (diameter). It has recently been pointed out that the bacterial status of nanobacteria is still lacking satisfactory evidence, and the term "calcifying nanoparticles" has recently been adopted to describe such bodies [47]. Other authors have reported on similar tiny corpuscles, but generally it is not clear whether they are "living" in the common sense of bacterial life [48].

It is clear from this analysis that the question about the minimal size of a cell is still open – and that is a relevant question. The clarification of such a question would be important in the field of the origin of life, as it is conceivable that the origin of cells started with minimal protocells (<200 nm?) along the pathway of evolution – and it is then interesting to assess whether and to what extent small compartments may permit life. The question becomes particularly timely in the present era of synthetic biology, as it has become possible to construct molecular systems in the laboratory which display living (or living-like) properties, as in the case of the semi-synthetic minimal cells. In fact, a recent published experimental study (see below) was specifically designed in order to address the question of minimal size of cells.

11.5 CURRENT EXPERIMENTAL APPROACHES FOR THE CONSTRUCTION OF MINIMAL CELLS

Although the complexity of a minimal cell based on protein synthesis inside liposomes is certainly higher and probably less historically realistic than RNA (ribozyme)-based minimal cells (see previous section), it has the great advantage of being experimentally accessible. In fact, genes, ribosomes, enzymes, and all the molecular components needed are available to the experimenter, whereas the ribozymes indicated in the previous section are not. In other words, a scientist may design and construct in the laboratory a semi-synthetic cell based on the encapsulation of genes, ribosomes, and enzymes within lipid vesicles.

From the theoretical point of view, a DNA/protein cell cannot be considered "prebiotic." In fact, it is not. It must be clarified from the beginning within which framework the research on semi-synthetic minimal cells is placed.

Traditionally, people working in the area of prebiotic chemistry have been interested in reproducing, in the laboratory, all the steps that led to the origins of the first living cell by pursuing the so-called "bottom-up" approach, based on the notion that a continuous and spontaneous increase of molecular complexity transformed inanimate matter into the first self-reproducing cellular entities. The impressive work on reconstruction of such paths, done in the last 60 years, was aimed at achieving such a result, in a stepwise and allegedly prebiotic fashion. For a number of reasons, this approach has not yet been successful, and several experimental and conceptual problems (i.e. the historical nature of such path, more related to contingency than to strictly deterministic routes) somehow hinder this research. Nevertheless, a large number of scientists are actively involved in such an approach. There is, however, another way of looking at the same problem (i.e. the emergence of living cells) that has been proposed in the last few years, as indicated in Figure 11.4. This corresponds to a different approach that starts conceptually from modern cells and reduces their complexity in order to arrive at a *minimal*

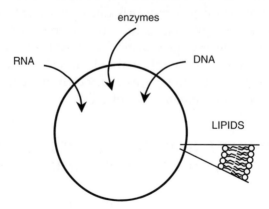

Figure 11.4 The semi-synthetic approach to the construction of the minimal cell. Reproduced, with permission from Luisi et al. [12]

living cell. From the practical viewpoint, extant nucleic acids, ribosomes, and enzymes are encapsulated in a suitable compartment (generally liposomes) so that a minimal cell is, therefore, *constructed*. The term "top-down" was initially used to define the engineering idea behind this approach. However, it should be noted that, from the experimental point of view, the construction of minimal cells starting from molecules (even if these are modern, complex molecules) goes from a lower complexity level (i.e. the molecules) to a higher one (i.e. the cell). In order to avoid confusion with other common uses of the term "top-down," the term "reconstruction" is perhaps more appropriate, making it clear that in this procedure one does not necessarily reach the construction of an extant cell or something that exists (or has existed) on Earth. The minimal cells obtained according to this strategy may be correctly called semi-synthetic, since natural macromolecules are used (modern genes, ribosomes, enzymes) that are synthetically/artificially assembled in a compartment such as a lipid or fatty acid vesicle (Figure 11.4).To date, studies on semi-synthetic minimal cells have a threefold relevance: (i) in the origins of life, they represent cell models and serve to demonstrate that life indeed emerges as a system property when a certain organization pattern is reached [49]; (ii) in the field of autopoietic systems, being designed and constructed in order to behave according to an autopoietic organizational scheme [50]; (iii) in synthetic biology, since the bioengineering approach ultimately aims to construct cells in order to accomplish specific functions [51]. It must finally remarked that the construction process is itself a source of biological knowledge; therefore, epistemological considerations are often associated with the goal of synthesizing living cells [52].

11.5.1 Reactivity in Vesicles

Liposomes have been used for many years as cellular models. Possible analogies between vesicles and cellular membranes have often been considered in terms of physico-chemical properties, such as stability, permeability, self-reproduction, and so on, to see whether (and to what extent) vesicles are close to cellular structures. The second area of inquiry considers the use of vesicles as hosts for complex molecular biological reactions, to see if the biochemistry of cellular life can be reconstructed inside vesicles.

Concerning the first area of study, it has been shown that vesicles are capable of multiplying themselves at the expense of surfactant precursors [53, 55, 74]; and under certain conditions this may happen with the retention of the original size distribution (the so-called matrix effect; see Refs [56–59]). It is not the aim of this chapter to review all these data, but it is important here to keep in mind that one of the most critical mechanisms of living cells can be simulated by vesicles based only on physical and chemical properties; that is, without the use of sophisticated biochemical machineries. This consideration is relevant if one focuses on the prebiotic scenario.

Another important, preliminary physico-chemical property is the membrane permeability to solutes. Here, things are more complicated, as lipid vesicles (liposomes) offer a considerable resistance to the uptake of simple biochemicals in their water pool. This is particularly true with phospholipid membranes, which are commonly used as models for the bilayer of modern membranes. Note, however, that phospholipids should be considered "modern" compounds in the evolutive scale, and most likely the first membranes and vesicles were constituted by other surfactants, such as fatty acids, which could offer higher permeability (although possibly less stability) by virtue of their different chemical structure and also presumable chemically heterogeneous composition. It is reasonable, in fact, that early cells might have been somehow more "permissive" in terms of boundary properties and functions. In a recent study, Szostak and coworkers investigated the permeability of nucleotides in mixed fatty acid vesicles [60]. It was shown that (i) myristoleate:glycerol monomyristoleate $(2:1)$ vesicles, (ii) decanoate:1-decanol:glycerol monodecanoate $(4:1:1)$, and (iii) myristoleate:farnesol $(2:1)$ allowed the entrance of an activated nucleotide into the vesicles. Template-directed (polymerase-free) oligomerization inside the vesicle followed the uptake. The use of membrane channels is an alternative. Alpha-hemolysis [61] and porines [62–65] have recently been used for this aim, but this method has not been standardized, although it can be considered successful. Within the framework of primitive cells, however, the effect of simple peptides on the membrane permeability appears much more interesting. The main conceptual problem is how to bring together several components into a lipid vesicles, so that the required concentration and composition can be achieved. It has been suggested that this difficulty might be partially circumvented if two or more liposomes, each containing a given substrate, could fuse together so as to produce liposomes containing all reagents [11]. In fact, fusion of vesicles is becoming an active area of research, and interesting results have already been obtained [66–69].

Table 11.5 Molecular biology reactions into liposomes

Description of the system	Main goal and results	Ref.
Enzymatic poly(A) synthesis	Poly(A) is produced inside simultaneously with the (uncoupled) self-reproduction of vesicles	[54]
Enzymatic poly(A) synthesis	Polynucleotides phosphorylase producing poly(A) from ADP	[70]
Oleate vesicles containing the enzyme Qβ replicase, an RNA template, and ribonucleotides. The water-insoluble oleic anhydride was added externally	A first approach to a synthetic minimal cell: the replication of a RNA template proceeded simultaneously with the self-replication of the vesicles	[71]
POPC liposomes containing all different reagents necessary to carry out a PCR reaction	DNA amplification by the PCR inside the liposomes; a significant amount of DNA was produced	[75]
T7 DNA within cell-sized giant vesicles formed by natural swelling of phospholipid films	Transcription of DNA and transportation by laser tweezers; vesicles behaved as a barrier preventing the attack of Rnase	[72]
DNA template and the enzyme T7 RNA polymerase microinjected into a selected giant vesicle; nucleotide triphosphates added from the external medium	The permeability of giant vesicles increased in an alternating electric field; mRNA synthesis occurred	[73]

Reprinted from [12] with kind permission from Springer Science+Business Media.
POPC: 1-palmitoyl-2-oleoyl-sn-glycero-3-phosphatidylcholine.

Let us now focus on compartmentalized biochemical reactions in liposomes. A significant amount of experimental work has paved the way to more recent and complex developments. Such initial research has been collected in Table 11.5, under the heading of "molecular biology reactions inside liposomes."

For example, the biosynthesis of poly(A) – a model for RNA – was reported independently by two groups [54, 70]. In both cases a polynucleotide phosphorylase (PNPase) was entrapped into vesicles, and the synthesis of poly(A), which remained in the aqueous core of such vesicles, was observed. In one case [54] the internal poly(A) synthesis proceeded simultaneously by the reproduction of the vesicle shell, due to external addition of a membranogenic precursor (oleic anhydride).

A more suggestive example was provided shortly after [71] with the use of Qβ replicase, an enzyme that replicates an RNA template. Also in this case, the replication of a core component was coupled with a replication of the vesicle shell. With an excess of Qβ replicase/RNA template, the replication of RNA could proceed for a few generations.

This system, as well as the previous one by Walde et al. [54], is interesting because it represents a case of "core and shell replication," in which both the inside of the core and the shell itself undergo duplication. However, the limitations of this analogy should be clear; in fact, a real core-and-shell reproduction should be synchronous, which was not the case.

In particular, even if the RNA template and vesicle shell are both replicating, the Qβ replicase is not continuously produced in the process; thus, the system undergoes a "death by dilution." After a while the new vesicles will not contain either enzyme or template; therefore, the construct cannot reproduce itself completely.

A more recent example of exploitation of Qβ replicase has been provided by the group of Yomo [76]. The system is composed of an RNA-containing vesicle that behaves as follows. The first messenger RNA strand (sense sequence) codifies for the Qβ replicase, which is synthesized in situ by ribosomes. Once formed, Qβ replicase replicates the sense RNA strand, to give the antisense strand, and, in the next cycle, a new sense RNA chain. In this way, a genetic polymer (the RNA) produces an enzyme (Qβ replicase) that produces the polymer, and so on. Clearly, antisense RNA strands accumulate in the vesicle. Yomo and coworkers took advantage of this fact to produce a second enzyme (β-galactosidase), coded by the antisense strand. This elegant work shows that genetic polymers and their products may replicate together (in a coordinated way) inside a synthetic cell.

Another complex biochemical reaction implemented in liposomes is the polymerase chain reaction (PCR) [75]. Liposomes were able to endure the hardship of the PCR conditions, with several temperature cycles up to 90 °C (liposomes were practically unchanged at the end of the reaction). In addition, nine different chemicals had to be encapsulated in each liposome for the reaction to occur. Depending on the liposome formation mechanism and on the concentration of the chemicals, the entrapment efficiency can be different from what is expected on a statistical basis. In particular, it is not obvious that all nine chemicals are simultaneously trapped within one liposome. A more recent, and prebiotically oriented, study on template-directed DNA elongation is the above-mentioned work by Szostak and coworkers [60].

Table 11.5 also reports the work of Fischer *et al.* [73] on mRNA synthesis inside giant vesicles utilizing a DNA template and T7 RNA polymerase, and the transcription of DNA by Tsumoto *et al.* [72]. Further considerations on polymerase activity inside vesicles have been reported by Monnard [77].

Apart from the above-mentioned studies, there are not many other data on the formation of biooligomers or biopolymers inside vesicles. Clearly, such studies have great relevance in origin-of-life scenarioes, since the formation of the first biopolymers inside vesicles may have some functional advantage when compared with the formation in bulk solution. In fact, vesicles may protect the newly formed biopolymer from chemical attack by all those compounds which are present in the external medium, as well as maybe furnishing a compartmentalized milieu for the occurrence of otherwise difficult reactions.

An important additional aspect of the systems described in this chapter is that these studies were a prelude to more advanced cell models, explicitly pointing toward the laboratory construction of minimal *living* cells. With this goal being very ambitious and probably approachable only by a stepwise strategy, many of the current approaches focus on the synthesis of liposome-based bioreactors which are able to express functional proteins and carry out some simple function.

11.5.2 Protein Expression in Liposomes

In the previous section we reviewed the occurrence of rather complex biochemical reactions in liposomes. Where the construction of a minimal cell is concerned, however, it is easily recognized that the machinery of protein synthesis represents the main challenging function to be reconstructed within liposomes. In fact, according to the composition of a minimal genome, protein synthesis plays a key role in the metabolism of a hypothetical minimal cell based on about 200 genes. Proteins, which are synthesized by the ribosomal machinery, may then accomplish all other minimal functions such as replicating DNA, synthesizing RNA (messenger, transfer, ribosomal), and creating channels in the lipid membrane. Moreover, proteins may translate electrochemical gradients into biochemical energy (ATP) and synthesize or degrade chemical messengers. From the point of view of cell organization, protein synthesis recursively stems from earlier protein synthesis, and it is really a core function of minimal cells that can be synthesized in the laboratory; that is, minimal cells based on modern biochemical mechanisms.

How many genes are conceptually involved in a minimal DNA/protein cell that performs protein synthesis? Well, this question is also not easy to elicit from current data, as a calculation of the genes/enzymes involved has not been fully discussed in the literature. Often, commercial cell extracts are used for protein expression, and these are notoriously black boxes where the number of enzymes is not made known. Quite recently, however, Takuya Ueda (Tokyo University) reported on the first totally reconstituted cell-free protein-expression system, called the PURE (protein synthesis using reconstituted elements) system [78], which is now considered the standard for the constructive/synthetic biology of minimal cells [79, 80]. The system, now commercially available with the trade name PURESYSTEM®, consists of a mixture of enzymes, translation factors, ribosomes, t-RNAa, and other energy-recycling enzymes, plus (of course) low molecular weight compounds. The composition of the PURESYSTEM is give in Table 11.6. When we focus only on proteins, we count 36 different components, which correspond to 36 genes. However, if the 55 prokaryotic ribosomal proteins are considered, this number becomes 91. When three rRNAs and 46 tRNAs are included [81], the overall number of independent genes that codify for the whole PURESYSTEM reaches 140. An up-to-date overview of the work done, limited to the expression of proteins in liposomes, is presented in Table 11.7. The common strategy adopted in these studies consists in the entrapment of all ingredients required for the *in vitro* protein expression in the aqueous core of liposomes.

The first work dates back to 1999, at the ETH in Zurich, where Oberholzer *et al.* [82] demonstrated the production of the polypeptide poly(Phe) starting from poly(U) as RNA template. By co-entrapping poly(U), phenylalanine, ribosomes, tRNA^{Phe}, and translational factors inside lecithin vesicles, it was shown that poly(Phe) was synthesized with an efficiency only 20 times lower than the control experiment in bulk solution (without liposomes). The authors argued that the yield was actually surprisingly high, considering that the liposomes occupy only a very small fraction of the total volume, and that only a very few of them would contain all the ingredients by the statistical entrapment.

Two years later, Yomo, Urabe and coworkers [83] reported for the first time the expression of a *functional* protein (green fluorescent protein, GFP) inside lecithin liposomes. The synthesis of fluorescent, well-folded GFP inside large vesicles was proved by flow cytometry and confocal laser microscopy.

Table 11.6 Composition of the PURESYSTEM [78]

No. of different macromolecules	Type	Specification	Source
1	RNA polymerase	T7 RNA polymerase	Recombinant[a]
10	Translation factors	IF1; IF2; IF3; EF-G; EF-Tu; EF-Ts; RF1; RF2; RF3; RRF.	Recombinant[a]
20	aa-t-RNA synthetases	AlaRS; ArgRS; AsnRS; AspRS; CysRS; GlnRS; GluRS; GlyRS; HisRS; IleRS; LeuRS; LysRS; MetRS; PheRS; ProRS; SerRS; ThrRS; TrpRS; TyrRS; ValRS	Recombinant[a]
1	Others	Methionyl-tRNA formyltransferase	Recombinant[a]
4	Energy recycling enzymes	creatine kinase; myokinase; nucleoside diphosphate kinase; pyrophosphatase	Recombinant[a]
46[b]	t-RNAs		56 AU_{260}/mL t-RNA mix (Roche)
3 r-RNA and 55 proteins	Ribosomes		Purified from *E. coli*
	Low molecular weight compounds	20 amino acids; ATP; GTP; CTP; UTP; creatine phosphate; potassium glutamate; magnesium acetate; spermidine; DTT; 10-formyl-5,6,7,8-tetrahydrofolic acid; HEPES potassium salt	

[a]His$_6$-tagged recombinant proteins.
[b]Calculated on the basis of Dong et al. [81].

Table 11.7 Protein expression in vesicles

Description of the system	Main goal and results	Ref.
POPC liposomes incorporating the ribosomal complex together with the other components necessary for protein expression	Ribosomal synthesis of polypeptides can be carried out in liposomes; synthesis of poly(Phe) was monitored via trifluoroacetic acid precipitation of the (14)C-labelled products	[82]
Liposomes from EggPC, cholesterol, DSPE-PEG5000 used to entrap cell-free protein synthesis	Expression of a mutant GFP, determined with flow cytometric analysis	[83]
Small liposomes prepared by the ethanol injection method	Expression of EGFP evidenced by spectrofluorimetry	[10]
Gene-expression system within cell-sized lipid vesicles	Encapsulation of a gene-expression system; high expression yield of GFP inside giant vesicles	[84]
A two-stage genetic network encapsulated in liposomes	A genetic network in which the protein product of the first stage (T7 RNA polymerase) is required to drive the protein synthesis of the second stage (GFP)	[85]
E. coli cell-free expression system encapsulated in a phospholipid vesicle, which was transferred into a feeding solution containing ribonucleotides and amino acids	The expression of the α-hemolysin inside the vesicle solved the energy and material limitations; the reactor could sustain expression for up to 4 ays	[61]
PURESYSTEM was encapsulated within lecithin vesicles and analyzed by flow cytometry	First example of PURESYSTEM usage for expressing proteins inside liposomes	[86]
PURESYSTEM was encapsulated within lecithin vesicles and analyzed by laser scanning microscopy	An explicit attempt to apply synthetic biology concepts to minimal cell construction	[79]
GFP was synthesized inside giant vesicles by entrapped PURESYSTEM. Vesicles were prepared by centrifugation of a pre-formed w/o emulsion	Further example of PURESYSTEM-based bioreactor based on giant vesicles	[87]

Table 11.7 (*Continued*)

Description of the system	Main goal and results	Ref.
β-Glucuronidase was expressed within multilamellar and multivesicular vesicles prepared by dehydration–rehydration method	A quantitative study of protein expression, carried out by flow-cytometric analysis	[88]
A minimal system for enzymatic lipid production inside liposomes was designed and realized by entrapping PURESYSTEMS inside lipid vesicles of proper composition. The reaction under study was the synthesis of phosphatidic acid from glycerol-3-phosphate and acyl-CoAs	G3PAT and LPAAT, two membrane proteins (membrane enzymes) can be expressed in functional way inside lipid vesicles. Proper substrates were converted into products by the two freshly synthesized enzymes, although in low yields	[89]
Vesicles of minimal size (radius 100 nm) were used as cell model in order to assess the minimal cell size	EGFP was synthesized in lipid vesicles. Results can be explained by considering special entrapment mechanism and a corresponding enhanced concentration and activities within small lipid vesicles	[90]

In the procedure utilized by Oberholzer and Luisi [10], all ingredients required for protein expression were first mixed in water solution, and then liposomes were formed by the ethanol injection method. The increase of the fluorescence signal over time indicated that enhanced GFP (EGFP) was properly synthesized inside liposomes.

The direct observation of protein expression was accomplished by the procedure utilized by Nakatani and Yoshikawa's team [84], by using giant vesicles. The progress of the reaction, observed by laser-scanning microscopy, indicates that the synthesis of rsGFP (red-shifted GFP) takes place inside giant vesicles with higher efficiency than in the bulk water reaction. The authors could also show that vesicles protect entrapped enzymes from proteinase K, added externally.

More recently, based on their earlier report [83], Yomo and coworkers were able to design and produce experimentally a two-level cascading protein expression [85]. A plasmid containing the T7 RNA polymerase (with SP6 promoter) and a mutant GFP (with T7 promoter) genes was constructed and entrapped into liposomes, together with an *in vitro* protein expression mixture – including SP6 RNA polymerase. Under these conditions, SP6 RNA polymerase drives the production of T7 RNA polymerase, which in turn induces the expression of the detectable GFP.

Of particular interest is the work by Noireaux and Libchaber [61]. A plasmid encoding for two proteins was entrapped inside giant vesicles, together with the transcription–translation machinery. The plasmid sequence coded for EGFP and α-hemolysin genes. In contrast to the cascading network described above, now the second protein (α-hemolysin) does not have a direct role in the protein expression, but is involved in a different task. In fact, although α-hemolysin is a water-soluble protein, it is able to self-assemble as heptamer in the bilayer, generating a pore of 1.4 nm in diameter (cut-off ~3 kDa). It follows that externally added low molecular weight compounds may enter in the liposome, whereas enzymes, ribosomes, and genes could not escape from it. In this way, it was possible to feed the inner aqueous core of the vesicles, realizing a long-lived bioreactor, where the expression of the reported EGFP was prolonged up to 4 days. This work certainly represents an important milestone in the road map to the minimal cell, because the α-hemolysin pore permitted the uptake of small metabolites from the external medium and thus solved the energy and material limitations typical of the impermeable liposomes.

The next generation of systems based on cell-free protein expression machinery entrapped within liposomes utilizes the PURESYSTEM kit [78, 91]. In 2006, Yomo and coworkers first reported on the entrapment of PURESYSTEM inside liposomes, which were able to express functional (fluorescent) GFP [86]. Similar results were reported by Murtas *et al.* [79], who also greatly emphasized the principle of protein expression with a minimal number of *known* components, a concept that is proper to synthetic biology. It is, in fact, very important to remark that the use of reconstituted kits (instead of cell-extracts) is not a mere technical advancement. It represent a novel approach to the construction of semi-synthetic cells, by foreseeing that the investigator use a known mixture of biochemical elements, such as enzymes, ribosomes, tRNA, lipids, DNA, and small molecule in order to "build" a cell. This approach represents a paradigm shift from discovery-driven to hypothesis-driven biology. In the latter case, the experimenter formulates hypotheses about a molecular system and tests/verifies/refutes them directly by constructing the system (assembling the parts). It is noteworthy that the creation of a Registry of Standard Biological Parts (http://

parts.mit.edu) by synthetic biologists, who adopt the parallelism between metabolic and genetic circuits with electrical/electronic ones.

Saito *et al.* [87] did similar work on the expression of GFP by PURESYSTEM, but giant vesicles were used instead. A quantitative study of protein expression in multilamellar and multivesicular vesicles has been recently reported, again from Yomo and coworkers [88].

To conclude this part on protein expression inside liposomes, we must mention a very recent study done by us [90], and aimed to assess experimentally the minimal cell size by employing the complex protein expression network as a paradigm of cell metabolism. In particular, we demonstrated that 200 nm (diameter) liposomes were surprisingly able to express a fluorescent protein (EGFP) in functional form. The study was done by using *E. coli* extracts as well as the PURESYSTEM kit. In addition to this important primary outcome, the experiments on protein synthesis in very small compartments gave a clue on a possible self-organizing mechanism operating during the formation of liposomes. In fact, although the expected probability for the formation of a small liposome containing all macromolecular elements of the PURESYSTEM is negligibly small ($\sim 10^{-26}$), the observation of functional liposomes challenges the classical view of stochastic entrapment, by suggesting that solutes may be entrapped within liposomes at a concentration that is one order of magnitude greater than their nominal concentration in the bulk. Studies are currently devoted to investigating the mechanism of such an effect (similar aspects on entrapment of solutes in giant vesicles have been reported by Keating and coworkers [92, 93]).

Summarizing, in the last few years a handful of pioneering studies have appeared on protein expression within liposomes, and some of these reports evidenced the effect of "compartmentation"; that is, a higher yield of protein expression in the vesicles compared with the bulk buffer [84, 90] – a very interesting fact deserving further investigation.

11.5.3 Future Developments towards Minimal Cells

Keeping in mind the notion of a minimal cell, analysis of the data presented in this chapter reveals what is still missing in order to proceed in this field.

For example, protein expression, as outlined in most salient experiments of Table 11.7, has been carried out without checking the number of enzymes/genes utilized in the work. Only recent developments, made possible by the use of PURESYSTEM, have explicitly considered the number of (DNA-coded) macromolecules involved in a minimal cell model. Another essential element that is still missing to reach the ideal case of a minimal living cell is self-reproduction. In fact, none of the systems of Table 11.6 after having produced GFP is capable of reproducing itself and giving rise to a new population of minimal cells.

A very interesting goal would be the achievement of vesicles self-reproduction by the endogenous synthesis of vesicle lipids. Two strategies can in principle be pursued: (i) incorporating first the enzymes that synthesize the lipids or (ii) starting from the corresponding genes; that is, expressing those enzymes within the vesicles.

Early attempts have been focused on the enzymatic production of lecithin in lecithin liposomes [94]. The metabolic pathway was the so-called salvage pathway, which converts glycerol-3-phosphate to phosphatidic acid, then diacylglycerol, and finally phosphatidylcholine. The four enzymes needed to accomplish these reactions were simultaneously inserted in liposomes by the detergent depletion method, and the synthesis of new phosphatidylcholine (10% yield) was followed by radioactive labeling. Liposome transformation, followed by dynamic light scattering, showed that vesicles changed their size distribution during the process.

This was indeed a complex system, and it was realized later that one could theoretically stop at the synthesis of phosphatidic acid, as this compound also forms stable liposomes. Further studies [95] were oriented to characterizing the process by means of overexpression in *E. coli* and reconstitution in liposomes of the first two enzymes of the phospholipid salvage pathway, in order to obtain self-reproducing vesicles with only two enzymes.

Production of the cell boundary (as depicted in Figure 11.5) from within corresponds to the notion of autopoiesis [50, 96, 97].

The internal synthesis of lecithin in lecithin-liposomes would be a significant step forwards. In particular, it will be very interesting to see, given a certain excess of the two enzymes, how many generations the cell self-reproduction could go on for.

First of all, clearly, a lipid-synthesizing liposome is needed. A very recent study, carried out in our laboratory, shows that the first two enzymes of the above-mentioned lipid-salvage pathway can be synthesized in functional form within liposomes [89]. Glycerol-3-phosphate acyltransferase and lysophosphatidic acid acyltransferases have been successfully synthesized in functional (catalytically active) form. The significance of this study concerns two aspects. The first is technical: for the first time, integral membrane enzymes have been produced in active form *inside* lipid vesicles. Second, the two enzymes could work together

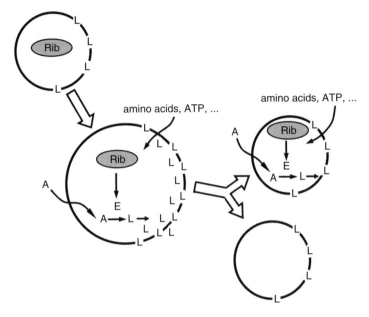

Figure 11.5 A cell that makes its own boundary. The complete set of biomacromolecules needed to perform protein synthesis (genes, RNA polymerases, and ribosomes) is indicated as Rib. The product of this synthesis (indicated as E) is the complete set of enzymes for lipid (L) synthesis. After some growth and division cycles, some of the "new" vesicles might undergo "death by dilution." (Reprinted from [12] with kind permission from Springer Science+Business Media)

– although in different redox conditions – to convert the precursors glycerol-3-phosphate and palmitoyl/oleoyl-CoAs, into 1-palmitoyl-2-oleoyl-*sn*-glycerol-3-phosphate, a compound that forms bilayer membranes [98, 99].

Suppose now we create a lipid-synthesizing minimal cell and that such a system synthesizes enough lipid molecules to allow membrane growth and division. Cleary, this cell is not yet "alive," since the shell is reproduced by an internal mechanism, but all the components of the internalized core are not. Clearly, after a certain number of generations, the system would undergo "death by dilution," because new "daughter" cells will form without all the required components to perform protein synthesis, for example. Moreover, since there is no way to control the division of solutes among the daughter cells during the act of division, it is possible that optimal core composition can be lost very early.

Finally, in order to get closer to the real minimal cell, there is the problem of further reduction of the number of genes. In all the systems of Table 11.7 we are still dealing with ribosomal protein biosynthesis, and this implies 100–200 genes. We are still far from our ideal picture of a minimal cell and we can pose, once more, the question of how to devise actual experiments so as to reduce this complexity.

As a way of thinking, we must resort to the conceptual knock-down experiments; for example, those outlined in the work by Islas et al. [20] and Luisi et al. [29] – a simplification that also corresponds to a movement towards the early cell. Simplification of the ribosomal machinery and of the enzyme battery devoted to RNA and DNA synthesis have been seen as the necessary steps.

Is this experimentally feasible? For example, can simple forms of rigid support for reactions and in particular protein biosynthesis be developed that are operative *in vitro* as ribosomes? Think of protein-free ribosomal RNA first. Can one operate, at costs of specificity, with only a very few polymerases? Similarly, it might not be necessary, at first, to have all possible specific tRNAs, but a few nonspecific ones instead. One might even conceive experiments with a limited number of amino acids.

It is also possible to imagine a hierarchy of "minimal cells." Some members of this hierarchy might require extensive resources from the environment, such as high-energy compounds. Others might be able to survive in a nutrient-poor environment, presumably more compatible with the "primordial soup." In fact, it would be quite interesting and revealing to analyze the differences between different members.

11.6 CONCLUDING REMARKS

The definition of a minimal cell, as given at the beginning of this review, appears simple and provided with its own elegance. Conversely, the experimental implementations of minimal cells may not appear equally satisfactory and elegant. We have outlined the main difficulties possibly encountered in the construction of an ideal minimal cell and pointed out, for example, that, in the best of the hypotheses, death by dilution is one limitation; and self-reproduction is already one target that has not yet been accomplished.

One problem with the present literature on minimal cells is that the link between a "minimal genome" and the minimal cell is rather weak. It would, of course, be advisable for researchers working on minimal cells to "count" the genes that are active under their conditions and compare the figures with the figures given by researchers on the minimal genome. Even within these limitations, experimental attempts to build a minimal cell are of great value in order to evaluate the specific simplification of the minimal genome. But not only this, the use of liposomes as a sophisticated "reaction vessel" is certainly instrumental in the technical realization of the minimal cell, but also has the added value of representing a possible route to the origin of early cells, emphasizing the manifold consequences of compartmentation as a possible role in self-organizing solutes so that functional cells can spontaneously form.

Constructs produced in the laboratory still represent poor models of full-fledged biological cells. This distance from fully biologically active cells makes it indeed premature to question possible hazards and bioethical issues in the field of minimal cells.

But there is still another very important topic that has not yet been discussed in due light by workers studying the minimal cell: the interaction with the environment. Of course, the feeding of the minimal cell is taken somehow into consideration, but only as a passive reservoir of nutrients and/or energy. In fact, we believe that the next generation of studies on the minimal cell should incorporate more actively such interactions with the surroundings, questioning, in particular, under which environmental conditions the minimal cell is able to perform its three basic functions.

Yet, these forms of "limping life," in our opinion, represent a very interesting part of this ongoing research. In fact, these approximations to life, such as a cell that produces proteins and does not reproduce itself, or one that does reproduce for a few generations and then dies out of dilution, or a cell that reproduces only parts of itself, and/or one characterized by a very poor specificity and metabolic rate.

All these constructs are important because, most probably, similar constructs were intermediates experimented with by nature to arrive at the final goal: the full-fledged biological cell. Thus, the creation of these partially living minimal cells in the laboratory may be of fundamental importance to understanding the real essence of cellular life, as well as the historical evolutionary pathway by which this target may have been reached. In addition, the construction of semi-synthetic living cells in the laboratory would be a demonstration – if still needed – that life is indeed an emergent property. In fact, in this case, cellular life would be created from non-life, since single genes and or single enzymes are non-living per se.

Generally, whereas the minimal cell can teach us a lot about early cellular life and evolution, it may not necessarily shed light on the origin of life. The reasons for this have already been expressed, and lie mostly in the fact that for the approach to the minimal cell we start with extant enzymes and genes – where life is already in full expression.

All this is very challenging, and perhaps for this reason, as already mentioned, there has been an abrupt rise of interest in the minimal cell. It appears that one additional reason for this rise of interest lies in a diffused sense of confidence, that the minimal cell is indeed an experimentally accessible target.

ACKNOWLEDGMENTS

This work has been funded by the SYNTHCELLS project (Approaches to the Bioengineering of Synthetic Minimal Cells, EU FP6 Grant #043359); by the Human Frontiers Science Program (RGP0033/2007-C); by the Italian Space Agency (Grant Nr. I/015/07/0); and by the Italian PRIN2008 program (Grant Nr. 2008FY7RJ4). It is also developed within the COST Systems Chemistry CM0703 Action.

REFERENCES

1. Morowitz, H.J. (1992) *Beginning of Cellular Life. Metabolism Recapitulates Biogenesis*, Yale University Press, New Haven, CT.
2. Jay, D. and Gilbert, W. (1987) Basic protein enhances the encapsulation of DNA into lipid vesicles: model for the formation of primordial cells. *Proceedings of the National Academy of Sciences of the United States of America*, **84**, 1978–1980.
3. Woese, C.R. (1983) The primary lines of descent and the universal ancestor, in *Evolution from Molecules to Man* (ed. D.S. Bendall), Cambridge Universiy Press, Cambridge, pp. 209–233.
4. Dyson, F.J. (1982) A model for the origin of life. *Journal of Molecular Evolution*, **18**, 344–350.
5. Pohorille, A. and Deamer, D. (2002) Artificial cells: prospects for biotechnology. *Trends in Biotechnology*, **20**, 123–128.
6. Szathmáry, E., Santos, M., and Fernando, C. (2005) Evolutionary potential and requirements forminimal protocells. *Topics in Current Chemistry*, **259**, 167–211.
7. Forster, A.C. and Church, G.M. (2006) Towards synthesis of a minimal cell. *Molecular Systems Biology*, **2**, 45.
8. Zhang, Y., Ruder, W.C., and LeDuc, P.R. (2008) Artificial cells: building bioinspired systems using small-scale biology. *Trends in Biotechnology*, **26**, 14–20.
9. Luisi, P.L. (2002) Toward the engineering of minimal living cells. *Anatomical Record*, **268**, 208–214.

10. Oberholzer, T. and Luisi, P.L. (2002) The use of lipsomes for constructing cell models. *Journal of Biological Physics*, **28**, 733–744.
11. Luisi, P.L. (2006) *The Emergence of Life: From Chemical Origin to Synthetic Biology*, Cambridge University Press.
12. Luisi, P.L., Ferri, F., and Stano, P. (2006) Approaches to semi-synthetic minimal cells: a review. *Naturwissenschaften*, **93**, 1–13.
13. Stano, P., Ferri, F., and Luisi, P.L. (2006) From the minimal genome to the minimal cell: theoretical and experimental investigations, in *Life as We Know It* (ed. J. Seckbach), Cellular Origin, Life in Extreme Habitats and Astrobiology, Vol. 10, Springer, Berlin, pp. 181–198.
14. Luisi, P.L., Chiarabelli, C., and Stano, P. (2006) From never born proteins to minimal living cells: two projects in synthetic biology. *Origins of Life and Evolution of the Biosphere*, **36**, 605–616.
15. Luisi, P.L., Stano, P., Murtas, G. *et al.* (2007) En route to semi-synthetic minimal cells, in *Proceedings of the Bordeaux Spring School on Modelling Complex Biological Systems in the Context of Genomics. April 3rd–7th 2006* (eds P. Amar, F. Képès, V. Norris et al.), EDP Sciences, Les Ulis (Paris), pp. 19–30.
16. Stano, P., Murtas, G., and Luisi, P.L. (2009) Semi-synthetic minimal cells: new advancements and perspectives, in *Protocells: Bridging Nonliving and Living Matter* (eds S. Rasmussen, M.A. Bedau, L. Chen *et al.*), MIT Press, Cambridge, MA, pp. 39–70.
17. Glass, J.I., Assad-Garcia, N., Alperovich, N. *et al.* (2006) Essential genes of a minimal bacterium. *Proceedings of the National Academy of Sciences of the United States of America*, **103** (2), 425–430.
18. Gibson, D.G., Benders, G.A., Andrews-Pfannkoch, C. *et al.* (2008) Complete chemical synthesis, assembly, and cloning of a *Mycoplasma genitalium* genome. *Science*, **319** (5867), 1215–1220.
19. Gibson, D.G., Benders, G.A., Axelrod, K.C. *et al.* (2008) One-step assembly in yeast of 25 overlapping DNA fragments to form a complete synthetic *Mycoplasma genitalium* genome. *Proceedings of the National Academy of Sciences of the United States of America*, **105** (51), 20404–20409.
20. Islas, S., Becerra, A., Luisi, P.L., and Lazcano, A. (2004) Comparative genomics and the gene complement of a minimal cell. *Origins of Life and Evolution of the Biosphere*, **34**, 243–256.
21. Shimkets, L.J. (1998) Structure and sizes of genomes of the archaea and bacteria, in *Bacterial Genomes: Physical Structure and Analysis* (eds F.J. De Bruijn, J.R. Lupskin and G.M. Weinstock), Kluwer Academic Publishers, Boston, pp. 5–11.
22. Moya, A., Gil, R., Latorre, A. *et al.* (2009) Toward minimal bacterial cells: evolution vs. design. *FEMS Microbiology Reviews*, **33**, 225–235.
23. Mushegian, A. and Koonin, E.V. (1996) A minimal gene set for cellular life derived by comparison of complete bacterial genomes. *Proceedings of*

the *National Academy of Sciences of the United States of America*, **93**, 10268–10273.
24. Fraser, C.M., Gocayne, J.D., White, O. *et al.* (1995) The minimal gene complement of *Mycoplasma genitalium*. *Science*, **270**, 397–403.
25. Itaya, M. (1995) An estimation of the minimal genome size required for life. *FEBS Letters*, **362**, 257–260.
26. Mushegian, A. (1999) The minimal genome concept. *Current Opinion in Genetics & Development*, **9**, 709–714.
27. Hutchison, C.A., Peterson, S.N., Gill, S.R. *et al.* (1999) Global transposon mutagenesis and a minimal *Mycoplasma* genome. *Science*, **286**, 2165–2169.
28. Koonin, E.V. (2000) How many genes can make a cell: the minimal-gene-set concept. *Annual Review of Genomics and Human Genetics*, **1**, 99–116.
29. Luisi, P.L., Oberholzer, T., and Lazcano, A. (2002) The notion of a DNA minimal cell: a general discourse and some guidelines for an experimental approach. *Helvetica Chimica Acta*, **85**, 1759–1777.
30. Kolisnychenko, V., Plunkett, G., III, Herring, C.D. *et al.* (2002) Engineering a reduced *Escherichia coli* genome. *Genome Research*, **12**, 640–647.
31. Gil, R., Sabater-Munoz, B., Latorre, A. *et al.* (2002) Extreme genome reduction in *Buchnera* spp: toward the minimal genome needed for symbiotic life. *Proceedings of the National Academy of Sciences of the United States of America*, **99**, 4454–4458.
32. Koonin, E.V. (2003) Comparative genomics, minimal gene-sets and the last universal common ancestor. *Nature Reviews Microbiology*, **1**, 127–136.
33. Gil, R., Silva, F.J., Peretó, J., and Moya, A. (2004) Determination of the core of a minimal bacteria gene set. *Microbiology and Molecular Biology Reviews*, **68**, 518–537.
34. Gabaldon, T., Pereto, J., Montero, F. *et al.* (2007) Structural analyses of a hypothetical minimal metabolism. *Philosophical Transactions of the Royal Society B: Biological Sciences*, **362**, 1751–1762.
35. Zhang, B. and Cech, T.R. (1998) Peptidyl-transferase ribozymes: trans reactions, structural characterization and ribosomal RNA-like features. *Chemistry & Biology*, **5**, 539–553.
36. Nissen, P., Hansen, J., Ban, N. *et al.* (2000) The structural basis of ribosome activity in peptide bond synthesis. *Science*, **289**, 920–930.
37. Calderone, C.T. and Liu, D.R. (2004) Nucleic-acid-templated synthesis as a model system for ancient translation. *Current Opinion in Chemical Biology*, **8**, 645–653.
38. Weiner, A.M. and Maizels, N. (1987) tRNA-like structures tag the 3' ends of genomic RNA molecules for replication: implications for the origin of protein synthesis. *Proceedings of the National Academy of Sciences of the United States of America*, **84**, 7383–7387.

39. Lazcano, A., Guerriero, R., Margulius, L., and Oró, J. (1988) The evolutionary transition from RNA to DNA in early cells. *Journal of Molecular Evolution*, 27, 283–290.
40. Lazcano, A., Valverde, V., Hernandez, G. *et al.* (1992) On the early emergence of reverse transcription: theoretical basis and experimental evidence. *Journal of Molecular Evolution*, 35, 524–536.
41. Frick, D.N. and Richardson, C.C. (2001) DNA primases. *Annual Review of Biochemistry*, 70, 39–80.
42. Suttle, D.P. and Ravel, J.M. (1974) The effects of initiation factor 3 on the formation of 30S initiation complexes with synthetic and natural messengers. *Biochemical and Biophysical Research Communications*, 57, 386–393.
43. Szostak, J.W., Bartel, D.P., and Luisi, P.L. (2001) Synthesizing life. *Nature*, 409, 387–390.
44. Knoll, A. (ed.) (1999) *Size Limits of Very Small Microorganisms*, National Academic Press, Washington, DC.
45. Boal, D. (1999) *Size Limits of Very Small Microorganisms*, National Academic Press, Washington, DC, pp. 26–31.
46. Kajander, E.O. and Ciftcioglu, N. (1998) Nanobacteria: an alternative mechanism for pathogenic intra- and extracellular calcification and stone formation. *Proceedings of the National Academy of Sciences of the United States of America*, 95, 8274–8279.
47. Kajander, E.O. (2006) Nanobacteria – propagating calcifying nanoparticles. *Letters in Applied Microbiology*, 42, 549.
48. Ciftcioglu, N., McKay, D.S., Mathew, G., and Kajander, E.O. (2006) Nanobacteria: fact or fiction? Characteristics, detection, and medical importance of novel self-replicating, calcifying nanoparticles. *Journal of Investigative Medicine*, 54, 385.
49. Luisi, P.L. and Oberholzer, T. (2001) Origin of life on Earth: molecular biology in liposomes as an approach to the minimal cell, in *The Bridge Between the Big Bang and Biology* (ed. F. Giovanelli), CNR Press, pp. 345–355.
50. Luisi, P.L. (2003) Autopoiesis: a review and a reappraisal. *Naturwissenschaften*, 90, 49–59.
51. De Lorenzo, V. and Danchin, A. (2008) Synthetic biology: discovering new worlds and new words. *EMBO Reports*, 9, 822–827.
52. Bich, L. and Damiano, L. (2007) Question 9: theoretical and artificial construction of the living: redefining the approach from an autopoietic point of view. *Origins of Life and Evolution of the Biosphere*, 37, 459–464.
53. Bachmann, P.A., Luisi, P.L., and Lang, J. (1992) Autocatalytic self-replicating micelles as models for prebiotic structures. *Nature*, 357, 57–59.
54. Walde, P., Goto, A., Monnard, P.A. *et al.* (1994) Oparin's reactions revisited: enzymatic synthesis of poly(adenylic acid) in micelles and

self-reproducing vesicles. *Journal of the American Chemical Society*, **116**, 7541–7544.
55. Luisi, P.L., Stano, P., Rasi, S., and Mavelli, F. (2004) A possible route to prebiotic vesicle reproduction. *Artificial Life*, **10**, 297–308.
56. Bloechliger, E., Blocher, M., Walde, P., and Luisi, P.L. (1998) Matrix effect in the size distribution of fatty acid vesicles. *Journal of Physical Chemistry*, **102**, 10383–10390.
57. Lonchin, S., Luisi, P.L., Walde, P., and Robinson, B.H. (1999) A matrix effect in mixed phospholipid/fatty acid vesicle formation. *Journal of Physical Chemistry B*, **103**, 10910–10916.
58. Berclaz, N., Mueller, M., Walde, P., and Luisi, P.L. (2001) Growth and transformation of vesicles studied by ferritin labeling and cryotransmission electron microscopy. *Journal of Physical Chemistry B*, **105**, 1056–1064.
59. Rasi, S., Mavelli, F., and Luisi, P.L. (2003) Cooperative micelle binding and matrix effect in oleate vesicle formation. *Journal of Physical Chemistry B*, **107**, 14068–14076.
60. Mansy, S.S., Schrum, J.P., Krishnamurthy, M. *et al.* (2008) Template-directed synthesis of a genetic polymer in a model protocell. *Nature*, **454** (7200), 122–125.
61. Noireaux, V. and Libchaber, A. (2004) A vesicle bioreactor as a step toward an artificial cell assembly. *Proceedings of the National Academy of Sciences of the United States of America*, **101**, 17669–17674.
62. Winterhalter, M., Hilty, C., Bezrukov, S.M. *et al.* (2001) Controlling membrane permeability with bacterial porins: application to encapsulated enzymes. *Talanta*, **55**, 965–971.
63. Vamvakaki, V., Fournier, D., and Chaniotakis, N.A. (2005) Fluorescence detection of enzymatic activity within a liposome based nano-biosensor. *Biosensors and Bioelectronics*, **21**, 384–388.
64. Graff, A., Winterhalter, M., and Meier, W. (2001) Nanoreactors from polymer-stabilized liposomes. *Langmuir*, **17**, 919–923.
65. Yoshimoto, M., Wang, S., Fukunaga, K. *et al.* (2005) Novel immobilized liposomal glucose oxidase system using the channel protein OmpF and catalase. *Biotechnology and Bioengineering*, **90**, 231–238.
66. Stamatatos, L., Leventis, R., Zuckermann, M.J., and Silvius, J.R. (1988) Interactions of cationic lipid vesicles with negatively charged phospholipid vesicles and biological membranes. *Biochemistry*, **27**, 3917–3925.
67. Marchi-Artzner, V., Jullien, L., Belloni, L. *et al.* (1996) Interaction, lipid exchange, and effect of vesicle size in systems of oppositely charged vesicles. *Journal of Physical Chemistry*, **100**, 13844–13856.
68. Pantazatos, D.P. and MacDonald, R.C. (1999) Directly observed membrane fusion between oppositely charged phospholipid bilayers. *Journal of Membrane Biology*, **170**, 27–38.
69. Thomas, C.F. and Luisi, P.L. (2004) Novel properties of DDAB: matrix effect and interaction with oleate. *Journal of Physical Chemistry B*, **108**, 11285–11290.

70. Chakrabarti, A.C., Breaker, R.R., Joyce, G.F., and Deamer, D.W. (1994) Production of RNA by a polymerase protein encapsulated within phospholipid vesicles. *Journal of Molecular Evolution*, **39**, 555–559.
71. Oberholzer, T., Wick, R., Luisi, P.L., and Biebricher, C.K. (1995) Enzymatic RNA replication in self- reproducing vesicles: an approach to a minimal cell. *Biochemical and Biophysical Research Communications*, **207**, 250–257.
72. Tsumoto, K., Nomura, S.M., Nakatani, Y., and Yoshikawa, K. (2001) Giant liposome as a biochemical reactor: transcription of DNA and transportation by laser tweezers. *Langmuir*, **17**, 7225–7228.
73. Fischer, A., Franco, A., and Oberholzer, T. (2002) Giant vesicles as microreactors for enzymatic mRNA synthesis. *Chembiochem*, **3**, 409–417.
74. Walde, P., Wick, R., Fresta, M. et al. (1994) Autopoietic self-reproduction of fatty acid vesicles. *Journal of the American Chemical Society*, **116**, 11649–11654.
75. Oberholzer, T., Albrizio, M., and Luisi, P.L. (1995) Polymerase chain reaction in liposomes. *Chemistry & Biology*, **2**, 677–682.
76. Kita, K., Matsuura, T., Sunami, T. et al. (2008) Replication of genetic information with self-encoded replicase in liposomes. *Chembiochem*, **9**, 2403–2410.
77. Monnard, P.A. (2003) Liposome-entrapped polymerases as models for microscale/nanoscale bioreactors. *Journal of Membrane Biology*, **191**, 87–97.
78. Shimizu, Y., Kanamori, T., and Ueda, T. (2005) Protein synthesis by pure translation systems. *Methods*, **36**, 299.
79. Murtas, G., Kuruma, Y., Bianchini, P. et al. (2007) Protein synthesis in liposomes with a minimal set of enzymes. *Biochemical and Biophysical Research Communications*, **363**, 12–17.
80. Luisi, P.L., Chiarabelli, C., and Stano, P. (2006c) From the never born proteins to the minimal living cell: two projects in synthetic biology. *Origins of Life and Evolution of the Biosphere*, **36**, 605–616.
81. Dong, H., Nilsson, L., and Kurland, C.G. (1996) Co-variation of tRNA abundance and codon usage in *Escherichia coli* at different growth rates. *Journal of Molecular Biology*, **260**, 649–663.
82. Oberholzer, T., Nierhaus, K.H., and Luisi, P.L. (1999) Protein expression in liposomes. *Biochemical and Biophysical Research Communications*, **261**, 238–241.
83. Yu, W., Sato, K., Wakabayashi, M. et al. (2001) Synthesis of functional protein in liposome. *Journal of Bioscience and Bioengineering*, **92**, 590–593.
84. Nomura, S.M., Tsumoto, K., Hamada, T. et al. (2003) Gene expression within cell-sized lipid vesicles. *Chembiochem*, **4**, 1172–1175.
85. Ishikawa, K., Sato, K., Shima, Y. et al. (2004) Expression of a cascading genetic network within liposomes. *FEBS Letters*, **576**, 387–390.
86. Sunami, T., Sato, K., Matsuura, T. et al. (2006) Femtoliter compartment in liposomes for *in vitro* selection of proteins. *Analytical Biochemistry*, **357**, 128.

87. Saito, H., Yamada, A., Ohmori, R. *et al.* (2007) Towards constructing synthetic cells: RNA/RNP evolution and cell-free translational systems in giant liposomes, in International Symposium on Micro-NanoMechatronics and Human Science, MHS '07, pp. 286–229.
88. Hosoda, K., Sunami, T., Kazuta, Y. *et al.* (2008) Quantitative study of the structure of multilamellar giant liposomes as a container of protein synthesis reaction. *Langmuir*, **24**, 13540–13548.
89. Kuruma, Y., Stano, P., Ueda, T., and Luisi, P.L. (2009) A synthetic biology approach to the construction of membrane proteins in semi-synthetic minimal cells. *Biochimica et Biophysica Acta*, **1788**, 567–574.
90. Pereira de Souza, T., Stano, P., and Luisi, P.L. (2009) The minimal size of liposome-based model cells brings about a remarkably enhanced entrapment and protein synthesis. *Chembiochem*, **10**, 1056–1063. DOI: 10.1002/cbic.200800810.
91. Shimizu, Y., Inoue, A., Tomari, Y. *et al.* (2001) Cell free translation reconstituted with purified components. *Nature Biotechnology*, **19**, 751–755.
92. Dominak, L.M. and Keating, C.D. (2007) Polymer encapsulation within giant lipid vesicles. *Langmuir*, **23**, 7148–7154.
93. Dominak, L.M. and Keating, C.D. Macromolecular crowding improves polymer encapsulation within giant lipid vesicles. *Langmuir*, 2008, **24**, 13565–13571.
94. Schmidli, P.K., Schurtenberger, P., and Luisi, P.L. (1991) Liposome-mediated enzymatic synthesis of phosphatidylcholine as an approach to self-replicating liposomes. *Journal of the American Chemical Society*, **113**, 8127–8130.
95. Luci, P. (2003) Gene cloning expression and purification of membrane proteins. ETH-Z Dissertation Nr. 15108, Zurich.
96. Varela, F., Maturana, H.R., and Uribe, R.B. (1974) Autopoiesis: the organization of living system, its characterization and a model. *Biosystems*, **5**, 187–196.
97. Luisi, P.L. and Varela, F.J. (1990) Self-replicating micelles – a chemical version of minimal autopoietic systems. *Origins of Life and Evolution of the Biosphere*, **19**, 633–643.
98. Hauser, H. and Gains, N. (1982) Spontaneous vesiculation of phospholipids: a simple and quick method of forming unilamellar vesicles. *Proceedings of the National Academy of Sciences of the United States of America*, **79**, 1683–1687.
99. Hauser, H., Gains, N., and Mueller, M. (1983) Vesiculation of unsonicated phospholipid dispersions containing phosphatidic acid by pH adjustment: physicochemical properties of the resulting unilamellar vesicles. *Biochemistry*, **22**, 4775–4781.
100. Gibson, D.G. *et al.* (24 authors), (2010) Creation of a bacterial cell controlled by a chemically synthesized genome. *Science*, **329**, 52–56.

Part Four
General Problems

12

Replicators: Components for Systems Chemistry

Olga Taran and Günter von Kiedrowski

Department of Organic Chemistry I, Ruhr University Bochum, Universitätstraße 150, 44780 Bochum, Germany

12.1 The Need for Systems Chemistry 290
12.2 Self-Replicating Reactions 293
    12.2.1 First Self-Replicating Systems 294
    12.2.2 Minimal Replicator Theory 297
    12.2.3 Artificial Replicators 302
    12.2.4 Networks of Replicators 306
12.3 Tools for Systems Chemistry 311
    12.3.1 Online Kinetic Measurement 311
    12.3.2 NMR Kinetic Titrations 312
    12.3.3 SimFit Fitting Program 314
References 315

The chapter emphasizes first the relevance of system chemistry to complement system biology and synthetic biology in particular, with keywords such as supramolecular self-organization, autocatalysis, molecular information processing, and possibly nano- to meso-systems. Then we examine the challenges arising from the synthesis of prebiotic molecules, discriminating between the endogenous compounds and those originating from astrobiology. The question of homochirality is particularly emphasized. Generally, the art of synthesizing coupled autocatalytic systems points to the future of chemical research inspired by the origin of life problem. Coupling will necessarily involve not only one class of molecules, but at least two – for example, nucleotides *and* peptides, nucleotides *and* lipids, sugars *and* peptides, etc. The chapter then dwells on chemical replicators, with a historical overview of the work carried out by different schools on the subject, and discussing in particular a few examples of chemical reactions displaying self-replication capability. There is then a theoretical section on minimal replicators, discussing various types of autocatalytic growth, then artificial replicators, to pass to networks of replicators. Here again some examples are discussed in detail based on current literature. In the last part of this review we dwell on the tools for system chemistry, from online kinetic measurements to nuclear magnetic resonance titration techniques, terminating with an analysis of the SimFit fitting program.

12.1 THE NEED FOR SYSTEMS CHEMISTRY

At the beginning of the twenty-first century biology has matured to become a leading science, in the sense that it generates questions, frontiers, and challenges to many different fields. But in spite of the formidable success of the Human Genome Project, biology itself has not been able so far to answer the most fundamental question concerning the origin of its subject. Biologists today have a wealth of information dealing with the parts making up a living system. The nature of this knowledge does not, however, allow one to draw any good conclusion about how life originated on this planet from simpler chemical precursor systems, or how life may have originated elsewhere in the universe, or whether new forms of life can be synthesized *de novo* in the laboratory. A new approach is required that needs to combine the "classical" knowledge of chemistry, namely the language of molecules, their structures, their reactions and interactions, together with the "classical" knowledge derived from existing forms of life. One key component of this approach, acting both as a translator and abstractor between these languages should come from the fields of theoretical biology and complex systems research; the other key component should come from a chemistry that is the offspring of both, supramolecular and prebiotic chemistry, and adds a new dimension that has not been sufficiently addressed so far. Over the past decades more and more chemists have learned to design and implement simple self-replicating and self-reproducing systems, and today we even have the first examples on the issue of chiral symmetry breaking in autocatalytic reactions. What seems to be missing is a kind of generalization of "synthetic methods" based on the principles of supramolecular self-organization, autocatalysis, molecular information processing, and, moreover, ranging to be applicable from small molecules via nano- to meso-systems. Systems chemistry is the name of a field in which chemists, theoretical biologists, and complex systems researchers interact to find the chemical roots of biological organization [1–3].

This approach complements the emerging field of "synthetic biology." Whereas the latter aims at the utilization of intracellular regulatory networks as biological building blocks to construct synthetic cells, the former aims to find the chemical roots of such regulation in nonevolved systems. "Protocells," "chemotons," and "minimal cells" are just different words for the same issue of research that may be biology driven as a top-down approach or chemistry driven as a bottom-up approach. In any case, the disassembly and reconstruction approach of synthetic biology means the existence of biological building blocks as the products of biological evolution. Even if synthetic biology will be able to generate a more primitive form of cellular life, this does not necessarily answer the question of how such a thing could emerge when starting from scratch. Unfortunately, the same fundamental problem underlies our current vision that directed evolution will help to reconstruct a preceding biochemistry such as the RNA world. Finding a set of self-replicating RNA molecules that cooperate in a vesicle to constitute a minimal cell will be a remarkable scientific achievement of the twent-first century. Nevertheless, one will stay puzzled by the question of how such things could have developed in the absence of evolved "tools" such as polymerases.

The origin of Darwinian evolvability is one of the central challenges. The other, equally challenging frontier is the origin of a sufficiently complex chemically organized system embodying a minimal living, namely a self-containing and self-sustaining entity. All life as we know it today is based on cells as the unit of life. The distinction between a unit of life and a unit of evolution may be made for present-day life; however, it is questionable whether any reasonable definition on the origin of life can be made without equating the living state of matter with an evolvable state of matter. If so, the task is to find answers to what sets of molecular structures, reactions, and interactions are required to arrive, finally, at a system that fulfills both the criteria of minimal life and minimal evolvability. The transition from limited to unlimited heredity may, however, be a later issue of research dealing with life's origin.

Life today is based on proteins and nucleic acids, lipids, sugars, amino acids, and other molecular building blocks, where almost all molecular components are in a homochiral (i.e. enantiomerically pure) state. The creation of minimal life and evolvability is thus necessarily connected to the question at which level in the transition from small molecules to minimal living and evolving systems the amplification of homochirality took place. From today's perspective it seems very likely that this process might be deeply linked to the emergence of self-replication and was even indistinguishable from this in the beginning. Whether an autocatalytic transformation of racemizing building blocks, a mutual annihilation of autocatalytic products of opposite handedness, or a process in which chiral templating proceeds as a higher order autocatalytic process played a leading role, or whether the building blocks were racemic (like in the formose reaction) or prochiral is an open question to which chemistry must find more answers in the future.

A primary problem here is that we do not even know the repertoire of organic molecules delivered from space or endogeneous sources and what was the initial set of chemical reactions that started the long transformation from space molecules to the first living systems. So long as this question is not fully answered by astrobiological research, any "primitive" organic or inorganic chemicals may be employed in the design and exploration of chemical systems that hopefully express dynamic signatures of the living. The good news is that research in autocatalytic systems currently gains much attention in various areas of chemistry and that the acceptance to work with "complex mixtures" of molecules has increased since the advent of combinatorial chemistry. What is still a challenge for present-day chemistry is to filter out the signal of self-organization in the noise of lots of side reactions. Gaining chemical control to a whole network of reactions having an autocatalytic "core" at a low level of information content but otherwise just producing "diversity," namely a rich mixture of constitutional and stereoisomers, is clearly a challenge.

Generally, the art of synthesizing coupled autocatalytic systems points to the future of chemical research inspired by the origin of life problem. Coupling will necessarily involve not only one class of molecules, but at least two. So, not only nucleotides or peptides or sugars or lipids, but nucleotides *and* peptides, nucleotides *and* lipids, sugars *and* peptides, peptides *and* lipids, to name only a few possible combinations here. The exploration of couplings between an autocatalytic core like the formose reaction, a self-reproducing micelle, or a self-replicating template does not necessarily require that the set of reactions "talking with the core" are autocatalytic by themselves, namely constituting an independently running autocatalytic cycle. Theory gives the insight that any reaction triggered by the core autocatalyst (e.g. in the sense of heterocatalysis) will also cause multiplication and growth of those compounds generated due to the presence of the autocatalyst. Thus, the outcome of such coupling manifests in a stoichiometric relationship between the number of "core autocatalysts" and the number of molecules coming up in the "periphery" of the reaction network. If the whole process is now selectively generating a specific set of molecules, different from the mixture in the absence of coupling, then informational "harvesting" takes place. As self-replication can be defined as autocatalysis *plus* information transfer, selecting specific products from an autocatalytic network means self-replication. Of course, templating is one of the best proven ways for establishing a mechanism for information transfer, but the term "template" may have a much broader meaning in the future than it has had in the past. The same extrapolation may be foreseeable to the issues of evolvability and "cellular compartmentation."

12.2 SELF-REPLICATING REACTIONS

Self-replication is one of the basic characteristics of the living system. In chemistry, self-replication means autocatalysis with information transfer. Chemical self-replicating systems have been designed in order to identify the minimal requirements for molecular replication, to translate the principle into synthetic supramolecular systems and to derive a better understanding of the scope and limitation of self-organization processes which, among others, are believed to be relevant to the origin of life on Earth.

In the simplest case of a self-replicating system a product works as a template for its own formation. The general reaction equation can be seen as follows:

$$A + B + nC \rightarrow (n+1)C \qquad (12.1)$$

The reaction mechanism can be visualized as a simple three-step process (Figure 12.1). Here, a template molecule C is self-complementary and thus able to catalyze its own formation from corresponding building blocks A and B. In the first step, the precursors A and B bind reversibly to the template C and form the termolecular complex ABC. Reactive ends of A and B molecules are brought in close spatial proximity to

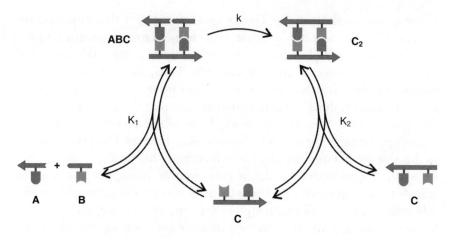

Figure 12.1 A schematic representation of a self-replicating system

each other, which facilitates the formation of a covalent bond between them. After an irreversible covalent ligation reaction the termolecular complex ABC is transformed to the duplex C_2. The duplex can reversibly dissociate in order to give two template molecules C that both can start new replication cycles.

This minimal scheme has served as a successful aid for the development of nonenzymatic self-replicating systems. Current implementations make use of oligonucleotide analogues, peptides [4], and other molecules [5] as templates and are based on autocatalytic, cross-catalytic, or collectively autocatalytic pathways for template formation.

12.2.1 First Self-Replicating Systems

Nucleic acids carry the inherent ability for complementary base-pairing and replication. Probably, they were the first reproducing molecules under primitive Earth conditions. Kühn and Waser [6] and others (Crick [7], Orgel [8], Eigen and Schuster 1979 [9]; for a review, see Joyce [10]) have drawn the picture of an "RNA world" that might have existed before mechanisms of translation had evolved. The hypothesis of the "RNA-world" states that, before formation of the first cells, sets of catalytic RNA molecules capable of performing replication or self-replication existed. This theory was supported by discovery of RNA (and also DNA) that can act as an enzyme-like catalyst [11–14]. A set of cross-catalytic ribozymes that can undergo self-sustained exponential replication in the absence of any enzyme were described by Lincoln and Joyce in 2009 [14b]. This system presents not only exponential growth, but also a Darwinian selection, when serial transfer experiments are realized with the population of similar RNA strands. That type of selection promises opportunities of open-ended evolution in non-natural systems.

The search for minimal self-replicating reactions started from the study of templated ribonucleotide synthesis more than 40 years ago. Since 1968, condensation of the chemically activated nucleotides (monoribonucleoside-5′-phosphorimidazolide) on ribonucleotide template have been studied by Orgel's group [8, 15, 16]. Pyrimidine-rich templates were found to be more efficient for the polycondensation than purine-rich oligonucleotides. Templates up to 14 base pairs long were successfully transcribed using monomeric building blocks to perform the reaction, but a complete catalytic cycle was not achieved. Regioisomers with a 2′–5′ bond between nucleotides have been observed as side products. Despite these problems, it was shown for the first time that transcription without enzyme participation is possible.

In 1986 von Kiedrowski [17] reported a first self-replicating system. Ribonucleotides were substituted by desoxyribonucleotides, which avoided regioselectivity problems of undesired 2′–5′ linkage formation between nucleotides. Instead of a condensation reaction with activated phosphorimidazolides, chemical ligation in the presence of the water-soluble carbodiimide EDC (1-ethyl-3-(3-dimethylaminopropyl)carbodiimide) [18] was applied. Mononucleotide building blocks on a long template were substituted for a restricted system of two trinucleotides that have to form one single bond on a hexanucleotide template. A 5′-terminally protected trideoxynucleotide 3′-phosphate d(Me-CCG-p) (1) and a complementary 3′-protected trideoxynucleotide d(CGG-p′) (2) react in the presence of EDC to yield the self-complementary hexadeoxynucleotide (Me-CCG-CGG-p′) (3) with natural phosphodiester linkage as well as the 3′–3′-linked pyrophosphate of 1. The chosen sequences were such that the product 3 could act as a template for its own formation (Figure 12.2).

The reaction was performed using milimolar concentrations of the reactants with initial template concentration varied from 0 to 8% of the concentration of trioligonucleoties. Formation of 3′–3′ product was not affected, but the rate of formation of hexanucleotide increased on template addition. The initial reaction rate was fitted to the empirical law:

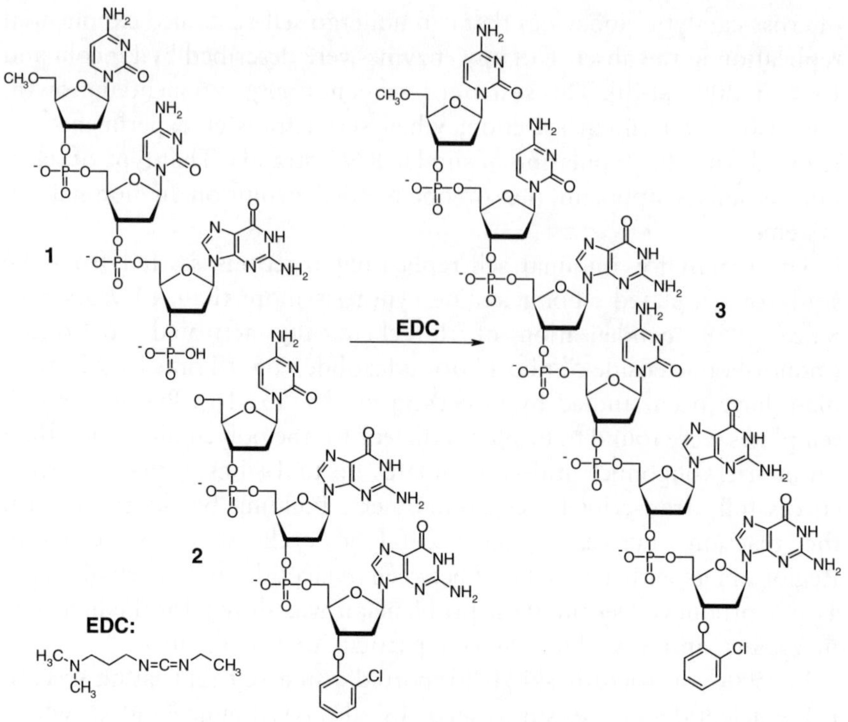

Figure 12.2 First non-enzymatic self-replicating system

$$\frac{dc}{dt} = k_a a_0 b_0 c^p + k_b a_0 b_0 \qquad (12.2)$$

Least-squares fitting showed the best results for $p = 0.48$. This finding was later explained by the formation of a nonreactive double-stranded complex of hexaoligonucleotide product and was called "the square-root law" of autocatalysis. Analysis of the initial reaction rates showed two pathways for the formation of hexanucleotide: non-autocatalytic and autocatalytic. Over the concentration range studied, the non-autocatalytic pathway is the dominant one. That is the reason why sigmoidal, or "S"-shaped, growth of the product, expected for autocatalysis, is not observed under experimental conditions.

In 1987 a replication of ribonucleotide analogues GNHpCNH2 and pGNHpCN3 in the presence of EDC was reported by Zielinsky and Orgel [19]. It was the first case of replication of nucleotides with artificial backbone. The reaction also shows the square-root dependence of initial reaction rates on concentration of added template. The non-autocatalytic reaction is still predominant and again no sigmoidal growth is observed.

Further change in the backbone and substitution of a 3'-amino-5'-phosphate bond to 5'-amino-3'-phosphate bond elevated the ratio of autocatalytic pathway over the non-autocatalytic one [20]. Two new systems that finally exhibit sigmoidal growth were found: von Kiedrowski's system based on 5'-amino-3'-phosphate oligonucleotide analogues [20] and Rebek's [21] first artificial system that used amide bond formation instead of phosphorus chemistry. Rebek's work led to the development of artificial replicators; that is, systems that use other organic molecules instead of RNA or DNA analogues.

12.2.2 Minimal Replicator Theory

Accumulated data needed a theoretical basis [22]. Experimentally observed replicators were different from autocatalytic reactions described before [23]. The initial reaction rates do not increase linearly with the concentration of a template added at the beginning of the reaction. Instead, the dependency presented on an initial template concentration is described by the following empirical equation:

$$\frac{dc}{dt} = k_a a_0 b_0 c^p + k_b a_0 b_0 = \alpha c^p + \beta \qquad (12.3)$$

where k_a is the reaction rate constant of the autocatalytic reaction and k_b is the reaction rate constant of spontaneous ligation; a_0, b_0 and c are the initial concentrations of species A, B, and C respectively. The order of autocatalytic reaction is p. For the real DNA-based reactions the value of p, calculated from the dependence of initial rates on added template, was 0.5. In that particular case, Equation (12.3) becomes

$$\frac{dc}{dt} = \alpha \sqrt{c} + \beta \qquad (12.4)$$

This finding has been called "the square-root law" of autocatalysis. It is an empirical dependence, and actually replicators with a value of p between 0.48 and 1 have been found. To understand how a product's growth is affected by reaction order we can examine two border cases, namely when $p = 1$ or 0.5. Considering the case where the noncatalytic pathway is negligible and integrating Equation (12.3) over time for both values of p, the following change in concentration of C at the beginning of the reaction can be observed:

$$p = 1: \qquad c = c_0 e^{\alpha t} \qquad (12.5)$$

$$p = 0.5: \qquad c = \left(\sqrt{c_0} + \frac{\alpha}{2} t\right)^2 \qquad (12.6)$$

Equation (12.5) corresponds to exponential growth. In the case when c_0 is infinitesimally small, Equation (12.6) becomes

$$c = \frac{\alpha^2 t^2}{4} \qquad (12.7)$$

Equation (12.7) represents a parabola. For this reason, autocatalytic growth based on the square-root law has been termed parabolic. Parabolic replicators are expected to present sigmoidal growth of the product concentrations. Sigmoidal growth is an "S"-shaped curve, where the initial induction period corresponds to formation of the product C, which gradually elevates the reaction rate. The rate of the reaction achieves its maximum at the so-called inflection point, after which saturation happens, either because of reagents consumption or inhibition by product (Figure 12.3). This type of growth was not observed in the first self-replicating reactions due to high non-autocatalytic reaction

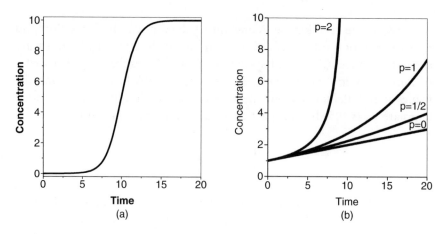

Figure 12.3 (a) Sigmoidal growth presented by parabolic replicator. (b) Typical growth laws: linear, parabolic, exponential and hyperbolic for $\alpha = c_0 = 1$. In both pictures the time and concentration are given in arbitrary units

rate. In order to characterize experimental replicators, the autocatalytic efficiency ε has been defined as a relationship between autocatalytic rate constant k_a and non-catalytic rate constant k_b:

$$\varepsilon = \frac{k_a}{k_b} \quad (12.8)$$

Calculations have shown that sigmoidal behavior can be first observed if $\varepsilon > 10^2$ for parabolic replicators and $>10^4$ for the exponential. At larger values of ε, reactions presenting an autocatalytic pathway approach the growth curve of the pure autocatalytic synthesis. In the case of exponential replicators, with increasing value of ε the reaction induction times are constantly increased without coming to any limit case.

Exponential and parabolic growths are two special cases of kinetic behavior. Other types of initial reaction rates are known. One is a linear mode, where $p = 0$ (Equation (12.9)). This corresponds to an initial rate of any simple noncatalytic reaction. Another one is a hyperbolic growth when $p = 2$ (Equation (12.10) and Figure 12.3b).

$$p = 0: \quad c_t = c_0 + \alpha t \quad (12.9)$$

$$p = 2: \quad c_t = \frac{c_0}{1 - c_0 \alpha t} \quad (12.10)$$

The ability of these systems to undergo Darwinian evolution has been explored theoretically under the constraint of constant organization (namely in a flow reactor in which the total concentration of the competing template is kept constant) [24]. A Darwinian selection in this case would be considered a situation when several species consisting of the same building blocks are produced in the solution. The "fittest" one (a molecule presenting a higher autocatalysis rate constant) could overgrow and "win" over the less efficient ones.

All four growing modes have a different selection range when several competing replicators are present. Different types of distribution of product concentrations arise in those cases. It has been shown that if the growth is sub-exponential then a "survival of everybody" takes place. In a system with linear growth, a relationship between product concentrations will be maintained constant and defined by the ratio of the rate constants. In the case of parabolic replicators, the selectivity will be enhanced and depend on the squares of the rate constants. In the case of exponential growth, the most efficient replicator can overgrow others. However, in the case of hyperbolic replicators, the species presented in higher concentration at the beginning of the reaction will "win." The inefficient selection problem of parabolic replicators can be solved, at least theoretically, if the decay of products and template over time is taken into account [25].

For a better understanding of the factors that rule the order of the reaction, a minimal model has been proposed and a resulting set of differential rate equations has been solved analytically. The minimal model is derived from the reaction mechanism from Figure 12.1. The nonautocatalytic pathway is ignored here for the reason of simplicity. The model describes a system in terms of first-order rate constant k and two equilibrium constants: K_1, a constant for the formation of termolecular complex, and K_2, the dissociation constant of the inactive duplex.

$$A + B + C \underset{}{\overset{K_1}{\rightleftharpoons}} ABC \quad (12.11)$$

$$ABC \overset{k}{\longrightarrow} C_2 \quad (12.12)$$

$$C_2 \underset{}{\overset{K_2}{\rightleftharpoons}} 2C \quad (12.13)$$

In this case the mechanism of the formation of the termolecular complex was not considered in order to have the possibility of analytical solution for the proposed set of reactions. The reaction order can be derived from Equation (12.3):

$$p = \frac{d\ln(dc/dt)}{d\ln(c)} = \frac{d\ln[ABC]}{d\ln(c)} \quad (12.14)$$

Finding the ABC mass balance equation from a reaction mechanism plotted with Equations (12.11)–(12.13) and substituting it into Equation (12.14) gives the expression for the reaction order p:

$$p = \frac{4K_2 q^2 c}{(q+1)^2 + 8cK_2 q^2 - (q+1)\sqrt{(q+1)^2 + 8cK_2 q^2}} \quad (12.15)$$

$$q = \frac{1}{K_1 ab} \quad (12.16)$$

Analysis of the growth of c in time as a function of p shows three different reaction types: with parabolic, strong exponential, and weak exponential growth (Figure 12.4 and Table 12.1). In systems with parabolic growth, the template mostly exists as a part of the inactive duplex

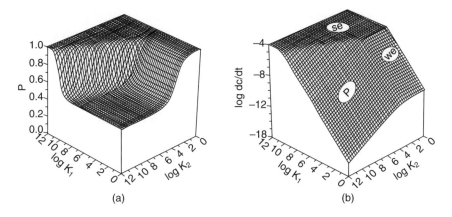

Figure 12.4 (a) Surface plot for the reaction order p as a function of constants K_1 and K_2. (b) Triple logarithmic plot of the autocatalytic rate dc/dt as a function of K_1 and K_2. In both cases $a_0 = b_0 = 10^{-3}$ M and $c = 10^{-4}$ M

Table 12.1 Types of autocatalytic growth

Autocatalytic growth type	First condition	Second condition	Limit rate law	Description
Parabolic	$\sqrt{2K_2c} \gg K_1ab$	$\sqrt{2K_2c} \gg 1$	$\dfrac{dc}{dt} = \dfrac{kK_1ab\sqrt{c}}{\sqrt{2K_2}}$	Most of the template exists as inactive duplex C_2
Weak exponential	$\sqrt{2K_2c} \ll 1$	$K_1ab \ll 1$	$\dfrac{dc}{dt} = kK_1abc$	A template molecule exists as termolecular complex ABC or in a free-form C
Strong exponential	$\sqrt{2K_2c} \ll K_1ab$	$K_1ab \gg 1$	$\dfrac{dc}{dt} = kc$	Limiting case when all the template exists as termolecular complex ABC and thus is independent of equilibrium constants

C_2. The dissociation of the inactive duplex is the rate-limiting step. In cases where a termolecular complex is more stable than the duplex, most of the template would exist as part of the catalytically active ABC complex. Here, the rate-limiting step would be the catalytic activity of ABC. In this case, strong exponential growth should be observed. In conditions where both termolecular complex and duplex are not stable and most of the product is present in its free form C, a weak exponential growth can be observed with the rate-limiting step being the formation of ABC complex. Parabolic growth has already been described. As an analogue of weak and strong exponential growth, an enzyme reaction kinetics far from saturation and at saturation of the enzyme with substrate respectively can be considered.

It can be seen (Equation (12.15)) that the order of the reaction is determined only by the relationship between the equilibrium constants of termolecular complex and duplex formation. The catalytic efficiency of the ABC complex does not alter the value of p. On closer examination it can be also noticed that the replicator's rate end efficiency will present a maximum with the change of the temperature. The ligation rate constant increases at higher temperatures. Values of K_1 and K_2 will change in opposite directions in this case. At low temperatures, K_2 decreases and inhibition by the product becomes stronger. At higher temperatures, the stability of the inactive product duplex decreases the same as the stability of the catalytically active termolecular complex. At some point, supramolecular interaction stops ruling the system and autocatalysis disappears. Thus, autocatalytic activity arrives at its maximum near the melting temperature of the duplex.

The minimal model of autocatalysis presented here usually provides necessary information for the characterization of the replicators. However, this reaction mechanism has several simplifications. In the studies of artificial replicators, more complex modeling is necessary. Complete modeling of all possible reactions was done first by Reinhoudt *et al.* [26] in 1996, by analyzing Rebeck's replicator [21].

12.2.3 Artificial Replicators

Another new field in bioorganic chemistry is the development of non-nucleotidic self-replicating molecules – so-called artificial replicators. Since supramolecular recognition and catalysis are common features among various organic molecules, it seems feasible to develop new replicators based on totally synthetic precursors. The relative simplicity of these molecules (compared with the oligonucleotides) makes them good candidates for theoretical studies that can help to understand better self-replicating reactions. The diverse geometries of molecules which organic chemistry can provide help in solving the square-root law problem, which seems unsolvable for oligonucleotide-based replicators, by designing a reaction with different conformations of reactive complex ABC and template dimer C_2. This can lead to replicators with a programmable reaction order. Autocatalysis of small organic chiral molecules can also be a key answer to the question of the origin of homochirality on the early Earth [27]. In recent years several self-replicating organic reactions have been described and an extensive review has been published [5].

The first work in this area was done by Rebek and coworkers in 1990 [21]. They designed a replicator consisting of an adenosine derivative as the natural component and a derivative of Kemp's acid as the artificial part. As shown in Figure 12.5, the replicator performs recognition via weak interaction of adenosine derivative and forms a new amide bond by nucleophilic attack of the amine that comes into close proximity with the activated carboxyl ester. Dissociation of the self-complementary template duplex closes the replication cycle.

Soon, it became clear that nucleic-acid-based recognition is not the only way to transmit information at the molecular level. In 1992, von Kedrowski presented replicators based on the condensation of amines

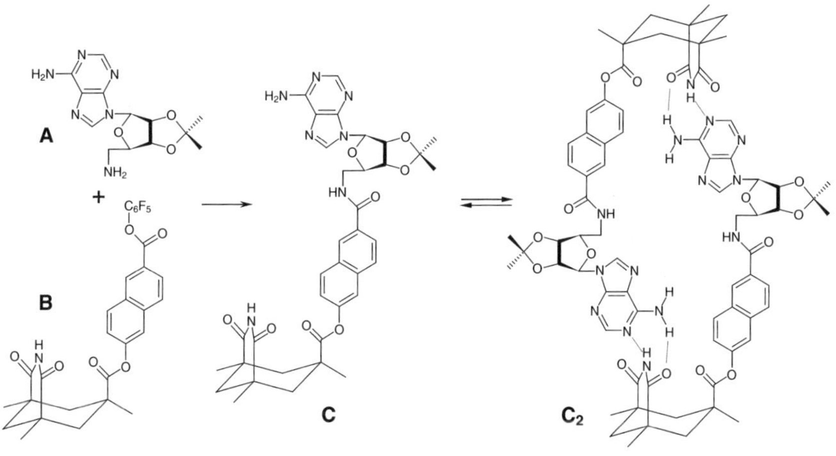

Figure 12.5 Rebek's replicator

with aldehydes with recognition realized by amidinium-carboxylate salt bridges [28].

Amide bond formation presents a high background reaction and is sensitive to acid–base catalysis. The ligation step was changed to Diels–Alder cycloaddition. It was used in a fully artificial replicator presented by Wang and Sutherland [29] in 1997, where the chemical ligation was based on a Diels–Alder reaction between a cyclohexadiene derivative and an N-substituted maleimide as dienophile. Analysis revealed that the autocatalytic reaction order in this system lies between parabolic and exponential growth (reaction order of 0.8). This reaction presented a complicated mechanism not fully addressed in 1997. Among others, a stereogenic center on diene gives a racemic mixture as a product.

This system was simplified by Kindermann et al. in 2005 [1,30]. In their work, R- and S-enantiomers are used as starting material that can produce enantiomeric products via autocatalytic or cross-catalytic cycles. The Diels–Alder reaction was analyzed by nuclear magnetic resonance (NMR). This method not only gives integrals related to the concentrations of the species involved in reaction, but can provide complete information about possible side products or different configurations of the products. The background reaction was measured directly instead of extrapolating observed reaction rates to zero template concentration. To measure the background reaction a proton was substituted by CH_3, destroying the hydrogen bond formation responsible for recognition. The proposed reaction mechanism involved full modeling with all the possible pre- and post-equilibrium fitted to the reaction curves. Later, quantum chemical calculations of energies of reactants, products, and transition states at the B3LYP/6-31G* level allowed one to explain high reaction order (with $p = 0.9$ this reaction gives an example of a strong exponential replicator) and showed no significant energy differences between autocatalytic and cross-catalytic reaction pathways. In these systems, the product inhibition seems to be reduced due to the fact that the association of the template molecules is sterically hindered, whereas the close spatial proximity in the termolecular complex is suitable for ligation. By this stabilization of the termolecular complex compared with the template duplex, the observed reaction order follows the theoretical requirements for near-exponential growth. This was a first attempt at a full characterization of a dynamic supramolecular system, which is now usually described as systems chemistry (Figure 12.1).

Philp and coworkers studied self-replication based on cycloaddition of maleimide and furan [31]. They performed a systematic study of replicator efficiency and conformation selectivity based on different

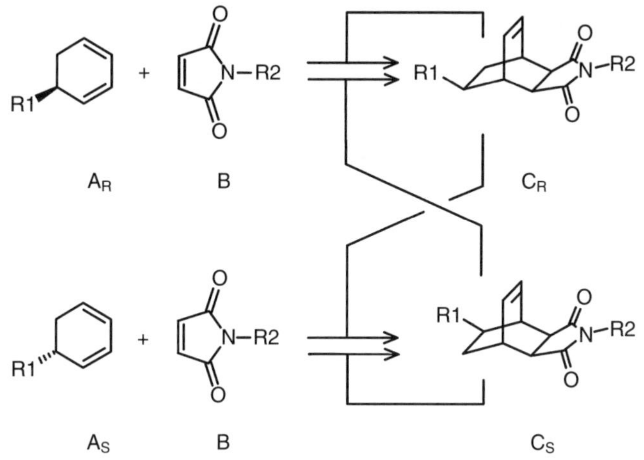

Figure 12.6 Kindermann's replicator presents autocatalytic and cross-catalytic reaction pathways [30]

positions of the substituent on the furan ring and the number of atoms between recognition and ligation units.

Philp and coworkers also explored other chemical systems: a 4π 1,3-dipole and a 2π dipolarophile addition, such as the reaction between maleimide and azide, with the recognition by amidopyridine and a carboxylic acid ($p = 0.4$, $\varepsilon = 20$, no turnover observed due to large duplex stability) [32]. 1,3-Dipolar cycloadition between maleimide and nitrone resulted in a more efficient self-replicating system ($p = 0.9$, $\varepsilon = 5000$) [33]. An interesting feature of the system is the possibility for chiral amplification. It presents the formation of two stereoisomers, *cis* and *trans*. The *trans* stereoisomer is an efficient autocatalyst, whereas the *cis* product presents no catalytic activity at all. Optimization of the system by change of temperature and reactive concentrations gives a *trans* to *cis* product ratio of 250:1. They also explored a combinatorial approach for the search of autocatalytic replicators in the reaction where maleimide reacts with a pool of reversibly formed nitrones [34]. Self-replication and cross-catalysis in peptide-based systems were discovered by Ghadiri in 1996 [4a]. These systems present high reaction orders ($p = 0.63$, reaction proceeds via quaternary template ABTT to give T_3 complex). Homochiral amplification of the products is possible [4d], and a rational design of large networks of 81 peptides interconnected by autocatalytic and cross-catalytic reactions was also achieved recently [4f].

12.2.4 Networks of Replicators

The simple self-replicating system can be seen as a building block for the creation of a network of reactions with complex dynamical behavior. Significant progress in this area has been made so far, including self-replicating oligonucleotide analogues, peptides, and small organic molecules. An extensive review that covers the topic has recently been published [35]. Here, we describe the systems formed by nucleotide analogue replicators.

The system described by Achilles and von Kiedrowski [36] was the first molecular network with a rudimentary information transfer by the feedback reaction pathways. The network consisted of three building blocks: 5′-methylthiomethyl-trideoxynucleotide-3′-phosphate A (^{MMT}CCG^p, MTM = methylthiomethyl), 5′-aminodimer-3′-phosphate B (^{NH2}CG^p), and 5′-amino-2′,5′-dideoxyguanosine-3′-(o-chlorophenyl)-phosphate C (^{NH2}Gp*). Five products, AC, AB, BB, BC, and ABC, were detected by reverse-phase high-performance liquid chromatography (RP-HPLC). The self-complementary tetramer AC is the major product, followed by tetramer BB. AB and BC present a maximum in concentration over the time curve, because they are also intermediates of the ABC product formation and are consumed over the course of the reaction. In order to find the major catalytic pathways in the network formed, template molecules were added at the beginning of the reaction. It was shown that externally added template with a phosphate diester bond has the same templating effect as the phosphoramidate-bonded template formed during the course of the reaction. The templates ^{HO}CCGG (AC″), ^{HO}CCGCG (AB″), and $^{M'}$CCGCGG^{p*} (ABC″) used in this work have been synthesized separately, which greatly simplifies the experiments. Six different pathways with feedback have been identified: autocatalytic synthesis of AC, BB, and AB, autocatalytic formation of ABC either from AB and C or A and BC building blocks, and cross-catalytic reaction between ABC and AB. In cases where different replicators are competing for the same building block, enhanced selectivity has been noticed when corresponding template is added at the beginning of the reaction. However, it is still not a Darwinian selection, due to square-root law limitations.

In 1994, Sievers presented a study where a network formed by four different building blocks was fully characterized [37] (Figure 12.7). In this system, the first complete cross-catalytic cycle was achieved. It can be considered as the simplest example of collective autocatalysis defined

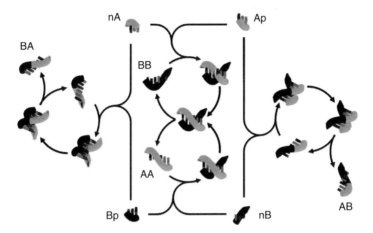

Figure 12.7 Overview of the Sievers network: starting materials are four trimers, two of them acting as nucleophilic aminotrimers (nA, nB) and two of them as electrophilic trimer 3'-phosphates (Ap, Bp). Black means A and gray means B. Reversible reactions are those with arrows in gray and irreversible reactions are those with arrows in black. The four products AA, AB, BA, and BB are formed in the presence of a water-soluble carbodiimide (not shown). Two of them (AB and BA) feed back into their synthesis as autocatalysts (left and right), while the other two (AA and BB) constitute a cross-catalytic replication system (center [37])

by Kauffman as the basis of life [38]. Trinucleotides Ap (CCG^p), nA (^{NH2}CCG), Bp (CGG^p), and nB (^{NH2}CGG) react to give four different hexamers involved in two autocatalytic cycles and one cross-catalytic cycle. The effect of the four possible templates (autocatalytic AB and BA and mutually cross-catalytic AA and BB) was studied for every mixture of Xp and nY building blocks, as well as for the complete set of building blocks constituting the whole system. The network was modeled as follows:

background reactions

$$Ap + nA \rightarrow AA \quad (12.17)$$

$$Bp + nB \rightarrow BB \quad (12.18)$$

$$Ap + nB \rightarrow AB \quad (12.19)$$

$$Bp + nA \rightarrow BA \quad (12.20)$$

template-directed reactions

$$Ap + nA + BB \rightarrow AA \cdot BB \qquad (12.21)$$

$$Bp + nB + AA \rightarrow AA \cdot BB \qquad (12.22)$$

$$Ap + nB + AB \rightarrow AB \cdot AB \qquad (12.23)$$

$$Bp + nA + BA \rightarrow BA \cdot BA \qquad (12.24)$$

reversible duplex formation

$$AA + BB \rightleftharpoons AA \cdot BB \qquad (12.25)$$

$$AB + AB \rightleftharpoons AB \cdot AB \qquad (12.26)$$

$$BA + BA \rightleftharpoons BA \cdot BA \qquad (12.27)$$

For the sake of simplicity, the template-directed reactions were treated as pseudo third-order reactions, comprising the network of reversible association and dissociation steps leading to the respective termolecular complexes as well as the latter's irreversible transformation into template duplexes. The rationale behind this simplification was a lack of good estimates for the equilibria involved. Without this knowledge, a deconvolution of the third-order rate constants into ligation rate constants and equilibria constants would have resulted in quantities with high negative covariance.

What happens, if more than four building blocks are allowed to react? Of course, more than two autocatalytic and two cross-catalytic cycles are expected to show up. While the increase of complexity is less interesting in the case of short oligonucleotides, simply because the high selectivity of sequence–sequence interactions is expected to yield sets of independent replicators with a low level of cross-talk, peptide recognition based on hydrophobic contacts has more potential here. This is why the extension from cross-catalysis as in the Sievers example to larger networks of coupled replicators started to fly some years later in the peptide replicator world, today beautifully explored by the laboratories of Ghadiri, Chmielewski, and Ashkenasy [4].

In the oligonucleotide world, our approach of taking four pieces to create a cross-catalytic replicator nevertheless survived. The most recent example is a seminal piece of work from the Joyce laboratory [14b], which applied directed coevolution of four pieces of an RNA ligase-based cross-catalytic replicator to optimize the whole system for exponential growth (Figure 12.8).

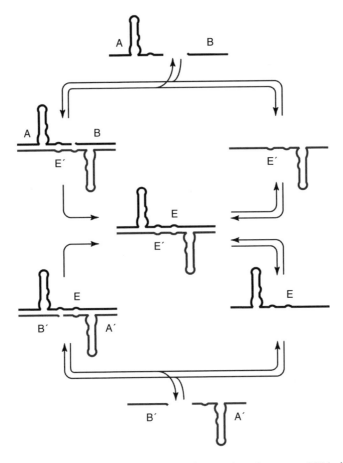

Figure 12.8 Cross-catalytic four-piece replicator based on an RNA ligase as described by Lincoln and Joyce [14]

Independently, another four-piece ribozyme-based replicator came from the Lehman laboratory. This work was based on the consideration that RNA not only has the templating option to express autocatalytic feedback, but also the option to express catalytic activity in a metabolic network leading to the ribozyme as product. A systems chemistry approach toward complex reaction mechanism analysis has been applied to the description of the autocatalytic network formed by self-reproducing ribozyme constructed from four RNA fragments of the Azoarcus group I intron [39]. Unlike other four fragments, W, X, Y, and Z preassemble in a noncovalent structure WXYZ and within this structure three covalent bonds are reversibly formed in order to obtain the final product W·X·Y·Z, which is a catalyst for covalent bond formation within the preassambled ribozyme. It also works as an autocatalyst for its own formation in three different reactions:

$$W + X \cdot Y \cdot Z \rightleftharpoons W \cdot X \cdot Y \cdot Z \tag{12.28}$$

$$W \cdot X + Y \cdot Z \rightleftharpoons W \cdot X \cdot Y \cdot Z \tag{12.29}$$

$$W \cdot X \cdot Y + Z \rightleftharpoons W \cdot X \cdot Y \cdot Z \tag{12.30}$$

When corresponding precursors have been synthesized it was noticed that the observed initial reaction rates of covalent product formation present a lineal increase when W·X·Y·Z is added at the beginning of the reaction. So, in this case, according to Equation (12.3) p is 1 and we have an example of an exponential growth. The autocatalytic efficiency ε is in the range 10^5–10^6 M^{-1}, showing that this system belongs to the most efficient self-reproducing reactions known so far. Simplification of the network resulted in a set of 13 reversible reactions, including self-assembly of the initial four fragments to the noncovalent structure WXYZ, the set of the reactions catalyzed by the ribozyme W·X·Y·Z to form direct precursors of the ribozyme shown in Equations (12.17)–(12.19), and the set of autocatalytic reactions shown by the same equations.

Even the simplified model gave a good fit to product formation profiles for different species. The constants estimated by fitting with the SimFit program resulted in a good agreement with the experiment. Though most of the catalysis described was unspecific and with low informational transfer, the rybozyme proved the possibility of the transfer from simple self-replicating templates to a formation of molecules of larger complexity via metabolic autocatalytic cycles.

Finally, beautiful examples for oligonucleotide replicator systems chemistry came from another completely different approach. The usual approach to a replication is a replication with a gain of information, where small molecules react to form larger ones. As more order is created, entropy is lost and a product inhibition happens. A rather creative way to fool this devil is to turn the whole machinery around: take large stuff of high complexity and cut it into smaller pieces, in this way winning entropy. In the protein world, Nature very often use this recipe to pulse (or shoot or fire) protease activity. Protease enzymes are very often formed from enzymogens, which are larger but inactive preproteases from which a short stretch has to be removed to activate the catalyst. Very often this process is autocatalytic. Some years ago the Ellington laboratory provided the first example that this picture is also an operation principle to make autocatalysis with oligonucleotides [40]. Ellington and Levy synthesized a cyclic single-stranded DNA as an inactive deoxyribozymogen. The active deoxyribozyme was obtained by a single phosphordiester cut to the cycle – catalyzed by the docking and action of the product. One single deoxyribozyme strand is enough to start a cascade hydrolysis of inactive circular substrate.

Similar ingenious schemes employ strand displacement and toehold intermediates to encode autonomous autocatalytic growth without involving covalent transformations [41, 42].

12.3 TOOLS FOR SYSTEMS CHEMISTRY

The first kinetic experiments with replicators were performed using RP-HPLC analysis. Experiments with large fragments of oligonucleotides are done by gel electrophoresis. These methods are time consuming, require large sample quantities, and return small amounts of data with relatively high dispersion between measurements. Several methods have been developed lately and can be relevant in further systems chemistry studies.

12.3.1 Online Kinetic Measurement

Autocatalytic reactions of fluorescent-dye-labeled molecules open the door to online kinetic measurement and allow at the same time the use of very low concentrations of the reagents. Two different FRET (fluorescence resonance energy transfer) approaches have been used in oligonucleotide replication experiments. In the first case, Cy5 or Cy3 dyes are attached to nonreactive 5'-ends of the otherwise identical building blocks [43]. During the reaction, ligation products with a dye at the 5'-end are formed and stay in equilibrium with a nonactive double-stranded species. These duplexes will have a Cy5 at one end and Cy3 at the other, in close enough proximity to enable energy transfer. Knowing the duplex dissociation equilibrium constant, a mass balance for every product can be found and the reaction rate calculated.

The second case is the Alexa-dabcyl quenching system [44] similar to the quenched autoligation (QUAL) probes described by Kool and coworkers for RNA sensing inside living cells [45]. The emerging fluorescence signal is caused by release of the quencher molecule and serves as probe for direct online monitoring of the ligation reaction (Figure 12.9). Here, the reaction rate can be measured directly without consideration of all the equilibria that are involved in the system.

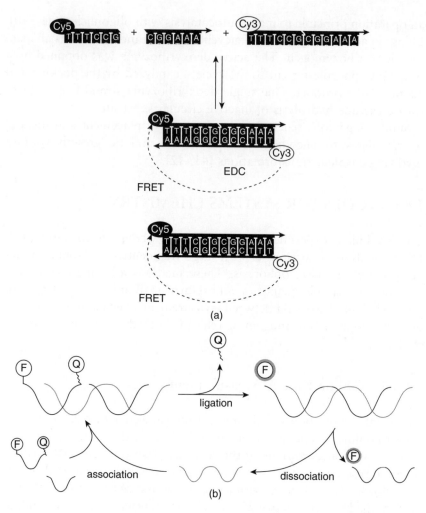

Figure 12.9 Two methods of fluorescence implementation for the online kinetic measurements: (a) FRET-based reaction monitoring and (b) a system with a quencher-labeled leaving group

12.3.2 NMR Kinetic Titrations

For the study of organic replicators based on Diels–Alder chemistry, NMR has been applied. That led us to the understanding of an NMR phenomenon, which resulted in a highly valuable source of kinetic and thermodynamic information (Figure 12.10) [46, 47].

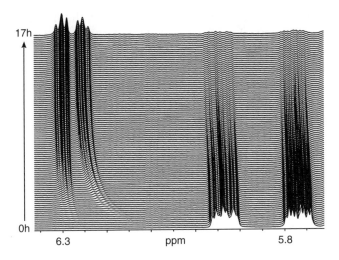

Figure 12.10 Stack plot of kinetic NMR titration. Temporal NMR shift-shifting of two observer protons in a self-replicating organic template

Briefly, if a reaction is monitored as a function of time, one observes just changes of NMR signal integrals. Growing signals are attributed to reaction products and decaying signals belong to the reaction precursors. For self-replicating systems one observes not only integral changes, but also characteristic changes of the signal position, called chemical shift. The reason for this phenomenon is that such systems involve supramolecular complexes, namely noncovalent reaction intermediates, which form and decompose rapidly on the NMR time scale. Single-stranded template molecules are in rapid equilibrium with hydrogen-bridged complexes such as productive termolecular complexes, template duplexes, and others. An observable proton in the template senses a difference in the chemical environment, and its chemical shift reflects whether the template is free in solution, has docked to its building blocks, or has docked to another template molecule. If the process of sensing is slow compared with the life-time of such complexes, an average chemical shift is observed whose value depends on the concentrations of interacting molecules and, thus, the population structure of the system. Since the latter changes in the course of the reaction, shift-shifting reflects the process of complex repopulation. In the beginning of the reaction the shift of a template signal is close to that of a single template molecule free in solution. At the end of the reaction we have high template concentrations but no precursors anymore. High template concentrations lead to the dimerization of templates and the NMR shift is now close to that for the template duplex. In the interim phase we have an involvement of complexes that are formed from both the template precursors and the template itself.

Methods to extract information on the thermodynamic stability of supramolecular complexes from chemical shift information have been developed by Wilcox and others [47]. They are known as NMR titrations. In a typical NMR titration one fills a series of NMR tubes with various mixtures of two interacting molecules. One concentration is usually kept constant while the other is varied. An NMR titration curve plots the observable chemical shift as a function of the concentration ratio. Fitting this curve to an equation derivable from the respective equilibrium and mass-balance equations yields the dissociation constant of the complex as well as its unknown chemical shift as the fit parameters. Repeating such titrations at various temperatures finally allows the calculation of the free energy and entropy of complex formation.

We have coined the means to read and understand reaction-caused shift shifting as "kinetic NMR titration." The term is, in a sense, both correct and misleading. It is correct because part of the underlying math is similar to that for classical NMR titrations. Also, both classical and kinetic NMR titrations are limited to exchange processes rapid enough for the NMR time scale and estimable from the Heisenberg uncertainty principle. The only difference is that the observable shift and the mole fractions are introduced as time-dependent quantities here, while they are time invariant in classical NMR titrations. It is misleading, however, because reaction-caused shift-shifting is not a titration. There is no experimental need to generate the various mixtures by pipetting. Instead, the system is able to produce these mixtures on its own. This, in turn, has the inherent advantage that kinetic titrations are in many cases much more accurate and reliable than classical NMR titrations.

We are confident that "kinetic NMR titrations" is an entry door to a whole field of applications in which advanced NMR techniques are combined with kinetic modeling to decipher complex dynamics in feedback networks. We are currently addressing the question how two-dimensional NMR techniques such as COSY or DOSY can provide both structural and colligative data for the analysis of complex systems in chemistry. "Systems NMR" may be on the doorstep soon.

12.3.3 SimFit Fitting Program

Kinetic data from self-replicating reactions are frequently analyzed using the SimFit program, written by GvK (1989–2008). Several other laboratories have also been using SimFit for complex kinetics analysis [48]. This program starts from a set of reaction equations introduced by the user, transforms them to differential equations, integrates them numerically, and then fits the proposed model to the experimental data. Mechanism is introduced to the program as a set of reaction equations with user-defined rate constants. The program finds an analytical Jacobian for all rates and concentrations of species declared by the user. The equations derived from the user-introduced reactions are integrated using solver for stiff differential equations. Then, the fitting takes place. SimFit uses the observable quantities (e.g. spectroscopic or HPLC integrals) that refer to concentrations or sum or differences of concentrations of reaction species. Observables are defined and then assigned to species by means of string expressions encoding the relationship between observables and species at run time. Kinetic data are read from external tables during the run time. A first approximation is done by a simplex optimization with the possibility of posterior refinement by a Newton–Raphson optimization. For the case of knowing almost nothing on rate parameters, a simple stochastic optimizer is also available. Rate parameters can be fixed, variable, or coupled. Initial concentrations can be optimized for cases where the temperature equilibrium has not been settled initially. SimFit yields a multi-window MDI-type output where one window holds the plots of fitted observables, residuals, fitted shifts, and species concentrations as a function of time, while the other contains all text-based output. The text-based output includes the encoded ordinary differential equation system, its Jacobian, the course of optimized rate parameters, their standard errors, and covariance. In addition, SimFit allows one to visualize the error-hypersurface in all $N \times (N - 1)/2$ possible two-dimensional projections of parameter pairs in another graphic window.

SimFitting of a whole temperature series of kinetic NMR titration data has allowed for the first time the construction of the energy profile for a parabolic replicator. At the level of an experimental energy profile, experimental and theoretical data derivable from quantum mechanics information become comparable.

REFERENCES

1. Kindermann, M., Stahl, I., Reimold, M. *et al.* (2005) Systems chemistry: kinetic and computational analysis of a nearly exponential organic replicator. *Angewandte Chemie, International Edition* **44**, 6750–6755.

2. Stankiewicz, J. and Eckardt, L.H. (2006) Chembiogenesis 2005 and Systems Chemistry Workshop. *Angewandte Chemie, International Edition* **45**, 342.
3. Ludlow, R.F. and Otto, S. (2008) Systems chemistry. *Chemical Society Reviews*, **37**, 101–108.
4. (a) Lee, D.H., Granja, J.R., Martinez, J.A. *et al.* (1996) A self-replicating peptide. *Nature*, **382**, 525–528; (b) Severin, K., Lee, D.H., Martinez, J.A., and Ghadiri, M.R. (1997) Peptide self-replication via template-directed ligation. *Chemistry: A European Journal*, **3**, 1017–1024; (c) Yao, S., Ghosh, I., Zutshi, R., and Chmielewski, J. (1997) A pH-modulated, self-replicating peptide. *Journal of the American Chemical Society*, **119**, 10559–10560; (d) Saghatelian, A., Yokobayashi, Y., Soltani, K., and Ghadiri, M.R. (2001) A chiroselective peptide replicator. *Nature*, **409**, 797–801; (e) Ashkenasy, G. and Ghadiri, M.R. (2004) Boolean logic functions of a synthetic peptide network. *Journal of the American Chemical Society*, **126**, 11140; (f) Ashkenasy, G., Jagasia, R., Yadav, M., and Ghadiri, M.R. (2004) Design of a directed molecular network. *Proceedings of the National Academy of Sciences of the United States of America*, **101**, 10872.
5. For a review on a subject: Vidonne, A., Philp, D. (2009) Making molecules make themselves – the chemistry of artificial replicators. *European Journal of Organic Chemistry*, **2009**, 593–610.
6. Kuhn, H. and Waser, J. (1981) Molekulare Selbstorganisation und Ursprung des Lebens. *Angewandte Chemie*, **93**, 495–515.
7. Crick, F.H.C. (1968) The origin of the genetic code. *Journal of Molecular Biology*, **38**, 367–379.
8. Orgel, L.E. (1968) Evolution of the genetic apparatus. *Journal of Molecular Biology*, **38**, 381–393.
9. Eigen, M. and Schuster, P. (1979) *The Hypercycle: A Principle of Natural Self-Organisation*, Springer, Berlin.
10. Joyce, G.F. (1989) RNA evolution and the origins of life. *Nature*, **338**, 217–224.
11. Sharp, P.A. (1985) On the origin of RNA splicing and introns. *Cell*, **42**, 397–400.
12. Cech, T.R. (1986) A model for the RNA-catalyzed replication of RNA. *Proceedings of the National Academy of Sciences of the United States of America*, **83**, 4360–4363.
13. Breaker, R. and Joyce, G.F. (1994) A DNA enzyme that cleaves RNA. *Chemistry & Biology*, **1**, 223–229.
14. (a) Paul, N. and Joyce, F.G. (2002) A self-replicating ligase ribozyme. *Proceedings of the National Academy of Sciences of the United States of America*, **99**, 12733; (b) Lincoln, T.A. and Joyce, G.F. (2009) Self-sustained replication of an RNA enzyme. *Science*, **323**, 1229–1232.
15. Orgel, L.E. and Lohrmann, R. (1974) Prebiotic chemistry and nucleic acid replication. *Accounts of Chemical Research*, **7**, 368–377.

16. Inoue, T. and Orgel, L.E. (1983) A nonenzymatic RNA polymerase model. *Science*, **219**, 859–862.
17. Von Kiedrowski, G. (1986) A self-replicating hexadeoxynucleotide. *Angewandte Chemie, International Edition in English* **25**, 932–935.
18. The first use of water-soluble diimide was reported a long time before by Naylor, R., and Gilham, P.T. (1966) Studies on some interactions and reactions of oligonucleotides in aqueous solution. *Biochemistry*, **5**, 2722–2728. Later, in Moscow, the group of Shabarova was using EDC as condensing agent with preparative means: Dolinnaya, N.G., Sokolova, N.I., Gryaznova, O.L., and Shabarova, Z.A. (1988) Site-directed modification of DNA duplexes by chemical ligation. *Nucleic Acids Research*, **16**, 3721–3738; Dolinnaya, N.G., Tsytovich, A.V., Sergeev, V.N. *et al.* (1991) Structural and kinetic aspects of chemical reactions in DNA duplexes. Information on DNA local structure obtained from chemical ligation data. *Nucleic Acids Research*, **19**, 3073–3080.
19. Zielinski, W.S. and Orgel, L.E. (1987) Autocatalytic synthesis of a tetranucleotide analogue. *Nature*, **327**, 346–347.
20. Von Kiedrowski, G., Wlotzka, B., and Helbing, J. (1989) Sequence dependence of template-directed syntheses of hexadeoxynucleotide derivatives with 3'–5' pyrophosphate linkage. *Angewandte Chemie, International Edition in English* **28**, 1235–1237.
21. Rotello, V., Hong, J.-I., and Rebek, J. (1991) Sigmoidal growth in a self-replicating system. *Journal of the American Chemical Society*, **113**, 9422–9423.
22. Von Kiedrowski, G. (1993) Minimal replicator theory. 1: Parabolic versus exponential growth, in *Bioorganic Chemistry Frontiers*, vol. 3, (ed. H. Dugas), Springer, pp. 113–146.
23. Nicolis, G. and Prigogine, I. (1977) *Self-Organization in Nonequilibrium Systems*, John Wiley & Sons, Inc., New York.
24. Szathmáry, E. and Gladkih, I. (1989) Sub-exponential growth and coexistence of non-enzymatically replicating templates. *Journal of Theoretical Biology*, **138**, 55–58.
25. Von Kiedrowski, G. and Szathmáry, E. (2000) Selection versus coexistence of parabolic replicators spreading on surfaces. *Selection*, **1**, 173–179.
26. Reinhoudt, D.N., Rudkevich, D.M., and de Jong, F. (1996) Kinetic analysis of the Rebek self-replicating system: is there a controversy? *Journal of the American Chemical Society*, **118**, 6880–6889.
27. (a) Frank, F.C. (1953) On spontaneous asymmetric synthesis. *Biochimica et Biophysica Acta*, **11**, 459–463; (b) Blackmond, D. (2004) Asymmetric autocatalysis and its implications for the origin of homochirality. *Proceedings of the National Academy of Sciences of the United States of America*, **101**, 5732–5736.
28. Terfort, A. and von Kiedrowski, G. (1992) Self-replication by condensation of 3-aminobenzamidines and 2-formylphenoxyacetic acids. *Angewandte Chemie, International Edition in English* **31**, 654–656.

29. Wang, B. and Sutherland, I.O. (1997) Self-replication in a Diels–Alder reaction. *Chemical Communications*, 1495–1496.
30. Stahl, I. (2005) Analysis and classification of minimal self-replicating systems based on Diels–Alder ligation chemistry, PhD thesis, Ruhr-Universität Bochum.
31. (a) Pearson, R.J., Kassianidis, E., and Philp, D. (2004) A completely selective and strongly accelerated Diels–Alder reaction mediated by hydrogen bonding. *Tetrahedron Letters*, 45, 4777–4780; (b) Pearson, R.J., Kassianidis, E., Slawin, A.M.Z., and Philp, D. (2004) Self-replication vs. reactive binary complexes – manipulating recognition-mediated cycloadditions by simple structural modifications. *Organic and Biomolecular Chemistry*, 2, 3434–3441; (c) Pearson, R.J., Kassianidis, E., Slawin, A.M.Z., and Philp, D. (2006) Comparative analyses of a family of potential self-replicators: the subtle interplay between molecular structure and the efficacy of self-replication. *Chemistry: A European Journal*, 12, 6829–6840.
32. Quayle, J.M., Slawin, A.M.Z., and Philp, D. (2002) A structurally simple minimal self-replicating system. *Tetrahedron Letters*, 43, 7229–7233.
33. Kassianidis, E. and Philp, D. (2006) Design and implementation of a highly selective minimal self-replicating system. *Angewandte Chemie, International Edition* 45, 6344–6348.
34. Sadownik, J.W. and Philp, D. (2008) A simple synthetic replicator amplifies itself from a dynamic reagent pool. *Angewandte Chemie*, 120, 10113–10118.
35. Dadon, Z., Wagner, N., and Ashkenasy, G. (2008) The road to non-enzymatic molecular networks. *Angewandte Chemie, International Edition* 47, 6128–6136.
36. Achilles, T. and von Kiedrowski, G. (1993) A self-replicating system from *three* starting materials. *Angewandte Chemie, International Edition* 32, 1198–1201.
37. Sievers, D. and von Kiedrowski, G. (1994) Self-replication of complementary nucleotide-based oligomers. *Nature*, 369, 221–224; Sievers, D. and von Kiedrowski, G. (1998) Self-replication of hexadeoxynucleotide analogues: autocatalysis versus cross-catalysis. *Chemistry: A European Journal*, 4, 629–641.
38. Kaufmann, S. (1996) *Investigations*, Oxford University Press.
39. Hayden, E.J., von Kiedrowski, G., and Lehman, N. (2008) Systems chemistry on ribozyme self-construction: evidence for anabolic autocatalysis in a recombination network. *Angewandte Chemie*, 120, 8552–8556.
40. Levy, M. and Ellington, A.D. (2003) Exponential growth by cross-catalytic cleavage of deoxyribozymogens. *Proceedings of the National Academy of Sciences of the United States of America*, 100, 6416–6421.
41. Seelig, G., Soloveichik, D., Zhang, D.Y., and Winfree, E. (2006) Enzyme-free nucleic acid logic circuits. *Science*, 314, 1585–1588.

42. Yin, P., Choi, H.M.T., Calvert, C., and Pierce, N.A. (2008) Programming biomolecular self-assembly pathways. *Nature*, **451**, 318–322.
43. Schöneborn, H., Bülle, J., and von Kiedrowski, G. (2001) Kinetic monitoring of self-replicating systems through measurement of fluorescence resonance energy transfer. *Chembiochem*, **2**, 922–927.
44. Patzke, V., McCaskill, J.S., and von Kiedrowski, G. *Angewandte Chemie, International Edition* in preparation.
45. Abe, H. and Kool, E.T. (2006) Flow cytometric detection of specific RNAs in native human cells with quenched autoligating FRET probes. *Proceedings of the National Academy of Sciences of the United States of America*, **103**, 263–268; Silverman, A.P. and Kool, E.T. (2005) Quenched autoligation probes allow discrimination of live bacterial species by single nucleotide differences in rRNA. *Nucleic Acids Research*, **33**, 4978–4986.
46. Stahl, I. and von Kiedrowski, G. (2006) "Kinetic NMR titration": including chemical shift information in the kinetic analysis of supramolecular reaction systems such as organic replicators. *Journal of the American Chemical Society*, **128**, 14014–14015.
47. Wilcox, C.S. (1991) *Frontiers in Supramolecular Organic Chemistry and Photochemistry* (eds H.-J. Schneider and H. Dürr), Wiley–VCH Verlag GmbH, Weinheim.
48. For other SimFit applications, see: (a) Sievers, D. and von Kiedrowski, G. (1994) Self-replication of complementary nucleotide-based oligomers. *Nature*, **369**, 221–224; (b) Sievers, D. and von Kiedrowski, G. (1998) Self-replication of hexadeoxynucleotide analogues: autocatalysis versus cross-catalysis. *Chemistry: A European Journal*, **4**, 629–641; (c) Schöneborn, H., Bülle, J., and von Kiedrowski, G. (2001) Kinetic monitoring of self-replicating systems through measurement of fluorescence resonance energy transfer. *Chembiochem*, **2**, 922–927; (d) Severin, K., Lee, D.H., Kennan, A.J., and Ghadiri, M.R. (1997) A synthetic peptide ligase. *Nature*, **389**, 706–709; (e) Yao, S., Gosh, I., Zutshi, R., and Chmielewski, J. (1998) Selbstreplikation eines Peptids unter Ionenkontrolle. *Angewandte Chemie*, **110**, 489–492; Yao, S., Gosh, I., Zutshi, R., and Chmielewski, J. (1998) A self-replicating peptide under ionic control. *Angewandte Chemie, International Edition* **37**, 478–481.
49. (a) Antsypovich, S.I. and von Kiedrowski, G. (2005) A novel versatile phosphoramidite building block for the synthesis of 5′- and 3′-hydrazide modified oligonucleotides. *Nucleosides, Nucleotides and Nucleic Acids*, **24**, 211–226; (b) Achilles, K. and Kiedrowski, G.V. (2005) Kinetic model studies on the chemical ligation of oligonucleotides via hydrazone formation. *Bioorganic & Medicinal Chemistry Letters*, **15**, 1229–1233.

13

Dealing with the Outer Reaches of Synthetic Biology Biosafety, Biosecurity, IPR, and Ethical Challenges of Chemical Synthetic Biology

Markus Schmidt[1], Malcolm Dando[2], and Anna Deplazes[3,4]

[1]*Organisation for International Dialogue and Conflict Management (IDC), Biosafety Working Group, Kaiserstr. 50/6, 1070 Vienna, Austria*
[2]*University of Bradford, Department of Peace Studies, Pemberton Building, Bradford, West Yorkshire BD7 1DP, UK*
[3]*University of Zurich UFSP Ethik, Zollikerstr. 117, 8008 Zurich,Zollikerstr. 117, 8008 Zurich, Switzerland*
[4]*Institute of Biomedical Ethics, University of Zurich, Pestalozzistr. 24, 8032 Zurich, Switzerland*

13.1 Introduction 322
      13.1.1 Let's Give Life a Second Chance 324
13.2 Societal Issues in Chemical Synthetic Biology 326
      13.2.1 Biosafety: Avoiding Unintended Consequences 327
            13.2.1.1 Dealing with Extraterrestrial/Unnatural Biological Agents 327
      13.2.2 Biosecurity: Preventing Hostile Misuse of Chemical Synthetic Biology 329

Chemical Synthetic Biology, First Edition. Edited by Pier Luigi Luisi and Cristiano Chiarabelli.
© 2011 John Wiley & Sons, Ltd. Published 2011 by John Wiley & Sons, Ltd.

    13.2.3   New Forms of Life, Ethical Aspects of Chemical Synthetic Biology 331
              13.2.3.1   What is Life? 332
              13.2.3.2   Value of Life 334
    13.2.4   Intellectual Property Rights 335
13.3   Conclusions 336
    13.3.1   Biosafety 336
    13.3.2   Biosecurity 337
    13.3.3   Bioethics 337
    13.3.4   Intellectual Property Rights 338
    References 339

Home is where one starts from. T. S. Eliot.

The chapter starts by recalling the possible analogy between astrobiology and synthetic biology, in the sense that in both cases we may be facing the encounter with new types of microorganism – with the observation that we are more likely to see "alien" species here on Earth. A brief analysis is then made on the work carried out in the field of synthetic minimal cells, and of alternative DNA forms, to then consider the societal issues in synthetic biology, with questions arising in biosafety and biosecurity, as well as ethical issues and intellectual property questions. A discussion on these challenges is presented here. Problems arising with sample return missions from outer-space objects are considered, with observations made by NASA studies. Concerning biosecurity, one problem is then recognized in preventing hostile misuse of chemical synthetic biology, and the chapter dwells on measures that are or could be taken to face this challenge. Further, the ethical aspects of chemical synthetic biology creating new forms of life are considered and discussed. This brings to the question "what is life" and to "the value of life." Finally, we discuss how chemical synthetic biology might challenge the current intellectual property rights regime.

13.1   INTRODUCTION

Many people will have heard media reports about the Search for Extra-Terrestrial Intelligence (SETI) project in the universe: the search for signals from extraterrestrial life-forms capable of sending them. Meanwhile, there is another lesser known aspect of astrobiology. In this second field of activity, called exobiology, the aim is to search the solar system for evidence of nonintelligent life-forms (such as microbes). Attention has been paid for some time to the question of what to do if

ETHICAL CHALLENGES OF CHEMICAL SYNTHETIC BIOLOGY   323

intelligent life-forms are detected [1], and similar consideration is now being given to the identical question in regard to nonintelligent life-forms [2].

The Outer Space Treaty of 1967 has provisions that require space-faring nations to conduct space exploration so as to avoid harmful contamination of the Earth and celestial bodies. The Committee on Space Research (COSPAR), a permanent committee of the International Council of Scientific Unions (ICSU), provides recommendations on how these aims should be achieved. Then, for example, in the USA, NASA's Planetary Protection Office is charged with assuring that missions are planned and carried out in accord with such laws and policies.

Box 13.1   PPO

Planetary protection is the term given to the practice of protecting solar-system bodies (i.e. planets, moons, comets, and asteroids) from contamination by Earth life, and protecting Earth from possible life forms that may be returned from other solar-system bodies. Planetary protection is essential for several important reasons: to preserve our ability to study other worlds as they exist in their natural states; to avoid contamination that would obscure our ability to find life elsewhere – if it exists; and to ensure that we take prudent precautions to protect Earth's biosphere in case it does. (Source: [3]).

This it does after consultation with both international and internal bodies, such as the Space Studies Board of the National Academy of Sciences. Additionally, for the United States' planned 2014 sample return mission to Mars, NASA will have to prepare a detailed environmental impact statement and this will be subject to public scrutiny. This tight regulation clearly indicates a cautious attitude towards the appearance of novel forms of life on Earth.

Theoretically, a Mars sample might contain microbes similar in many ways to those found on Earth, and these might be dangerous to Earth life-forms. However, the samples could also be very different, and there is clearly a range of other chemical possibilities to maintain living systems [4]. The amino acids in the alien life-form's proteins, for example, could be different from those which have evolved on Earth. The structure or mechanism of operation of its information-storing molecule could be different from our DNA. On some celestial bodies, even more "alien" life-forms may have developed, say through the use of a solvent other than water or the use of very different chemical elements – say silicon rather than carbon [4]. Of course, there are other possibilities, such as variations in the tripartite DNA–RNA–protein architecture found in Earth life-forms. One such possibility would be a dual architecture with, say, just RNA and proteins. Again, some of these possibilities might be dangerous if they arrived on Earth.

Astrobiology is a science with an understanding of the principles and processes involved in the development of the universe and tools such as spacecraft and large budgets to put this understanding to use; for example, in sample return missions. But while there is every reason to avoid complacency about the possible dangers, at least there are the rudiments of an international control system in place. Such assurance cannot be given in regard to all aspects of synthetic biology, especially those working on the design of biological systems based on an alternative biochemistry. This, again, is a new science with an understanding of how life operates and the tools and money to put that understanding to use; for example, in the design and creation of new life forms.

13.1.1 Let's Give Life a Second Chance

"To understand life, it is necessary to build it from scratch," is the motto of synthetic biologists. Those who think of naturally evolved DNA as an unalterable biological axiom will be surprised by recent efforts to release life (as we know it) from its evolutionary constraints. So far, perceptions of synthetic biology are often dominated by the idea that engineers have taken over this part of biology and are busy working on the production of a library of standardized biological parts of natural systems (biobricks) which they will then be able to combine in different ways for various design purposes [5]. Such work is certainly being carried out by engineers, but synthetic biology is a diverse field of activity and also includes biologists and chemists who are trying to produce unnatural molecules and architectures [6] in order, eventually, to create artificial microbes of the kind of concern in the exobiology field of astrobiology. The most prominent research areas dealing with the creation of unnatural (alien) biological systems are protocells and xenobiology [7] (see Table 13.1).

Scientists working on protocells try to create life from the bottom up, by assembling relevant and necessary biochemical subunits. Many dif-

Table 13.1 Characteristics of the main research fields in chemical synthetic biology [6, 7, 8, 59]

	Protocells	Xenobiology
Aims	To construct viable approximations of cells; to understand biology and the origin of life	Using atypical biochemical systems for biological processes, creating a parallel form of life
Method	Theoretical modeling and experimental construction	Changing structurally conservative molecules such as the DNA
Techniques	Chemical production of cellular containers, insertion of metabolic components	Searching for alternative chemical systems with similar biological functions
Examples	Containers such as micelles and vesicles are filled up with genetic and metabolic components	DNA with different set of base pairs, nucleotides with different structural molecules

ficulties accompany this endeavor, but step-by-step small successes have been achieved (e.g. see Ref. [9]). These protocells show some but not all of the characteristics of life, and they can be considered as "limping cells" (Luisi, 2006, personal communication).

Once life is built from scratch, why not try something new? Based on the idea that life could have evolved differently scientists are now trying to design xenobiological systems. The focus of their efforts has been to come up with alternative biomolecules to sustain living processes. Areas of research include the chemical modification of DNA, polymerases, amino acids and proteins. One area of research is the identification of amino acid sequences (proteins) that have a stable architecture but do not occur in nature. Actually, there is only a tiny fraction of theoretical possible proteins occurring naturally, with many more theoretically possible but not-yet-assembled proteins. These so-called never-born proteins could provide a lot of useful novel functions for molecular biology [8, 10, 11]. Changing the translational mechanism from mRNA to proteins via tRNA and the ribosome is another focus of interest. A mutant *Escherichia coli* aminoacyl-tRNA synthetase was evolved to selectively aminoacylate its tRNA with an unnatural amino acid and site-specifically incorporate the unnatural amino acid into a protein in mammalian cells in response to an amber nonsense codon [7, 12].

Yet another area of work consists of modifying DNA by replacing its chemical building blocks, the (desoxy) ribose molecules, and the base pairs. The attempt to come up with an unnatural nucleic acd consisting of a different backbone was the more difficult one, but resulted in novel informational biopolymers such as:

TNA: threose nucleic acid [13, 14];
GNA: glycol nucleic acid [15];
HNA: hexitol nucleic acid [16, 17];
PNA: peptide nucleic acid [18, 19]; and
LNA: locked nucleic acid[1] [20, 21].

On the other hand, experiments replacing or enlarging the genetic alphabet of DNA with unnatural base pairs lead to a genetic code that instead of four bases ATGC had six bases ATGCPZ [22, 23, 58]. In a recent study, 60 candidate bases (resulting in 3600 base pairs) were tested for possible incorporation in the DNA [24].

In respect to these recent efforts and further possibilities, the activities of some in the outer reaches of synthetic biology need careful examination. It is early days, perhaps, but the literature on chemical synthetic biology contains examples with obvious signs of success and promise of further constructive developments. Indeed, it is not unreasonable to suggest that we are much more likely to see "alien" species produced here on Earth before we have to deal with those brought from outer space!

13.2 SOCIETAL ISSUES IN CHEMICAL SYNTHETIC BIOLOGY

The successful design of unnatural biological systems or even "alien" species will definitely not go unnoticed outside the scientific community, and a number of societal issues might be triggered. On the one hand, the scientific results will be well received as a further important step towards understanding what life is and how it could have begun on Earth almost 4 billion years ago. It will also be seen as a powerful way to design new beneficial tools for molecular biology. The design of

---

[1] The LNA is a nucleic acid analogue containing one or more LNA nucleotide monomers with a bicyclic furanose unit locked in an RNA-mimicking sugar conformation.

unnatural biochemical systems or even life forms, however, also raises several critical questions in the areas of:

- biosafety – concerns regarding the prevention of unintended consequences;
- biosecurity – dealing with potentially harmful misuse of unnatural biological systems;
- ethical, philosophical and religious questions – reflecting the moral implications of creating life; and
- intellectual property rights – whether this new form of life can be owned by someone.

A first discussion of these challenges is presented here.

13.2.1 Biosafety: Avoiding Unintended Consequences

The handling of (potentially dangerous) biological agents is regulated through existing guidelines and laws covering microorganisms and viruses (including those which have been genetically modified). The biological material is classified into four risk groups, and the risk assessment that provides the basis for the classification depends on factors such as pathogenicity, severity of disease, individual worker and community risk, host range, availability of treatment or prophylaxis, and endemicity[2] [25–29]. The challenge of risk assessment, however, lies in those cases where a serious health risk is suspected and full information on these factors is not available. In such a case the material should be treated as potentially hazardous (application of standard universal precautions).

13.2.1.1 Dealing with Extraterrestrial/Unnatural Biological Agents

Precautions are also advised when sample-return missions from outerspace objects (e.g. Mars, Titan, asteroids) are carried out. As quoted in Rummel *et al.* [30], the Space Studies Board (SSB) of US National Research Council concluded that:

---

[2] Endemicity means if the biological agent is already present in the environment.

samples returned from Mars by spacecraft should be contained and treated as potentially hazardous until proven otherwise

and further on:

rigorous physical, chemical, and biological analyses (should) confirm that there is no indication of the presence of any exogenous biological entity.

In the recommendations by NASA a distinction is made between life detection and biohazard testing, as nonliving samples could also pose a hazard to Earth-life. According to the SSB, the initial evaluation of samples returned (from Mars) will focus on whether they pose any threat to the Earth's biosphere. The only potential threat posed by returned samples is the possibility of introducing a replicating biological entity of nonterrestrial origin into the biosphere. Only replicating entities (but not necessary living entities) pose a potential widespread threat, especially if they defy the natural, evolved defense mechanisms of Earth organisms. Nonreplicating entities can be considered a toxin and represent "only" a real threat to scientists or people who may be directly exposed to them, as the toxin would be diluted below a toxic concentration when released from the sample.

Similar considerations should be made with respect to unnatural biological systems (see Table 13.2). Non-self-replicating, rather simple agents can be considered and treated as new toxins (or pharmaceuticals); for example, third-type nucleic acids (e.g. HNA, LNA) that can act as steric blockers by duplex formation with mRNA [16, 20]. Greater caution, however, is necessary for self-replicating agents.

A (utopian?) worst-case scenario, for example, would be the design of a novel type of virus based on a different nucleic acid and using an unnatural reverse transcriptase. So far, however, unnatural nucleic acids

Table 13.2 Attempt to classify unnatural biochemical systems with respect to biosafety

Complexity	Ability to self-replicate	No	Yes
Low		Novel toxins; e.g. steric blocker	Unnatural virus
High		Unnatural biological system, protocells	Unnatural life forms

cannot be recognized by natural polymerases, and one of the challenges is to find/create novel types of polymerase that will be able to read the unnatural constructs. At least on one occasion a mutated variant of the HIV-reverse transcriptase was found to be able to PCR-amplify an oligonucleotide containing a third-type base pair. Only two amino acids must be substituted in this natural polymerase optimized for the four standard nucleotides to create one that supports repeated PCR cycles for the amplification of an expanded genetic system. It is without doubt surprising to find a useful polymerase to be so close in "sequence space" to that of the wild-type polymerase [22]. Finding such altered but working polymerases in the evolutionary neighborhood clearly raises the necessity to ask what should be done once an unnatural replicating system has been created in the laboratory.

Sample return missions from non-Earth space objects must place their samples in special sample receiving facilities (SRFs) that can manage to prevent contamination of terrestrial material from the sample and that can maintain a strict biological containment for the sample. Requirements for such an SRF are even higher than that for high-risk biosafety Level 3 and 4 facilities, representing the strictest forms of biological containment. NASA concluded that a facility that meets the strict requirements of such an SRF is not available anywhere in the world [30]. In other words, a sample-return mission with potential biological material would not have an adequate place to deposit and investigate its samples. But while the strictest containment rules are foreseen for extraterrestrial unnatural biological agents, this is not the case for terrestrial unnatural biological agents. "Sample-return missions from Earth," such as synthesis of third-type nucleic acid, for example, are carried out in BSL 1 or 2 laboratories as current regulations do not foresee a stricter handling of this material. After all, unnatural biological systems are not mentioned by the approved list of biological agents/select agent list (e.g. see Refs [26, 31]).

13.2.2 Biosecurity: Preventing Hostile Misuse of Chemical Synthetic Biology

A major difference between astrobiology and chemical synthetic biology is that it is very unlikely that anyone involved in astrobiology would yet have in mind the hostile use of any life-forms found in space, but there is a long history of major state-level offensive biological weapons programs ever since the microbial nature of infectious diseases was discovered by scientists like Koch and Pasteur towards the end of the nineteenth century [32, 33], and it is probable that some such programs persist today [34]. In looking forward to a malign future in which there is an offensive–defensive arms race based on the new biotechnology, US military analysts envisaged three overlapping phases [35]. The first would involve the classical agents, such as anthrax, used in previous programs. However, there are few such agents with ideal properties for biological warfare; thus, the defense would eventually be able to cope. For that reason, the offense would move to modify the agents; for example, by using genetic engineering to make them resistant to antibiotics or difficult to detect in standard tests. Again, however, there are only a limited number of modifications that can be made; so, theoretically, the defense would again catch up.

As this century progresses, however, more and more of life's fundamental processes will become understood and then, these analysts suggest, the offense will turn its attention not to the agent but to the target that they wish to attack. The analysts envisage "an entirely new class of fully engineered agents ... advanced biological warfare (ABW) agents" and suggest that:

> ... Emerging biotechnologies likely will lead to a paradigm shift in BW agent development; future biological agents could be rationally engineered to target specific human biological systems at the molecular level ...

As there are a very large number of physiological processes that could be targeted to cause incapacitation or death, and many ways in which each could be attacked, it seems probable that, if we allow such an arms race to proceed, there will be a long period of offensive supremacy.

Synthetic biologists involved in efforts to bring engineering disciplines into biology have not been ignorant of such dangers [36] and have suggested a range of possible new controls, such as the systematic checking of orders for DNA sequences by manufacturing companies. It is also clear that the States Parties to the Biological and Toxin Weapons Convention (BTWC) have agreed that all biological entities, whatever their mode of production, are covered by the prohibitions in the convention [37]. However, the situation may not be so clear if synthetic biologists create alien species that some might not regard as biological entities.

Article I of the BTWC states that:

> Each State Party to this Convention undertakes never in any circumstances to develop, produce, stockpile or other wise acquire or retain:
>
> (1) Microbial or other biological agents, or toxins whatever their origin or methods of production, of types and in quantities that have no justification for prophylactic, protective or other peaceful purposes.

As Jürgen Altmann argued in his analysis of the military implications of nanotechnology, while a fully artificial toxin (not known in nature) would not be covered by Article I of the BTWC, it would be captured by the provisions of the recent Chemical Weapons Convention [38]. More seriously, he suggested that a fully or partially artificial microbe – for example, one not based on the usual biochemistry including the usual DNA coding system – might not be universally regarded as being covered by the BTWC because it was not a natural microbe. Such an artificial "microscopic" organism would also not be covered by the Chemical Weapons Convention, as it would be much more complex than a chemical. On this reading, an arms race in ABW would be unconstrained in regard to such artificial microscopic organisms.

Such differences of interpretation in regard to the prohibition of chemical and biological weapons would not be unique in the historical record. As Mark Wheelis has argued in regard to German biological sabotage in World War I [39], the German General Staff probably regarded the prohibition at that time as only covering anti-human biological warfare and their substantial anti-animal sabotage campaign as, therefore, being quite legal. Again, today, there are clearly differences of opinion as to the meaning of Article II.9(d) of the Chemical Weapons Convention and whether the peaceful exemption for "Law enforcement including domestic riot control" allows the development of new forms of incapacitating chemical weapons and, thus, provides a route by which the whole prohibition may be subverted [40].

13.2.3 New Forms of Life, Ethical Aspects of Chemical Synthetic Biology

Ethical issues in synthetic biology have been discussed at an unusually high rate, considering the early developmental stage of the technology [41]. The discussion is very often led by topics in bioengineering or synthetic genomics, resulting in debates on the methods applied, as well as different applications and distributions of the technology. However, since synthetic biology is a multi-approach technology, it is important to make clear which branch is addressed in the assessment. In this chapter we focus on ethical issues related to chemical synthetic biology as it has been defined at the outset. The protocell approach and projects on unnatural biochemical systems have been used before they have been understood as a part of synthetic biology, and certainly the integration into this emerging technology should not serve as a reason to look for ethical issues where none has been detected before. However, the context of synthetic biology does shed a new light on these approaches. The idea of designing new forms of life comes to the fore and combination with other synthetic biology approaches seems to be obvious, or at least thinkable, and might lead to the development of new types of synthetic biology products.

The aim of none of the other synthetic biology approaches is as close to the idea of "creating life from the scratch" as is the case for the chemical synthetic biology approaches. Chemical synthetic biology can result in fundamentally novel forms of life based on new types of molecular biology. Therefore, it challenges our concept of life indeed, similar to the idea of extraterrestrial life, by raising the question about the basic features of life and whether life with a completely altered biochemistry should be regarded in the same way as traditional forms of life. In this article we address in more detail what protocells or unnatural genomes can tell us about life and what consequence the establishment of these products of chemical synthetic biology might have on our ethical and philosophical understanding of life.

13.2.3.1 What is life?

Before addressing the question of what chemical synthetic biology can tell us about life, it is necessary to clarify what we mean by the term "life." Several interpretations are possible for instance from a biological, a philosophical or a religious point of view.

Even when restricted to its biological features, a definition of life is not easily established; questions such as whether the life of an individual or that of a population should be described and what features of life are the most important ones have been discussed extensively (e.g. as summarized in Ref. [42], pp. 17–23 and Ref. [43], pp. 197–205). Some authors doubt that it is possible to define life adequately because the conditions that are required to establish a definition of life, such as the necessary and sufficient features, are not available at our current state of knowledge [44]. Others point out that, on linguistic grounds, it is difficult to find a definition of life because there are different types of definitions that tend to, but should not be, mingled – for example, a lexical definition, which attempts to give the meaning of a word and an operational definition that sets out parameters to verify whether the term can be applied [45].

However, researchers working in the fields of synthetic biology or exobiology need to be able to decide under which conditions an object can be considered "alive." As suggested by Oliver and Perry [45], we start, therefore, from different "working descriptions" which do not necessarily claim to be exhaustive definitions. NASA uses a characterization of life which focuses on life as feature of a population:

> "Life is a self-sustained chemical system capable of undergoing Darwinian evolution" [46].

This description has been refined by P.L. Luisi as follows:

> "Life is a system which is self-sustaining by utilizing external energy/nutrients owing to its internal process of component production and coupled to the medium via adaptive changes which persist during the time history of the system" [47].

This understanding of life focuses on the individual organism and comprises the idea of autopoiesis, a notion that has been created by Maturana and Varela specially to describe life; it means self-production and self-organization [48].

Taken together, these working descriptions of life are based on empirically testable biological criteria that distinguish living from inert systems. However, are biological criteria sufficient to describe what we mean by the notion "life"? There is for instance a widespread notion that there are aspects of human life, such as human dignity, which cannot be explained in scientific terms [49, 50]. Human dignity asks for a special form of respect in the contact with others. Every person has certain rights with corresponding duties and responsibilities to other people. However, descending the phylogenetic tree of life, such features get more and more questionable. Does the life of a cat have meaning? Does a rose have dignity? Do we owe any kind of respect to living beings in general? And if we do, what would this respect be based on? The answers to these questions are related to our concept of life.

In Western culture, the concept of life is, for example, influenced by ideas of an immortal soul in some types of living organism (such as human beings). Furthermore, because life originally existed in parallel and not under control of human beings, it is very often directly related to "nature" and "environment." Finally, life is the feature we share with all the other organisms, which gives this property even in microorganisms a particular significance. In these meanings, the term "life" has a positive connotation; it is not purely descriptive, but comprises a normative aspect. Ethicists call such terms with a descriptive and a normative aspect "morally thick concepts";[4] "life" belongs to these concepts.

13.2.3.2 Value of life

The above-mentioned normative components are closely related to an intrinsic value in life.[3] There are many different theories of environmental ethics which argue for the assignment of intrinsic value at different phylogenetic levels of life. Many positions assign intrinsic value only to human beings (e.g. Kant) and others extend it to sentient beings (e.g. Bentham and Singer), but some do argue that all living organisms are carriers of intrinsic value (e.g. Schweitzer, Attfield, and Taylor). Interestingly, for some authors speaking of intrinsic value in lower forms of life, this value seems to be related to the naturalness of life. Paul Taylor, for example, distinguishes between life in the environment and life in bioculture. Whereas in the first category the intrinsic value[5] of living organisms is the only value that needs to be considered, in the second type this value has to be balanced against the instrumental value that living organisms have for human beings. Paul Taylor states [51], pp. 57–58:

> It becomes a major responsibility of moral agents in this domain of ethics to work out a balance between effectiveness in producing human benefits, on the one

---

[3] The term "intrinsic value" is used in different meanings. In this context we understand it in a very broad sense as the claim that certain entities are morally considerable (have moral standing). This means that we cannot deal arbitrarily with any carrier of intrinsic value, but instead we should ask how we are allowed and required to treat it, because we owe moral respect to such an entity.

[4] The term "morally thick concept" was introduced by Bernhard Williams; such terms describe a certain person or fact, but at the same time imply an evaluative or normative component; Williams gives coward, lie, brutality, and gratitude as examples of morally thick concepts [57], pp. 140–143.

[5] Paul Taylor speaks of "inherent worth" when he means what we are calling intrinsic value.

hand, and proper restraint in the control and manipulation of living things, on the other.

This constriction indicates that not even for a biocentrist[6] is the intrinsic value of microorganisms absolute.

13.2.4 Intellectual Property Rights

Little has been said about how current intellectual property rights (IPRs) will shape the development and use of unnatural biological systems, and how unnatural biological systems will shape the way IPR are applied. Articles 52 and 53 of the European Patent Convention (EPC) stipulate that European patents are granted for inventions that are new, involve an inventive step, and are susceptible of industrial application [52]. In a similar way, the US Patent and Trademark Office only grants a patent if the invention is new, not obvious (that means it cannot exist in a natural state like a plant, animal), and must be useful [53]. Different ownership regimes exist internationally when it comes to patenting fragments of DNA. While it is not possible to own random pieces of DNA, a DNA with a useful function can be owned in some countries [54]. While it is not possible to own random fragments of DNA without known function, we do not know if it would be possible to own random fragments of, let us say, TNA or HNA, or any other third-type nucleic acid. In other words, is it possible to "copy and paste" the complete genetic diversity of life from DNA onto a chemically different informational polymer and then patent it? Can artificial genetic alphabets undermine the exclusion of broad patents of life?

Going from DNA to species, the European Patent office states that plants and animal varieties are excluded from patentability; however, microbiological processes and products thereof are not excluded. It seems that a minimal life form, a bacterial chassis, can in principle be patented. In the USA the team of Craig Venter has already filed a patent application for a minimal bacterial genome (US Patent application number 20070122826). A similar treatment can be expected for simple protocells; in other words, they seem not to be excluded from patentability as long as their invention is not contrary to "ordre public" or morality.

---

[6] "Biocentrism" considers all forms of life as having intrinsic value.

13.3 CONCLUSIONS

In this chapter we have attempted to review some of the societal impacts and novel aspects of synthetic biology, particularly protocells and xenobiology. We have asked whether work in these fields is sufficiently guarded by measures of biosafety and biosecurity, whether we have a sufficiently shared understanding of the ethics of the creation of life, and what IPR issues might arise as technology matures. Even this initial review leads to some interesting implications and questions for those working in the fields and those who might be affected by the work.

13.3.1 Biosafety

Currently, no living organisms based on an unnatural nucleic acid are known to exist. But the combination of an extended genetic code and an adequate novel polymerase could certainly lead to the next step towards implementing an artificial genetic system in, for example, *E. coli* [22]. The creation of such unnatural organisms will be done in increments, giving us some, but not too much, time to find out how we could assess the potential risk that these alien organisms could bring and how we should contain them until they are understood well enough to release them to BSL 1 and 2 laboratories or even beyond.

Regulators and scientists are aware that the list and classification of the biological agents must be examined regularly and revised on the basis of new scientific data [25]. The scope and impact of the ongoing research, however, would require a more proactive anticipatory approach, comparable to what NASA has done in anticipating a (possible) sample-return mission of extraterrestrial life.

The probable response to such a suggestion is likely to be, first, that "alien" species are unlikely to be competitive against the highly evolved natural species here on Earth. It will also be argued that nothing should be done to restrict work that has already produced major benefits, such as the Bayer VERSANT branched DNA diagnostic assay for HIV and hepatitis viruses [6]. These are strong arguments, but they do not give enough weight to the possibility that these new species could well be able to survive on Earth, especially as very little information is available on them and scientific predictions on the fate of these organisms in the environment will hardly be possible.

13.3.2 Biosecurity

At the very least, synthetic biologists should look carefully at whether the BTWC can be strengthened with their support through, for example, an agreed understanding at the 7th Review Conference in 2011 that Article I does indeed cover all such artificially created organisms, not just natural or modified natural organisms [38].

Of course, nobody is arguing that scientists have the sole responsibility for preventing the hostile misuse of modern biology. A web of preventive policies has to be constructed by many different actors in many different dimensions [55]. Yet scientists do have a particular responsibility in regard to protecting what they are creating and need to be aware of their obligations under the BTWC and in particular the ongoing discussions among State Parties related to the generation of a culture of responsibility among scientists [37].

This will involve scientists becoming much more involved in discussions of biosafety and biosecurity, but also to be aware of the elements of a new culture of responsibility, such as oversight, codes of conduct, and a much greater focus in education and professional training on the problem of dual use – that is, hostile applications of the results of benignly intended work.

13.3.3 Bioethics

Chemical synthetic biology brings bioethics into chemistry. Bioethics in its broader sense is

> "the study of the moral, social and political problems that arise out of biology and the life sciences generally and involve, either directly or indirectly, human wellbeing" [56].

Independently of potential future applications of chemical synthetic biology in, for example, medicine, interesting bioethical discussions of this field concern the occurrence and meaning of "life" in products of chemical synthetic biology. The synthesis of living systems from scratch or the designing of fundamentally different forms of life raises questions about the meaning of the concept "life" in our society. It reveals that "life" is a multilayer concept implying descriptive, but also normative, aspects. It is an interesting question how and whether normative aspects of life, such as an intrinsic value in certain living organisms, is related to the biological criteria of life and whether or under which conditions such normative issues might apply to products of synthetic biology such as protocells.

In chemical synthetic biology, human creativity, human purposes, and human priorities are the principles deciding about the existence of living organisms. Synthetic organisms resulting from these technologies may fulfill the biological criteria of life; however, this does not necessarily mean that they also fulfill its normative criteria. Some of these features, related, for example, to an intrinsic value or the fact that life is something that exists in parallel to and not because of human beings, may not be found in synthetic organisms.

Therefore, it would be sensible to clarify what type of life ethicists, philosophers, biologists, synthetic biologists, and the public are talking about. Do they understand "life" as something related to an intrinsic value, to naturalness or to biological criteria? If these different interpretations of the term "life" are clearly separated, then a multi-stakeholder discussion of synthetic life might bypass several misunderstandings, which are currently part of the debate. On the one hand, representatives of positions arguing life was something special carrying intrinsic value that should or must *not* be "created" by humans might take into account that synthetic life may, per definition, belong to another category exactly *because* it is designed by human beings. Therefore, it may ask for evaluation based on other standards than natural life, similar to the difference between environmental ethics and the ethics of bioculture suggested by Paul Taylor. Furthermore, synthetic life at a single-cell level should be distinguished from synthetic life in higher organisms. Synthetic biology in higher organisms may raise new types of ethical issue; however, at this stage chemical synthetic biology is not attempting to design or create higher organisms. On the other hand, those scientists who argue that life can be fully explained as soon as one is able to build it may consider that there might be aspects of life that cannot be explained scientifically. In summary, the term "life" is loaded with many different meanings. New scientific developments, such as synthetic biology or exobiology, are adding additional interest and signification to this list. However, not every application of this term necessarily refers to all its different meanings, this may lead to a complicated but fascinating diversification of our concept of life.

13.3.4 Intellectual Property Rights

Chemical synthetic biology deals with biological systems and simple life forms, not (yet) with multicellular organisms. But looking into a possible future, how would plants and animals that are based on a different genetic alphabet be treated by the patent system? Would they be treated in a similar manner to natural life forms and excluded from patentability, or would they not be considered animals and plants and, therefore, not be excluded from patentability? Which definition of life will be used in order to make the decision: the biological, ethical, or philosophical definition?

Clearly, the ethical, legal, and social implications of the outer reaches of synthetic biology will require a great deal more attention before we have even the level of assurance that we now have in regard to the implications of exobiology – and the situation in regard to that field can hardly be considered to be satisfactory.

REFERENCES

1. Race, M.S. and Randolph, R.O. (2002) The need for operating guidelines and a decision making framework applicable to the discovery of non-intelligent extraterrestrial life. *Advances in Space Research*, 30 (6), 1583–1591.
2. Race, M.S. (2007) Societal and ethical concerns, in *Planets and Life: The Emerging Science of Astrobiology* (eds T.S. Woodruff III and J.A. Baross), Cambridge University Press, Cambridge, Chapter 24, p. 483.
3. NASA (2008) About planetary protection. http://planetaryprotection.nasa.gov/pp/about/ (last access 2 December 2008).
4. Ward, P.D. and Benner, S.A. (2007) Alien biochemistries, in *Planets and Life: The Emerging Science of Astrobiology* (eds T.S. Woodruff III and J.A. Baross), Cambridge University Press, Cambridge, Chapter 27, pp. 537–544.
5. Aldhous, P. (2006) Redesigning life: meet the bio-hackers. *New Scientist*, (2552), 43–48.
6. Benner, S.A. and Sismour, A.M. (2005) Synthetic biology. *Nature Reviews Genetics*, 6, 533–543.
7. Schmidt, M. (2010) Xenobiology: a new form of life as the ultimate biosafety tool. *BioEssays*, 32 (4), 322–331.
8. Luisi, P.L. (2007) Chemical aspects of synthetic biology. *Chemistry & Biodiversity*, 4 (4), 603–621.
9. Szostak, J.W., Bartel, D.P., and Luisi, P.L. (2001) Synthesizing life. *Nature*, 409, 387–390.
10. Luisi, P.L., Chiarabelli, C., and Santo, P. (2006) From never born proteins to minimal living cells: two projects in synthetic biology. *Origins of Life and Evolution of the Biosphere*, 36, 605–616.
11. Seelig, J. and Szostak, J.W. (2007) Selection and evolution of enzymes from a partially randomized non-catalytic scaffold. *Nature*, 448, 828–831.

12. Liu, W., Brock, A., Chen, S. et al. (2007) Genetic incorporation of unnatural amino acids into proteins in mammalian cells. *Nature Methods*, **4** (3), 239–244.
13. Schöning, K.-U., Scholz, P., Guntha, S. et al. (2000) Chemical etiology of nucleic acid structure: the α-threofuranosyl-(3′→32′) oligonucleotide system. *Science*, **290**, 1347–1351.
14. Chaput, J.C., Ichida, J.K., and Szostak, J.W. (2003) DNA polymerase-mediated DNA synthesis on a TNA template. *Journal of the American Chemical Society*, **125**, 856–857.
15. Zhang, L., Peritz, A., and Meggers, E. (2005) A simple glycol nucleic acid. *Journal of the American Chemical Society*, **127** (12), 4174–4175.
16. Vandermeeren, M., Préveral, S., Janssens, S. et al. (2000) Biological activity of hexitol nucleic acids targeted at Ha-ras and intracellular adhesion molecule-1 mRNA. *Biochemical Pharmacology*, **59**, 655–663.
17. Declercq, R., Van Aerschot, A., Read, R.J. et al. (2002) Crystal structure of double helical hexitol nucleic acids. *Journal of the American Chemical Society*, **124** (6), 928–933.
18. Wittung, P., Nielsen, P.E., Buchardt, O. et al. (1994) DNA-like double helix formed by peptide nucleic acid. *Nature*, **368**, 561–563.
19. Ng, P.S. and Bergstrom, D.E. (2005) Alternative nucleic acid analogues for programmable assembly: hybridization of LNA to PNA. *Nano Letters*, **5** (1), 107–111.
20. Vester, B. and Wengel, J. (2004) LNA (locked nucleic acid): high-affinity targeting of complementary RNA and DNA. *Biochemistry*, **43** (42), 13233–13241.
21. Kauppinen, S., Vester, B., and Wengel, J. (2005) Locked nucleic acid (LNA): high affinity targeting of RNA for diagnostics and therapeutics. *Drug Discovery Today: Technologies*, **2** (3), 287–290.
22. Sismour, A.M., Lutz, S., Park, J.-H. et al. (2004) PCR amplification of DNA containing non-standard base pairs by variants of reverse transcriptase from human immunodeficiency virus-1. *Nucleic Acids Research*, **32**, 728–735.
23. Yang, Z., Hutter, D., Sheng, P. et al. (2006) Artificially expanded genetic information system: a new base pair with an alternative hydrogen bonding pattern. *Nucleic Acids Research*, **34** (21), 6095–6101.
24. Leconte, A.M., Hwang, G.T., Matsuda, S. et al. (2008) Discovery, characterization, and optimization of an unnatural base pair for expansion of the genetic alphabet. *Journal of the American Chemical Society*, **130** (7), 2336–2343.
25. EC (2000) Directive 2000/54/EC of the European Parliament and of the Council of 18 September 2000.
26. HSE (2004) The Approved List of Biological Agents. http://www.hse.gov.uk/pubns/misc208.pdf (last access October 2010).
27. WHO (2004) Laboratory Biosafety Manual, 3rd edn. http://www.who.int/csr/resources/publications/biosafety/Biosafety7.pdf (last access October 2010).

28. Fleming, D.O. (2006) Risk assessment of synthetic genomics: a biosafety and biosecurity perspective, in *Working Papers for Synthetic Genomics: Risks and Benefits for Science and Society* (eds M.S. Garfinkel, D. Endy, G.L. Epstein and R.M. Friedman), J. Craig Venter Institute/CSIS/MIT, pp. 105–164.
29. ABSA (2008) Risk Group Classification for Infectious Agents. American Biological Safety Association. http://www.absa.org/riskgroups/index.html (last access October 2010).
30. Rummel, J.D., Race, M.S., DeVincenzi, D.L. *et al.* (2002) A Draft Test Protocol for Detecting Possible Biohazards in Martian Samples Returned to Earth, NASA/CP-2002-211842, Washington, DC. http://spacescience.nasa.gov/admin/pubs/marssamples/draft_protocol.pdf (last access 2 December 2008).
31. CDC (2008) HHS and USDA Select Agents and Toxins. http://www.selectagents.gov/resources/List%20of%20Select%20Agents%20and%20Toxins_111708.pdf (last access October 2010).
32. Geissler, E. and Van Courtland Moon, J.E. (eds) (1999) *Biological and Toxin Weapons: Research, Development and Use from the Middle Ages to 1945*, Oxford University Press, Oxford.
33. Wheelis, M.L., Rozsa, L., and Dando, M.R. (2006) *Deadly Culture: Bioweapons Since 1945*, Harvard University Press, Cambridge, MA.
34. Kerr, P.K. (2008) Nuclear, Biological, and Chemical Weapons and Missiles:Status and Trends. CRS Report for Congress. Congressional Research Service, United States.
35. Petro, J.B., Plasse, T.R., and McNulty, J.A. (2003) Biotechnology: Impact on biological warfare and biodefense. *Biosecurity and Bioterrorism: Biodefense Strategy, Practice, and Science*, **1** (3), 161–168.
36. Garfinkel, M., Endy, D., Epstein, G.L., and Friedman, R.M. (2007) Synthetic Genomics–Options for Governance. http://www.jcvi.org/cms/fileadmin/site/research/projects/synthetic-genomics-report/synthetic-genomics-report.pdf (last access October 2010).
37. United Nations (2006) Final declaration. Sixth Review Conference of the Convention on the Prohibition of the Development, Production and Stockpiling of Bacteriological (Biological) and Toxin Weapons and on their Destruction. Geneva.
38. Altmann, J. (2006) *Military Nanotechnology: Potential Applications and Preventive Arms Control*, Routledge, London.
39. Wheelis, M.L. (1999) Biological sabotage in World War I, in *Biological and Toxin Weapons: Research, Development and Use from the Middle Ages to 1945* (eds E. Geissler and J.E. Van Courtland Moon), Oxford University Press, Oxford, pp. 35–62 in Geissler and van Courtland Moon.
40. Pearson, A.M., Chevrier, M.I., and Wheelis, M.L. (2007) *Incapacitating Biochemical Weapons: Promise or Peril?* Lexington Books, Boulder, CO.
41. Presidential Commission for the Study of Bioethical Issues. (2010) New Directions: The Ethics of Synthetic Biology and Emerging Technologies. www.bioethics.gov

42. Luisi, P.L. (2006) *The Emergence of Life*, Cambridge University Press, Cambridge.
43. Popa, R. (2004) *Between Necessity and Probability: Searching for the Definition and Origin of Life*, Springer, Berlin.
44. Cleland, C.E. and Chyba, C.F. (2002) Defining "life". *Origins of Life and Evolution of the Biosphere*, **32** (4), 387–393.
45. Oliver, J.D. and Perry, R.S. (2006) Definitely life but not definitively. *Origins of Life and Evolution of the Biosphere*, **36** (5–6), 515–521.
46. Joyce, G.F. (1994) Foreword, in *Origins of Life: The Central Concepts* (eds D.W. Deamer and G.R. Fleischacker), Johnes and Bartlett, Boston, MA, pp. xi–xii.
47. Luisi, P.L. (1998) About various definitions of life. *Origins of Life and Evolution of the Biosphere*, **28** (4–6), 613–622.
48. Varela, F.G., Maturana, H.R., and Uribe, R. (1974) Autopoiesis: the organization of living systems, its characterization and a model. *Currents in Modern Biology*, **5** (4), 187–196.
49. Hollinger, D. (2004) Life, sanctity of, in *Encyclopedia of Bioethics*, Vol. 3 (ed. S. Post), Macmillan Reference USA, New York, pp. 1402–1406.
50. Metz, T. (2008) The meaning of life, in *The Stanford Encyclopedia of Philosophy* (ed. E.N. Zalta), Metaphysics Research Lab, Stanford, CA http://plato.stanford.edu/archives/fall2008/entries/life-meaning (last access October 2010).
51. Taylor, P. (1986) *Respect for Nature*, Princeton University Press, Princeton, NJ.
52. EPO (2008) About Patents. http://www.epo.org/patents/Grant-procedure/About-patents.html (last access October 2010).
53. USPTO (2008) General Information Concerning Patents. http://www.uspto.gov/web/offices/pac/doc/general/index.html (last access October 2010).
54. Oye, K.A. and Wellhausen, R. (2009) The intellectual commons and property in synthetic biology, in *Synthetic Biology. The Technoscience and Its Societal Implications* (eds M. Schmidt, A. Kelle, A. Ganguli-Mitra and H. de Vriend), Springer Academic Publishing, pp. 121–140.
55. International Committee of the Red Cross (2002) *Appeal of the International Committee of the Red Cross on Biotechnology, Weapons and Humanity*, ICRC, Geneva.
56. Frey, R.G. (1998) Bioethics, in *Routledge Encyclopedia of Philosophy* (ed. E. Craig), Routledge, London. http://www.rep.routledge.com/article/L008 (accessed 17 December 2008) (subscriber log in required).
57. Williams, B. (1985) *Ethics and the Limits of Philosophy*, Fontana, London.
58. Yang, Z., Sismour, A.M., and Sheng, P. et al. (2007) Enzymatic incorporation of a third nucleobase pair. *Nucleic Acids Research*, **35** (13), 4238–4249.
59. O'Malley, M., Powell, A., Davies, J.F., and Calvert, J. (2008). Knowledge-making distinctions in synthetic biology. *BioEssays*, **30** (1), 57.

14

The Synthetic Approach in Biology: Epistemological Notes for Synthetic Biology

Pier Luigi Luisi
University of Roma 3, Biology Department, Viale G. Marconi 446, 00146 Roma, Italy

14.1 Introduction 344
14.2 Setting the Framework 345
14.3 The Two Souls of Synthetic Biology 347
14.4 Life as Emergence, and as a Process of Collective Integration 349
14.5 The Way of Operation of Bioengineering SB 354
14.6 Concluding Remarks 358
Acknowledgments 359
References 360

This chapter is an attempt to provide a framework of epistemology to the growing field of synthetic biology (SB). It is first argued that there are two kinds of SB: one (bioengineering SB) clearly and purposely directed towards one goal set from the start; the other being more concerned with basic science and responding to the basic question "why this and not that?" They need to be considered separately from the point of view of epistemology.

Chemical Synthetic Biology, First Edition. Edited by Pier Luigi Luisi and Cristiano Chiarabelli.
© 2011 John Wiley & Sons, Ltd. Published 2011 by John Wiley & Sons, Ltd.

Some basic notions are necessary for this enterprise. One is the clarification between teleology and teleonomy; another is the clarification of the apparent dichotomy between reductionism and emergentism. Initially, one needs an operational definition of life, and the vision given here, based on system biology and autopoiesis in particular, is one in which life is seen as a dynamic integration of parts, which have all to interact with each other in order to give rise to an emergent whole – which is life. Within this framework, SB appears at first sight as reductionism, as most of its operations are based on assembly of biobricks, and "cut and paste" of genomic parts, seen often like the components of an electronic circuit. However, the necessary condition to arrive at a novel form of life (the goal of SB) is the integration of parts in the complete unity, which corresponds to life as an emergent property. Emergentism is then the real basis of SB, although the researchers in the field are not always conscious of that. It is also argued that emergence is somewhat linked to bio-ethical problems, since novel, unexpected and, in principle, harmful properties may arise from the genetic manipulations. This point is discussed, emphasizing that, in general, epistemic considerations should be brought more and more to the attention of students as an integral part of their understanding of life sciences.

14.1 INTRODUCTION

Synthetic biology (SB) has been developing and spreading both in the academic world and in the applied industrial research at a rate which, in my opinion, few of us were capable of predicting. Just count the international meetings which are offered every year on the subject in all parts of the world. It is easy to foresee that a large part of the life sciences and of pharmaceutical products – and, therefore, our daily life – will be more and more influenced by SB. Therefore, it is right to propose at this stage some basic questions concerning the conceptual framework of this discipline.

The epistemological questions one wishes to tackle are many. Is there something new from the epistemological point of view with respect to the classic genetic engineering, where finally SB is coming from? And is SB a homogeneous field in terms of epistemology, or should one discern different working directions each with its own philosophical charge? Is it correct to view SB, with its gene cut-and-paste procedure, a pure reductionist science? Where is the notion of emergent properties in all this? And how is it with the relation with the classic old items of epistemology, say teleonomy and teleology, or contingency and determinism? In addition, the question that comes to mind immediately with SB is about bioethics: isn't the claim of SB to pretend to synthesize new forms of life something highly arguable – and how to propose it to the mass media? And how is the idea of making life in the laboratory going to help us in understanding the old question: what is life?

There is another general and important consideration about the necessity of offering a discussion platform for the epistemology of SB: the necessity that epistemological concepts should be seen right away by the young researchers who are beginning to work with a comprehensible enthusiasm within the SB field – so as to have scientists who at least partly know, when doing experiments, in which conceptual and epistemological setting they are moving themselves.

In setting the framework of SB, we keep clear from artificial life (AL). I have to mention this, because in some of the literature and minds of workers there is confusion about the two. Aside from the general directive of dealing with forms of life which are not extant in our nature, they are different sciences. SB is basically experimental genetic engineering, whereas AL is basically computational or modelistic or robotics. AL, generally, is not concerned with experimental molecular biology and genetics, and in fact in the classic books of AL as edited by Langton and coworkers [1, 2] one does not find any experimental genetic engineering work; rather, the classic keywords of AL are computer simulation, chaos theory, robotics, alternative chemistry; modeling in terms of electronic circuitry and other forms of metaphoric design of life, and so on. If we design a metabolism based on silicon instead of carbon, this will be an exercise of AL. If we construct in the laboratory a new bacterium by synthesizing its genes and other biochemical parts, this will be a work of SB, not of AL. Of course, this is all matter of definitions, and there will be gray zones in between, but it is important to keep the two things conceptually clear from each other. This chapter deals with SB, not with AL.

14.2 SETTING THE FRAMEWORK

There are by now a few definitions of SB, and the common denominator describes a science that aims at synthesizing alternative forms of life, or, more generally, biological structures which do not exist in nature. If we want a more official definition, we can refer to the expert panel at the SB3 in Zurich (2007), which formulated that: *SB aims to engineer and study biological systems that do not exist as such in nature, and use this approach for achieving better understanding of life processes, generating and assembling functional modular components, or develop novel applications or processes.*

The term "alternative forms of life" above should be further qualified by referring simply – for the moment at least – to bacteria; in fact, the most popular version of SB is the construction of bacteria which are supposed to perform novel "useful" tasks, like the production of fuels, hydrogen, and other forms of energy. Examples of typical titles are: "Engineering a synthetic dual organism system for hydrogen production" [3]; "Metabolic engineering of microorganisms for biofuels production" [4]; and "Light-energy conversion in engineered microorganisms" [5]. The notion of alternative forms of life is also emphasized; for example, see the paper in Craig Venter's group titled "Genome transplantation in bacteria: Changing one species into another" [6]; or "Engineering microbes with synthetic biology frameworks" [7].

The term "bio-bricks" is used to indicate the genetic elements that can also be made available commercially and that can be used as elements for constructing the novel genetic circuits (e.g. see Ref. [8], and see also the Register of Standard Biological Parts at http://parts.mit.edu). In the same way as we used to buy chemicals to start the synthesis of more complex organic molecules, or electronic parts to make an electronic gadget, now there seems to be the possibility of buying genetic parts to build up synthetic life. Whereas the traditional approach to learn about life was to dissect the living and study the single parts one at a time, now, with SB, we can learn about life by way of synthesis. And this is, of course, an important conceptual difference – in the chemical sciences we used to say that we can fully understand only what we can synthesize. It is also interesting to notice that SB has been in full development in concomitance with system biology, namely with the capability of studying and in part conceiving the living in its entirety – actually even in its social context.

Also, the term "chassis" is used in the field of SB, like the naked chassis of an automobile on which you put all the parts – wheels, motor, brakes, and so on – in order to make the whole. And indeed, some of the research titles reflect this assembly endeavor, which sounds like the construction of electronic circuits; for example: "An integrated cell-free metabolic platform for protein production and synthetic biology" [9]; "Toward scalable parts families for predictable design of biological circuits" [45]; and "Principles of cell-free genetic circuit assembly" [10].

The notion of bio-bricks may give lay people the idea that, by buying all components, you can assemble a living organism. Is something like that really possible? The answer really depends on what you mean by the question and by the term "life". We will come back to this later on in this chapter.

Aside from the big question of making life, the bioengineering applications of SB also cover the field of medicine; for example, see the work of Benner *et al.* [11], or the work by Lee *et al.* [12]. Also, see the work by Fussenegger and coworkers [13] on the biotin-triggered genetic switch, which enabled dose-dependent vitamin H control in certain cell lines, the work by Chang and Keasling [14] on production of isoprenoid pharmaceuticals using engineered microbes, and that by Stephanopoulos and coworkers [15] on terpenoid synthesis from microorganisms.

14.3 THE TWO SOULS OF SYNTHETIC BIOLOGY

The statement given above from SB3 in Zurich emphasizes two different aspects of SB. One has to do with the engineering aspects – making new products; the other has to do with the development of basic science. This does not represent simply a generic description, but corresponds to two different ways of conceiving SB.

Let us deal first with the engineering aspect of SB. In fact, most of this work on SB follows a classic bioengineering approach, whereby a determined product is designed at the very beginning, and all routes and tools are bent and focused for obtaining that product – for example, bacteria to make hydrogen, or a particular drug.

Beginning with epistemic terminology, we can then say that we are dealing with a classic *teleological* enterprise, where teleology means that the purpose is defined at the very beginning – as engineers do when they have to construct a bridge or an airplane. It is worthwhile recalling that teleology is not the way by which nature and biological evolution in particular are supposed to proceed. The function (for example, sight, the spinning flagellum, wings, and so on) in nature's evolution is never set a priori, but is the result, the consequence of the contingent structure's development. There is no programmer or intelligent designer in the plans of nature: the ameba, the bees, or the ants move about and do what they do, obeying more or less blindly a genetic program, borne out on them by the laws of natural evolution. Yes, it may look like finality, but it is *teleonomy* instead of teleology, namely the more or less blind working of an implemented genetic program. The function is a consequence of the structure's development, and the structure is constructed according to the "bricolage" as described by the pioneers of contingency, Monod and Jacob, up to more recent authors like Stephen Jay Gould. Teleonomy at work.

Teleonomy, then, is not the procedure of SB, which, by constructing its alternative forms of life, operates completely on the basis of teleology – at least the bioengineers do so. In this case, the intelligent designer is the synthetic biologist at work. Teleology instead of teleonomy.

Having clarified that, let us now consider the other aspect of SB mentioned. It is, in fact, proper to say that SB has a double soul. One corresponds to the bioengineering approach outlined above. The other is instead rooted into basic science.

In this second case, the basic, underlying question is: Why did nature do things in a certain way, and not in another one? Why 20 amino acids, and not 15 or 55? Why do nucleic acids contain ribose instead of glucose? Must hemoglobin be constituted by four chains, why not six or twelve?

From one general, philosophical point of view, this kind of questioning (why this and not that?) links to the dichotomy between *determinism and contingency*: Are the things of nature the way they are simply because there was no other way to make them ("absolute determinism")? Or are they the way they are due to contingency – something that some time ago, less properly and less fashionably, we used to call "chance"?

SB possesses the tools that may permit one to tackle this kind of philosophical question. The way is conceptually simple: let us synthesize the alternative form and see whether there are some reasons why this route may have not been chosen by natural selection.

Take the example of the work by Albert Eschenmoser and collaborators at the Swiss Federal Institute of Technology (ETH-Z) on alternative DNA forms (a case of chemical SB *ante litteram*). He and his group synthesized DNA with pyranose instead of ribose in the chain [16, 17], arriving at important considerations about the working procedures of biological chemical evolution. And take the work in Yanagawa's group [18] on proteins with a reduced alphabet of amino acids, showing that in certain instances enzymes "remade" with only 10–12 amino acids instead of 20 may work rather well. And there is the field of the alternative proteins, with the question "why these proteins and not others?" which leads to the synthesis of proteins which do not exist in nature, the so-called "never-born proteins" [19, 20]. Or take the work of Benner and collaborators on nucleic acids built with bases different from the four canonical ones [46]. And why are unicellular organisms constituted by thousands of genes? Can't they work with a much lower complexity degree? And this is a question which leads the way to the field of the "minimal cell/minimal life," which is now being pursued by several groups [21–28].

Several other examples could be given – which, by the way, are examples of "clean" SB, in the sense that they are being carried out without genetic manipulations of the living typical of most engineered SB. In fact, the term "chemical synthetic biology" has been coined to represent this field [29]. In this case, the operation is also teleological, in the sense that we set a priori the construction of the new structure – for example, the DNA with pyranose. However, the function is not set a priori; actually, the functionality corresponding to this novel structure is not known – it is what we want to discover. In this sense, there is a significant difference with the bioengineering approach, where the function is the main prerequisite (e.g. production of hydrogen) set a priori at the very beginning.

However, it is the bioengineering approach that is more popular in SB, also because it is certainly the one which is more interesting from the applicative, and money-making, perspective.

It should also be noted, obviously, that the distinction between the two souls of SB is not always so clear, as you can spot in the literature examples of bioengineering work which bring important scientific advancements.

14.4 LIFE AS EMERGENCE, AND AS A PROCESS OF COLLECTIVE INTEGRATION

Since SB deals with the idea of making life in the laboratory, it is necessary to briefly dwell on the question, "what is life?", limiting the inquiry to the simple cellular constructs that SB is concerned with. For that, consider the simple cartoon of Figure 14.1. This represents the semipermeable membrane, which identifies the compartment in which many reactions and the many corresponding transformations take place.

The phenomenology of a cell is actually the procedure followed by Maturana and Varela to derive their autopoiesis (autopoiesis from the Greek, meaning self-production). Actually, we could say, with them [30–32], that the cell's main function is to maintain its own individuality despite the myriad of chemical transformations taking place in it. This apparent contradiction is explained by the fact that the cell regenerates from within those components which are consumed – be that ATP or glycogen, glucose, α-chymotrypsin, or t-RNA.

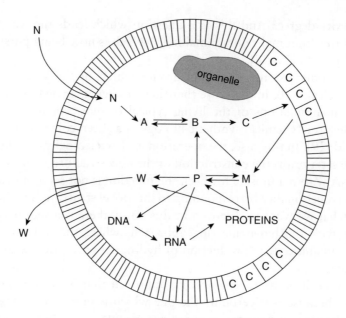

Figure 14.1 A cartoon representing the activity of a cell. Nutrients/energy (N) enter the cell, are integrated in the metabolism, which produces all its own components (A, B, M, P, W, DNA, RNA, proteins, organelles) as well as the membrane component (C). Waste material (W) is then released into the environment

Thus, the living cell is a system of processes which makes the components that assemble in the systems themselves – we are witnessing cyclic logic, by which the system produces itself (of course, at the expense of nutrients and energy coming from the medium). Autopoiesis, then, is characterized by this particular organization, which is the invariant property of life, whereas the structure may vary from cell to cell or within a cell depending on the actual function and metabolic changes.

Autopoiesis, then, according to the authors, is the blueprint of life, as all living cells and higher forms of life obey this principle of self-maintenance from within, due to a self-generating internal mechanism.

This we can simply learn from the phenomenology of a cell, and then give a first answer to the question "what is life?"

But we can learn something more from the cartoon of Figure 14.1. Consider the question: where is cellular life localized? There is an obvious and very important answer to this question: life is not localized, life is a global property, being given by the collective interactions of all molecular species, by the invariant self-organization mentioned above.

This is not true only for the cell, but for any other macroscopic form of life. The life of a large mammal is the organized, integrated interaction of heart, kidneys, lungs, brain, arteries, and veins. And each of these organs can be seen as the integrated and self-organized ensemble of different tissues and organelles (they can be seen as autopoietic systems of higher order); and each tissue or organelle is the integrated ensemble of different cells; and each cell is the organized integration of the molecular species, as already discussed from Figure 14.1.

There is another apparent contradiction in cellular life and life in general, which can be germane for our discourse here. This is the fact that a living cell, and by inference all living things (at least during the homeostasis period), must be a *thermodynamically open system*, as this allows the input of energy and nutrients through a semi-permeable membrane; conversely, the living cell, from the epistemic point of view, is an *operationally closed* system, in the sense that the information for all life activity is contained within the structure's self-organization. The aspect of being thermodynamically open links the living with the interaction with the environment, and we can say that the living is participating in a *cognitive interaction* with the environment. The terminology of cognition is taken again from the Maturana and Varela [31, 32] autopoiesis theory and indicates the specific coupled interaction between the living and the environment with the consequent constitution of a mutual cooperativity, or co-emergence into a mutual unit.

Then, for the time being, let us see life as an integrated system which is capable of self-sustainment due to the activity from within, and which is operationally closed and thermodynamically open, capable then of interacting with the environment in a specific manner. There is, however, another aspect of life which must be taken into consideration, and which is particularly important for the field of SB.

To say that life is a global property that depends on the interaction of parts is tantamount to saying that *life is an emergent property.*

It is important for our discourse here to briefly recall a few aspects of this notion, in particular those which may be important for SB, like, for example, the notion of downward causation.

First of all, emergence and emergent properties are notions that arise upon the formation of a higher hierarchic order starting from smaller parts or components, and indicate the arising of novel properties – novel in the sense that they are not present in the parts or components. Generally, it is accepted that the emergent properties cannot be predicted from, or reduced to, the properties of the components, and in the literature the notion of "strong emergence" is opposed to the notion of "weak emergence." The first term indicates that the prediction or interpretation from the parts is in principle (ontologically) impossible; the second term indicates that such prediction or interpretation is not in principle impossible, but simply technically too difficult. In the case of bioengineering SB, the properties to be expected are set a priori, whereas in the case of basic science SB ("why this and not that") the emergent properties, as we have already noted, are precisely the matter of investigation. There would be much to say about emergence, about predictability and deducibility, about the relation with reductionism, but we do not need all this here. However, there is another important element we should not miss for our present discussion on SB. This is the notion of *downward causation*.

To deal with this concept in detail would strain the chapter into a high degree of specialization, for which the reader is referred to the literature (e.g. see Refs [33, 34]). For the purpose of this chapter, it is enough to recall that this notion has to do with the fact that there is always a mutual interaction between the parts and the whole, in the sense that the formation of the whole affects and modifies the parts. Clearly, the amino acids that form a protein are modified by their mutual covalent and noncovalent bonding with respect to their free state of amino acids; and a tribe, formed by the ensemble of several families, modifies the behavior and rules of the family, while the family, formed by the ensemble of single people, modifies the properties and behavior of these constituent persons.

This ties with the notion of bio-bricks discussed previously. To understand this point better, consider that it does not make much sense to consider the heart by itself, as a separate abstract part. We may have it as a frozen organ, but it acquires its real meaning when it becomes an integrated part involved in the interactions with all other organs and parts of the body. It is, then, generally important to make clear that the "parts" of life cannot be considered as inert pieces as in a Lego-game or bricks that can be added mechanically; in reality, the parts (and, therefore, the bio-bricks also) should be seen as being substantially modified and functionalized by the making of the whole. Actually, and this is the important point, they will acquire a meaning only if they become a functional part of an integrated unity.

All this ties to the question we have raised in the introduction, whether by assembling all the appropriate bio-bricks you can build an organism in the laboratory. If by that we mean the construction of a living organism from scratch, it should first of all be pointed out that this has not been accomplished thus far by SB. The claim of some of the media, that Craig Venter had accomplished "the synthesis of life", were simply wrong. As already mentioned, what Venter's group had done was to substitute the original genome of a cellular organism with a synthetic one, a very remarkable job, but conceptually something like an organ transplant. The cell has not been synthesised, and the very idea that all that you need to make life is to make a genome, is simply fallacious: put that genome in a nutrient buffer solution and nothing would happen. Only if you put the genome in a cell, will things begin to move.

The idea of making an organism from scratch was not in the genetic SB agenda. But let us pursue for a short time the idea of trying to make an organism from scratch. In this respect, it is useful to recall the attempts to reconstitute living cells starting from their dismantled components: nucleus, cytoplasm and cellular membrane. The first thing to say is that the re-assembling is not a simple, spontaneous process, as for example in the case of TMV [35] or some ribosomal forms [36]. In the case of *Amoeba Proteus* [37], *Acetabularia Mediterranea* [38], and some mammalian cells [39], the re-assemblage was partially obtained, but always with the assistance of an operator, who with micro syringes or by other mechanical manipulation, has helped the reconstitution by hand.

Why the case of the cell is different from the case of TMV? Basically, because the self-organization of TMV is under thermodynamic control, whereas the construction of a cell is not; we are dealing with a lot of kinetic control, and all construction plans depend on a series of events in a very precise sequential order.

These considerations are very important also for proceeding with SB. Even supposing that one can dispose of all possible bio-bricks to make the entire organism, which is per se a fantastic assumption, how many do you need to make an amoeba, or a worm? What can you do? In fact, only if the biological system to be constructed by SB is under thermodynamic control can the simple mixing of the corresponding bio-bricks, in a mechanical or genetic way, bring about the complete structure and potentially all the corresponding functions. If, conversely, the system is under kinetic control, which means a lot of sequential order in the ontogenic pathway, then just mixing the various bio-bricks together would not work. The prerequisite to do something viable would then be to know a priori the exact construction order (the precise "cellular ontogenesis"), and simulate this ordered pathway. And even in this experiment, there would be problems. What to do with the cell membrane? At which point would the compartment form? How would all the components be put inside? More generally, what would be the experimental set-up to start this kind of experiment?

As we said, SB was not born to make organisms from scratch, but to modify the extant ones. Perhaps this will change in the near future: making organism from the scratch can really be seen as one of the challenges of future bio-engineering SB.

Things are somewhat easier with the "mechanical approach", namely starting from extant macromolecules(enzymes and genes) and basic bio-structures. Some progress in this direction has been made with the previously mentioned "minimal cell project", the mechanical approach of SB. In this project, one starts from already formed membrane models, the liposomes, into which single macromolecular components are introduced. The idea is to add the minimal but sufficient number of components to arrive at the first threshold of living cells.

From the above discussion, is also apparent that there are two ways of building up an organism by way of SB. One is to use the physical elements, the other is using the genetic bio-bricks, adding them to one another. Looking at these two ways, one could talk of a "mechanical SB" and of a "genetic SB" but there are already too many classifications and too much terminology.

Having clarified these general concepts, let us see more in detail how SB operates to reach its goals.

14.5 THE WAY OF OPERATION OF BIOENGINEERING SB

One major operational scheme of SB is based on modularity, which implies a "cut and paste" of genetic parts from one organism to another. Some of these parts can be "bio-bricks", originally belonging to a living organism, and then made commercially available, or can be genetic parts synthesized in the laboratory. Let us schematize the major modes of modularity operation, and this with the help of Figure 14.2.

Given a starting organism containing the four genetic elements A, B, C, and D, we can have the case in which one element (or more than one) is eliminated, as in the knock-out experiments carried out by Venter's group on *Mycoplasma genitalium* [40] – see Figure 14.2a. The result is a genetically simpler novel organism. And the question is, what is its performance? The goal of the investigators was to maintain the viability and "life" of the original organism, a goal that, according to the authors, was achieved.

THE SYNTHETIC APPROACH IN BIOLOGY 355

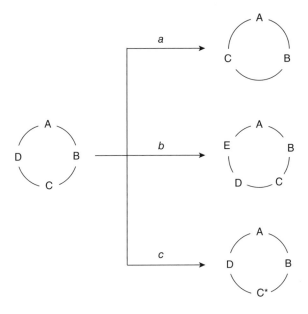

Figure 14.2 Basic operations in synthetic biology leading to non-extant forms of life

This procedure also epitomizes one possible way to the "minimal cell," a cell consisting of the minimal and sufficient number of genes. It is a kind of top-down approach and is conceptually quite different from the other kind of approach, to be discussed later on in this section, which offers a successive addition of genes to an empty scaffold.

The figure on the left represents an organism composed by four genetic elements (A, B, C, and D). In procedure "a" one such element is deleted; in procedure "b" a new element (E) is added instead; and in procedure "c" one of the original genetic elements (C) is modified.

In contrast to the mood of operation of Figure 14.2a, Figure 14.2b illustrates the case in which one gene, or one entire bio-brick, is added to a pre-existing organism. This is a more complex case, as there is a genetic addition: the organism A is now supposed to perform all its previous functions, plus the new ones due to the added B element.

The simplest case of this second manipulation is classic bioengineering: if you want *E. coli* to fabricate insulin, you add to its genome the pig insulin gene, and the *E. coli* genome is thus enriched and does the job of making insulin. However, SB offers more complex and more interesting cases. For example, in the work by Weber *et al.* [13], simplifying somewhat, the *E. coli* repressor of the biotin biosynthesis operon was fused to the *Herpes simplex* transactivation domain to generate a biotin-dependent transactivator; biotin-inducible transgene expression was functional in a variety of rodent, monkey, and human cell lines.

Let us consider now the case illustrated in Figure 14.2c. We have here – indicated with an asterisk – the replacement of a genomic part, in analogy with an organ transplant in humans. The new element added into A can also be synthetic; an example is the replacement of an entire gene of *M. genitalium* by the corresponding synthetic gene, a synthesis performed in Venter's group [41]. In all these cases we may obtain a more powerful functionality in the organism, and the integration and the consistency with the whole should be guaranteed by the fact that we have substituted similar or identical parts. And, more generally, this demonstrates that that life – at least at this level – can be seen in terms of a cut-and-paste of parts. A good example of this is the work by Lee and coworkers [12], who operate on the metabolic engineering for the production of drugs at the whole cell level (a notion germane to system biology) for enforcing or removing the existing metabolic pathways toward enhanced product formation.

Venter's group experiment deserves a further comment. The original bacterium, once deprived of its own genome, was not alive; the synthetic genome obtained in Venter's laboratory was also, obviously, per se, not alive. By putting these two non-alive parts together, a living organism came out. Actually, when this work was published, we all admired the labor endeavor, but nobody was surprised by the fact that life could be regained. Why not, if DNA is just a molecule?

Conceptually, this work reminds one of general cloning, where we have that the two parts (the enucleated ovocyte and the nucleus) each per se being not living, but once properly put together, the two nonliving things give rise to a living entity. This is a demonstration, then, that life is indeed an emergent property, as it can be obtained starting from nonliving parts?

Figure 14.2c may also illustrate the case in which the new genetic element has been created in situ in the A organism by genetic manipulation of the original organism. For example see the paper by Endy and coworkers [42], in which more than one single module has been genetically changed. Also in this case, as in the previous one, we are dealing at the end with a novel organism.

This is also in the example of the *iGEM* work by a young team from Peking University (2007) who modified the *E. coli* bacterium so that it changes color when light is switched on/off – see http://2009.igem.org/Main_Page). A novel operational unit is integrated in the original circuit, so that the final product (bacterium) is a new form of life.

What is missing in Figure 14.2 is the total synthesis from scratch, starting from the synthesis of the genome, as shown by Venter's group, and then making one by one what we have called A, B, C, and D moduli. I believe this is still well out of reach.

Let us now turn to epistemology again. Based on the schemes of Figure 14.2, life is seen in SB as a series of operational units, which are spatially and logically separated from each other and which can be added and taken away singularly. Under this viewpoint, we can say that there is a strong reductionism attitude. Do the SB schemes of Figure 14.2 point out a reductionism scenario?

Yes and no, we have to be careful. In fact, together with the cut-and-paste operation, even if implicitly, there is also the idea that all these operational units must be interconnected with each other to make possible the emergence of life. In other words, insofar as life is reduced to the sum of operational units, we are in reductionism. As soon as we say that life cannot be interpreted or explained with the sum of operational units, but we need their integration to form a novel emergent unit, then we abandon the narrow limits of reductionism.

In my opinion, this simple discriminating argument is not always clear in the mind of the workers in SB. Actually, the workers can be defined as reductionist or emergentists, depending upon the stand they take in the above discrimination.

Can these new forms of life be beautiful or monstrous, or be patented as a new gadget can? There is an ongoing discussion on that; fortunately, this is an issue which lies outside the scope of this chapter.

What about the other soul of SB, the basic science soul, that based on the question: why this and not that?

In this case, there is a structural design, set from the start – for example, the making of proteins with a reduced alphabet of amino acids, or of DNA having pyranose instead of ribose. The synthesis then follows a teleological perspective. However, contrary to the case of engineering SB, the function is not set a priori; namely, one does not know a priori what these novel proteins, or the DNA analogs, will be capable of doing. It is all open ended. The question, rather, is which kind of emergent properties are going to arise from such a novel biological structure? The fact that we do not know what the emergent properties are going to be sets this kind of science in the realm of basic science, rather than in the bioengineering area.

14.6 CONCLUDING REMARKS

The epistemology of SB can be highlighted in terms of a few keyword dichotomies. One is teleology versus teleonomy, and here things are relatively clear. The other is reductionism versus emergentism, and here things are subtler. We have seen that, according to a superficial point of view, SB appears as a reductionist science, as it is all based on interchangeable modules which can be assembled together, or dismantled. However, the underlying philosophy, even if it is not fully seen by all workers in the field, is one heavily based on a concept of life which is highly emergentist – life as an emergent property, as life emerges only from the dynamic integration of parts into a whole. More generally, one could say that these basic notions of epistemology help in giving a broader framework to the technical work of SB, and it would be good if the scientists in the field would foster more interest for philosophy of science, and possibly transmit this to their students.

The other thing that should be transmitted more clearly to students of SB and in the life science in general, is a first answer to the question "what is life?" I believe that the few common sense considerations outlined in this paper – based basically on system biology and autopoiesis – may be sufficient to see life in a broader perspective and perhaps dispel the confusing and partly false notion that all there is in life is DNA.

The notion of life as emergence might enjoy with SB its triumph, once SB succeeds in synthesizing life from scratch from nonliving parts. We have reasoned, however, that the synthesis of an entire organism is still far from reach.

The notion of emergence is worth a particular additional consideration: with the genetic manipulation of the genome of living bacteria, and within the teleological design of engineering SB, emergence may play some unexpected trick, in the sense that a novel combination of modules into a novel organism configuration may well give rise, in principle, to unexpected properties.

This argument links with the bioethical problems, which is a sore aspect of SB, particularly if we consider the increasing power of the mass media. In general, it must be said that SB has shown from the very beginning a great attention to bio-ethics, as witnessed by the considerable amount of contributions dealing with ethical problems in the major SB international conferences. It is as if the workers in the field have learned from the older field of genetic engineering, and defend themselves a priori from accusations of dealing too light heartedly with potentially socially scary items. Ethical problems are, of course, less of an issue for the second soul of SB, that of dealing with the basic science of chemical SB.

In the light of the present scientific production, the idea that by bioengineering SB one can produce novel and dangerous bacteria is far-fetched. Different might be the case in which somebody wants to produce purposely novel harmful bacteria – a consideration that may link with the suspicion that some military research agencies might precisely do that in their biowarfare programs. I believe SB scientists should be careful not to give a hand to such programs.

There is another level of the bioethical discourse that deals not with dangerous novel bacteria, but with the more subtle question of whether it is right to "play God", and produce in the laboratory forms of life which do not exist in nature – particularly when the enterprise may not stop at the level of bacteria, but has the ambition of creating new forms of life by mixing the genome of, say, worms with the genome of butterflies, or by screwing the homo-box during ontogeny so as to obtain monstrous living constructions.

I know that a few(?) colleagues see even this as a challenging enterprise and are not touched by the concern that the construction of biological monsters may be something unethical. Here, too, there is only the hope that the main substrate of all sciences, namely common sense, may prevail over all other ambitions.

A final consideration. I have purposely restricted the analysis to SB and avoided taking about AL, robotics, and the like. This does not mean, of course, that these fields do not necessitate an epistemological clearing. Actually, if one considers in general the "sciences of the artificial," then there are insights showing a growing interest towards basic science aspects, those which have more to do with emergent properties rather than solely engineering applications (e.g. see Refs [43, 44]).

Clearly, all this necessitates a deeper and differentiated study, for which the present article may only represent a starting point.

ACKNOWLEDGMENTS

I expresss my gratitude for the critical reading of the manuscript by Dr Luisa Damiano, Michel Bitbol, and Dr Pasquale Stano.

REFERENCES

1. Langton, C.G., Taylor, C., Doyne Farmer, J., and Rasmussen, S. (1992) *Artifical Life*, Addison-Wesley.
2. Langton, C. (ed.) (1995) *Artificial Life, An Overview*, The MIT Press.
3. Waks, Z. and Silver, P.A. (2009) Engineering a synthetic dual organism system for hydrogen production. *Applied and Environmental Microbiology*, 75, 1867–1875.
4. Lee, S.K., Chou, H., Ham, T.S. *et al.* (2008) Metabolic engineering of microorganisms for biofuels production. *Current Opinion in Biotechnology*, 19, 556–563.
5. Johnson, E.T. and Schmidt-Dannert, C. (2008) Light-energy conversion in engineered microorganisms. *Trends in Biotechnology*, 26, 682–689.
6. Lartigue, C., Glass, C., Alperovich, J.I. *et al.* (2007) Genome transplantation in bacteria: changing one species into another. *Science*, 317, 632–638.
7. Leonard, E., Nielsen, D., Solomon, K., and Prather, K.J. (2008) Engineering microbes with synthetic biology frameworks. *Trends in Biotechnology*, 26, 674–681.
8. Voigt, C.A. (2006) Genetic parts to program bacteria. *Current Opinion in Biotechnology*, 17, 548–557.
9. Jewett, M.C., Calhoun, K.A., Voloshin, A. *et al.* (2008) An integrated cell-free metabolic platform for protein production and synthetic biology. *Molecular Systems Biology*, 4, 220.
10. Noireaux, V., Bar-Ziv, R., and Libchaber, A. (2003) Principles of cell-free genetic circuit assembly. *Proceedings of the National Academy of Sciences of the United States of America*, 100, 12672–12677.
11. Benner, S.A., Hoshika, S., Sukeda, M. *et al.* (2008) Synthetic biology for improved personalized medicine. *Nucleic Acids Symposium Series*, 52, 243–244.
12. Lee, S.Y., Kim, H.U., Park, J.H. *et al.* (2009) Metabolic engineering of microorganisms: general strategies and drug production. *Drug Discovery Today*, 14, 78–88.
13. Weber, W., Lienhart, C., Daoud-El Baba, M., and Fussenegger, M. (2009) A biotin-triggered genetic switch in mammalian cells and mice. *Metabolic Engineering*, 11, 117–124.
14. Chang, M.C.Y. and Keasling, J.D. (2006) Production of isoprenoid pharmaceuticals by engineered microbes. *Nature Chemical Biology*, 2, 674–681.
15. Ajikumar, P.K., Tyo, K.E.J., Carlsen, S. *et al.* (2008) Terpenoids: opportunities for biosynthesis of natural product drugs using engineered microorganisms. *Molecular Pharmaceutics*, 5, 167–190.
16. Eschenmoser, A. (2005) *Chimia*, 59, 836–850.
17. Bolli, M., Micura, R., Pitsch, S., and Eschenmoser, A. (1997) *Helv. Chim. Acta*, 80, 1901–1951.

18. Doi, N., Kakukawa, K., Oishi, Y., and Yanagawa, H. (2005) *Protein Eng. Des. Sel.*, **18**, 279–284.
19. Chiarabelli, C., Vrijbloed, J.W., De Lucrezia, D. et al. (2006) Investigation of *de novo* totally random biosequences, Part II. *Chemistry & Biodiversity*, **3**, 840–859.
20. Chiarabelli, C., Vrijbloed, J.W., Thomas, R.M., and Luisi, P.L. (2006) Investigation of *de novo* totally random biosequences, Part I. *Chemistry & Biodiversity*, **3**, 827–839.
21. Pohorille, A. and Deamer, D. (2002) Artificial cells: prospects for biotechnology. *Trends in Biotechnology.*, **20**, 123–128.
22. Ishikawa, K., Sato, K., Shima, Y. et al. (2004) Expression of a cascading genetic network within liposomes. *FEBS Letters*, **576**, 387–390.
23. Noireaux, V. and Libchaber, A. (2004) A vesicle bioreactor as a step toward an artificial cell assembly. *Proceedings of the National Academy of Sciences of the United States of America*, **101**, 17669–17674.
24. Luisi, P.L., Ferri, F., and Stano, P. (2006) Approaches to a semi-synthetic minimal cell: a review. *Naturwissenschaften*, **93**, 1–13.
25. Murtas, G., Kuruma, Y., Bianchini, A. et al. (2007) Protein synthesis in liposomes with a minimal set of enzymes. *Biochemical and Biophysical Research Communications*, **363**, 12–17.
26. Stano, P. (2008) Approaches to the construction of the minimal cell. *Mining Smartness from Nature* (eds P. Vincenzini and S. Graziani), *Advances in Science and Technology*, **58**, 10–19.
27. Mansy, S.S., Schrum, J.P., Krishnamurthy, M. et al. (2008) Template-directed synthesis of a genetic polymer in a model protocell. *Nature*, **454**, 122–125.
28. Souza, T., Stano, P., and Luisi, P.L. (2009) The minimal size of liposome-based model cells brings about a remarkably enhanced entrapment and protein synthesis. *Chembiochem*, **10**, 1056–1063.
29. Luisi, P.L. (2007) Chemical aspects of synthetic biology. *Chemistry & Biodiversity*, **4**, 603–621.
30. Varela, F., Maturana, H., and Uribe, R. (1974) Autopoeisis: the organization of living systems, its characterization and a model. *BioSystems*, **5**, 187–195.
31. Maturana, H. and Varela, F. (1980) *Autopoiesis and Cognition: The Realization of the Living*, Reidel, Dordrecht.
32. Maturana, H. and Varela, F. (1998) *The Tree of Knowledge*, revised edition, New Science Library, Shambhala, Shambala.
33. Bitbol, M. (2007) Ontology, matter and emergence. *Phenomenology and the Cognitive Science*, **6**, 293–307.
34. Weber, A. and Varela, F. (2002) Life after Kant: natural purposes and the autopoietic foundations of biological individuality. *Phenomenology and the Cognitive Science*, **1**, 97–125.

35. Fraenkel-Conrat, H. and Williams, R.C. (1955) Reconstitution of active tobacco mosaic virus from its inactive protein and nucleic acid components. *Proceedings of the National Academy of Sciences of the United States of America*, **41**, 690–698.
36. Cohlberg, J.A. and Nomura, M. (1976) Reconstitution of *Bacillus stearothermophilus* 50 S ribosomal subunits from purified molecular components. *Journal of Biological Chemistry*, **251**, 209–221.
37. Jeon, K.W., Loech, I.J., and Danielli, J.F. (1970) Reassembly of living cells from dissociated components. *Science*, **167**, 1623–1626.
38. Pressman, E.K., Levin, I.M., and Sandakhchiev, L.S. (1973) Reassembly of an *Acetabularia mediterranea* cell from the nucleus, cytoplasm, and cell wall. *Protoplasma*, **76**, 327–332.
39. Veomett, G., Prescott, D.M., Shay, J., and Porter, K.R. (1974) Reconstruction of mammalian cells from nuclear and cytoplasmic components separated by treatment with cytochalasin B. *Proceedings of the National Academy of Sciences of the United States of America*, **71**, 1999–2002.
40. Fraser, C.M., Gocayne, J.D., White, O. *et al.* (1995) The minimal gene complement of *Mycoplasma genitalium*. *Science*, **270**, 397–404.
41. Gibson, D.G. *et al.* (2008) Complete chemical synthesis, assembly, and cloning of a *Mycoplasma genitalium* genome. *Science*, **319**, 1215–1220.
42. Chan, L.Y., Kosuri, S., and Endy, D. (2005) Refactoring bacteriophage T7. *Molecular Systems Biology*, **1**, 0018.
43. Dawson, M.R.W. (2002) From embodied cognitive science to synthetic psychology, in Proceedings. First IEEE International Conference on Cognitive Informatics, pp. 13–22.
44. Pfeifer, R., Lungarella, M., Sporn, O., and Gomila, T. (eds) (2008) *Handbook of Cognitive Science, An Embodied Approach*, Elsevier.
45. Lucks, J.B., Qi, L., Whitaker, W.R., and Arkin, A.P. (2008) Toward scalable parts families for predictable design of biological circuits. *Current Opinion in Microbiology*, **11**, 567–573.
46. Sismour, A.M. and Benner, S.A. (2005) Synthetic biology. *Expert Opinion on Biological Therapy*, **5**, 1409–1414.

Index

Page references given in italic type refer to figures. References given in bold type refer to tables.

adenine, 20, *89*
adenosine triphosphate (ATP) analogs, 216
advanced biological warfare (ABW), 229–330
AEGIS nucleotides, 91
affinity chromatography, 55–6
Agrobacterium tumefaciens, **190**
AIDS, 93, 101
alanine, 125, **182**
 polymer, 267–8
alanine synthetases, 235
Alexa-dabcyl quenching, 311, *312*
α-PNAs, 115
amino acid synthetases, 235–6, **236**
amino acids
 addition of novel to genetic code, 180–1
 evolution of, 123, 177–8
 prebiotic synthesis, 180
 random sequences of *see* random-sequence peptides
 replacement in genetic code, 190–1
aminoacyl tRNA synthase (aaRS), 187–8, **188**, **189**, 325
aminoadenine, 89
α-amino-*n*-butyric acid, **182**
α-aminoisobutyric acid, **182**
aminoethylglycine PNA (aegPNA), 108–9, *109*
p-aminophenylalanine (*p*-aminoPhe), 190
Amoeba proteus, 353
ANA, 21
anhydrohexitol nucleic acid (HNA), *21*, 211, 212–13, *212*
antibiotics, 183
antisense technology, 8–9, 83
 PNA, 110–11
aptamers, 55

Chemical Synthetic Biology, First Edition. Edited by Pier Luigi Luisi and Cristiano Chiarabelli.
© 2011 John Wiley & Sons, Ltd. Published 2011 by John Wiley & Sons, Ltd.

arabinopyranosyl RNA, 25–6
Arthrobacter luteus, **190**
artificial cells, 102, 324–6
   see also minimal cell
artificial genetics, 77–9
   see also minimal genome
artificial life, 345
   see also synthetic life
artificial replicators, 302–6
asparagine, 180
astrobiology, 323–4
   see also extraterrestrial biological agents
ATP analogs, 216
autocatalysis, 292–3, 293–4
   Darwinian evolution, 299–300
   online kinetic measurement, 311, *312*
   reaction rates, 297–300, **301**
   replicator networks, 306–11
automated combinatorial chemistry, 159–60
autopoiesis, 276, 350
p-azidophenylalanine (*p*-azidoPhe), 190
Azoarcus group I intron, 309

Bacillus polymyxa, 183
Bacillus stearothermophilus, **190**
Bacillus subtilis, 187, 190, **190**, 191, 251
bacitracin, 183
bacteria, RNA transcription, 236–7
base pairing, 9
   arabinopyranosyl RNA, 25–6
   DNA, 88–90
   glycerol DNA analog, 80–2
   homo-DNA, 20
   RNA, 60–1
   Watson-Crick model, 89–90
   see also nucleobases
base-stacking, 60–1
Beginnings of Cellular Life, 249
bicyclo-DNA, 20
bioavailability, 110–11

biobricks, 57, 324, 346, 354–5
biocatalysis, 179–80
bioengineering, 355
bioethics, 337–8, 344–5, 358–9
Biological and Toxic Weapons Convention (BTWC), 330–2, 337
biology, 8
biomimetic chemistry, 71
biosafety, 100, 202–3, 327–9, **328**, 336
biosecurity, 329–30, 337
bottom-up cell construction, 264
Buchnera aphidicola, 251
Buchnera spp., **252**

Caenorhabditis elegans, 241
calcifying nanoparticles, 263
L-canavanine, 186
cancer, 93, 111
cells
   artificial, 102
   membranes, see also liposomes
   minimal see minimal cell
   minimal size, 262–3
   spontaneous reassembly, 353
   synthetic see synthetic cells
cellular delivery assays, 110–11
central dogma of molecular biology, 82
cesium-induced DNA cleavage, 114
chassis, 346
chemical synthetic biology, 49, 157
Chemical Weapons Convention, 331
Chlamydia muridarum, **252**
Chlamydophilia pneumoniae, **252**
chromosome synthesis, 202
circular dichroism (CD) spectroscopy, 124, 129
Collins, Francis, 73
combinatorial chemistry, 159–60
   compound identification, 162–3
   library preparation, 160–1
   screening, 161–2
Committee on Space Research, 323

INDEX

comparmentalized self-replication, 220
computation, using DNA, 72–3
conformation
   DNA pyranosyl analogs, 13–19
   RNA, 60–1
   see also folding
contingency, 348
core and shell replication, 268
coupled autocatalysis, 292–3
cyclohexenyl nucleic acid (CeNA), 211, 213

Dabbs, Eric, 231–2
Darwinian evolution
   autocatalytic reactions, 299–300
   base-pairing and, 89–91
   nucleic acid chemical properties and, 86–8
   proteins, 150
   synthetic genomes and, 95–6
   synthetic nucleotides and, 96–7
   systems chemistry and, 291
   using synthetic nucleobases, 95–7
decanoate, 266
deconvolution, 162–3
Defense Threat Reduction Agency, 98–9
dehydroquinase, 202
deoxyadenosine monophosphate (dAMP), 216, *217*
deoxyhypusine synthase, 185
deoxyribose nucleic acid *see* DNA
determinism, 348
α,γ-diaminobutyric acid, 183
α,β-diaminopropionic acid, 183
diphtamide biosynthesis methyltransferase, 184–5
dipthine, 185
directed evolution (DE), 54–6, 59–60, 207
DNA, 79, *109*
   A-type, 36–8

analogs, 21
   pyranose, 13–14, *22*
   backbone charge, 87–8
   backbone phosphate replacement, 83–8
   backbone sugar replacement, 80–3, 348
   computation using, 72–3
   dimethylsulfone substituted, 83–4
   duplex inversion binding, 112
   duplexation thermodynamics, *21*
   encoding random-sequence proteins, 125–6
   RNA, TNA and self pairing energies, *30*
   transcription duplexes, **208–10**
DNA detection, 93–4
DNA libraries, random-sequence proteins, 123–4, 125–6, 133
DNA polymerases, 95–6, 212, 219, 257–60
E. coli, 213
downward causation, 352
duplex DNA-DNA analog confomational models, 16–19
dynamic libraries, 115

EDC, 295, *296*
elongation factors, 231
emergence, 351–2
enzyme-linked immunosorbent assay (ELISA), 162
epistemology, 343–4
Escherichia coli, 144, 168, 187, 188, 192, 213
   genome size, 251
   ribosome, 228
   ribosome protein mutants, 231–2
Eschermoser, Albert, 348
ethics, 337–8, 344–5, 358–9
1-ethyl-3-(3-dimethylaminopropyl) carbodiimide *see* EDC
Euplotes crassus, 181
European Patent Office, 335

evolution
  amino acids, 123, 177–8
  Darwinian *see* Darwinian evolution
  directed *see* directed evolution
  natural proteins, 150
  ribosome, 234–6
exponential growth, 298, **301**, 302
extraterrestrial biological agents, 327–9, 336
extraterrestrial life, 78, 88

farnesol, 266
Feynman, Richard, 48, 71
fitness space, 141–2
  ruggedness, 147
fluorescence resonance energy transfer (FRET), 311, *312*
fluorescence spectroscopy, 125
folding
  oligosulfone-substituted DNA and RNA, 84–5
  proteins, 172
  random RNA sequences, 50–3, 60–1
  see also conformation
functional proteins
  early evolution, 142–3
  frequency in protein sequence space, 143–7
functional selection, 132–3
funding, 98–9

gene repair, PNA-mediated, 113–14
genetic code, **126**, 177–8, *179*
  evolution, 123
  expansion, 179–80
    synthetic, 186–90
  proteome-wide expansion, 180–2
  redundancy, **252**
  synthetic, applications, 191–2
  turnover, 190–1
genetic complementation, 220
genetically modified organisms (GMO), 100
GeneX, 73

genome
  components for minimal self-maintenance, 256
  gene functions, 255–6, **255**
  minimal, 250–60, **253–4**, 278–9
  sizes, 251
glutamic acid (Glu), 125
glutamine, 180
glutamine synthetase, 150
glycerol monomyristoleate, 266
glycerol nucleic acid (GNA), 80, *81*, 211, 213, 326
glycerol-3-phosphate acyltransferase, 276
glycine, 125, **182**
GNHpCNH2, 296–7
GNW codon, 125
gramicidin S, 183
green fluorescent protein (GFP), 111, 270–3, **272**, 275
  red-shifted, 273
guanine, 20, *89*

Halmonas halmophila, 251
Haloarcula marismortui, 241
Halobacterium cutirubrum, **190**
hazard assessment, 100–101
health care, 91–5
HeLa cells, 110–12
helix invasion, 112
hepatitis detection, 95
hexitol nucleic acid (HNA), 326
high-throughput screening (HTS), 53, 54
histidine, 163
HIV, 93
  detection, 94–5
HIV-1 reverse transcriptase, 329
  M184V mutant, 212
HNA *see* anhydrohexitol nucleic acid
homo-DNA, 13–19, *15*, 22
  base-pairing, 20
  duplexation thermodynamics, *21*
  structure, 31–2

INDEX

homo-RNA, *14*, 23–9, *24*
homochirality, 291–2, 292
HTS *see* high-throughput screening

iGem, 356
influenza virus, 101
information content, 178–9
intellectual property rights, 335, 338–9
3-iodotyrosine (3-iodoTyr), 190
isocytonisine, 94
isoguanine, 94
isoleucine, 183
Isua Supercrustal Belt, 234
iterative deconvolution, 162–3

kasugamycin particle, 227–8, 233–40
Kindermann's replicator, 304, *305*
kinetic titration, 312–14
Knowles, Jeremy, 74

last universal common ancestor (LUCA), 184–6, 233–4
leaderless mRNA, 237–8
lecithin, 276
Lehn, Jean-Marie, 71
leucine, 183
life, 77–8
  definition, 78–9, 249, 332–4, 358
  early evolution, 150–1
  as emergence, 356, 358
  as emergent property, 349–54
  origins, 291–2
  synthetic, 97–9, 192–3
  value, 334–5
limping life, 279, 325
liposomes, 265–70, **267**, 279
  lipid synthesis, 276–7
  polymerase chain reaction, 268–9
  protein expression, 269–76, **272–3**
locked nucleic acid (LNA), 326
LUCA, 184–6, 233–4
luciferase genes, 111
Luisi, P. L., 333
lysine, 183

lysophosphaditic acid acyltransferases, 276

m-synthetic life, 193
mammals, 351
Mars, 323–4
Mars Climate Orbiter, 77–8
meaning, 333–4
Mendel, Gregor, 75
messenger RNA (mRNA)
  evolution, 236–7
  leaderless, 237–8
  synthetic base substitution, *92*
Methanobacterium thermoautotrophicum, 188
Methanococcus jannaschii, 188
Methanopyrus, 184
Methanosarcina acetivorans, 181
methylene phosphonate ATP analogs, 216
Micrococcus luteus, **190**
microreactors, 160
minimal cell, 248–50, 355
  construction, 263–5
  gene products, **258–9**
  genome size, 278
  see also minimal genome
  membrane permeability, 266
  nutrient availability, 257
  relationship with environment, 279
  RNA cell, 260–2
  size, 262–3
  vesicles as membrane, 265–9
minimal genome, 250–60, **253–4**, 278, 278–9
  bacterial, **255**
  components for minimal self-maintenance, 256
  gene functions, 255–6, **255**
  protein expression, 275–6
minimal replicator theory, 297–302
mitochondrial ribosome, 232, 241, *242*
modeling, 75
molecular design, RNA, 50–3

Morowitz, Harold J., 249
Murchison meteorite, 182
mutation, 96
Mycoplasma genitalium, 202, 251, 252, 354
Mycoplasma pneumoniae, 252
myristoleate, 266
Myxococcus xanthus, **190**

nanobacteria, 263
nanoengineering, 73–4
NASA, definition of life, 78–9
National Research Council (US), 327–9
NBP127, 171–2
never-born proteins, 165–6, 167–73, 348
  NBP127, 171–2
  see also random-sequence proteins
never-born RNA *see* random-sequence RNA
novel biological entities (NBE), 49–50
nuclear magnetic resonance (NMR) spectroscopy, 16, 304
  kinetic titration, 312–14
nucleic acids
  as autocatalysts, 294–5
  DNA *see* DNA
  RNA *see* RNA
  see also nucleic acid analogs
nucleic acid analogs, 9, 201–2, 326
  advantages, 202–3
  anhydrohexitol (HNA), 211, 212–13, *212*
  as autocatalytic system, 295–7
  backbone charge and water-solubility, 86–7
  backbone motif substitution, 80–3, 211–14
  base-pairing, 38
  bicyclo-DNA, 20
  co-periodicity with natural nucleic acids, 11–12, *12*
  conformation, 35–8
  cyclohexenyl (CeNA), 211, 213

  DNA sugar substitution, 80–3
  generalized *in vivo* synthesis scheme, 204, *205*, 207–11, **208–10**
  history, 202
  interaction with nucleic acids *see* orthogonal episomes
  leaving group substituted, 83–8, 215–17
  nucleobase modified, 90–1, 204–6, 217–19
  orthogonality, 31–2, 206–11
  peptide nucleic acids (PNA) *see* peptide nucleic acids
  phosphate-substituted DNAs, 83–8
  polymerase diversification, 219–20
  precursors, 216
  propagation in bacteria, **208–10**
  pyranosyl DNA, 13–19, *14*, *15*, 22, 31–2
  pyranosyl RNA, *15*, 22, 23–9, *24*
  replicator networks, 306–11
  threose (TNA), 28–33, 38–9, 211, 213, 326
nucleobase synthesis, 90–1
  in nucleic acid analogs, 217–19
  precursor supply, 218–19
  see also base pairing
nucleotide
  constitutional periodicity, 10–12, *12*
  delivery systems, 214, *215*
  substitution, 91–5
  synthesis, 295
  see also nucleic acid; nucleobases

o-synthetic life, 192
oligodipeptamidinium salts, 10–12
oligodipeptides, 36–8, *37*
oligonucleotides *see* nucleotides
oligopeptide permeases, 214
orthinine, 183
orthogonal episomes, 31–2, 202–3
  nucleic acid polymerases, 219–21
  nucleobase synthesis, 217–19

INDEX   369

replication duplexes, 207–11,
   208–10
requirements, 206–11
Outer Space Treaty (1967), 323

P nucleobase, 96, 97
paleogenetics, *78*
parabolic growth, 298–9, **301**
Patent and Trademark Office (US),
   335
Pauling, Linus, 85
PCR *see* polymerase chain reaction
peptide libraries, 157–9
   active compound identification,
      162–3
   library creation, 160–1, 164–5, *169*
   phage-display, 158, 168–70, 170–1
   random sequence generation, 163–7
   screening, 161–2
peptide nucleic acids (PNA), 9, 11–12,
      12, 38–9, 107–8, 326
   aminoethylglycine (aegPNA), 108–9
   analogs, 109
   DNA ligations, 115
   duplex DNA recognition, 112–13
   dynamic libraries, 115
   gene repair using, 113–14
   pLuc-HeLa cellular delivery assay,
      110–11
   sequence information transfer,
      113–14, 114–15
   thioester (tPNA), 115
peptide-bond formation, 240–1
peptides, effect on cell membrane
   permeability, 266
phage display libraries, 158, 168–70,
   169, 170–1
phosopholipid membranes, 266
phosphates, in DNA backbone, 83–8
phosphoramidate, 216
Pilibara, 234
planetary protection, 323
pLuc-HeLa cells, 110–12
polyalanine (poly(A)), 267–8

polymerase chain reaction (PCR), 95,
   329
   in liposomes, 268–9
   synthetic nucleobases, *97, 98*
polymerase substrates, 216
polymyxin B, 183
polynucleotide phosphorylase, 267–8
polyphenylalanine (poly(Phe)), 270
polyuracil (poly(U)), 270
positional scanning, 163
post-transcriptional modification, 184
pre-mix synthesis, 161
prebiotic chemistry, 264–5
pretran synthesis, 180
PRG, 167–8
prokaryotes, genome size, *251*
proline-arginine-glycine (PRG),
   167–8
protease inhibitors, 115
protein engineering, 73–4
protein sequence space, 141–2
   frequency of evolvable sequences,
      143–7
   structure of fitness landscape, 147–8
proteins
   as biocatalysts, 179–80
   Darwinian evolution, 150
   expression in liposomes, 269–75,
      272–3
   fitness space, 141–2
   functional evolution, 140–1,
      142–3
   information content, 178–9
   natural evolution, 150
   non-essential ribosomal, 231–2
   in primitive cells, 150–1
   random sequence, 122–3, 163–7
      binary code library, 133
      characterization, 127–31
      cloning, expression and
         purification, 124
      creation, 156–7
      evolution of other protein
         properties, 149–50

proteins (*Continued*)
  library construction, 123–4
  properties, **128**, *129*
  solubility, 126–8, *127*, 132, 149–50, 165
  sequence space, 141–8
  synthesis from artificial base-pair systems, 91
protocell *see* minimal cell
PURE system, 270, **271**, 274
PylS gene, 181
pyranosyl DNA analogs, 13–19, *15*, 22, 31–2
pyranosyl RNA, *24*
pyrimidine, 295
Pyrococcus horikoshii, 188

Qβ replicase, 268

random-sequence DNA, 123–4
random-sequence proteins and peptides
  binary code library, 133
  cloning, expression and purification, 124
  creation, 156–7
  DNA design, 125–6
  evolution
    protein functions, 143–7
    other protein properties, 149–50
  library construction, 123–4, 157–9, 163–7
  properties, **128**, *129*
  solubility, *127*, 132, 149–50, 165
  VADEG, 126–31, *128*
random-sequence RNA
  as modular synthetic scaffold, 57–61
  thermal stability, 52
rational design (RD), 53–4
reagent mixture synthesis, 161
Rebek's replicator, 303–4
Redesigning Life, 77–8
Registry of Standard Biological Parts, 274–5
1,5-repulsions, 13
resin beads, 160

Rhodopseudomonas spheroides, **190**
ribonucleic acid *see* RNA
ribonucleoside triphosphates (rNTP), 220
ribosomal RNA, 241
ribosome, 278
  E. coli, 228, 229, 241
  evolution, 234–6
  last universal common ancestor (LUCA), 236
  mitochondrial, 232
  nucleic acid analogs, 207
  peptide bond formation, 240–1
  proteins, 257
    non-essential, 231–2
  size reduction, 241
  structure and function, 228–31, *230*
ribozyme, minimal RNA cell, 260–2
Rickettsia spp., **252**
risk assessment, 100–2, *102*
  see also safety
RNA
  DNA, TNA and self pairing energies, *30*
  messenger (mRNA), 92, 236–8
  pyranosyl analogs, *14*, 23–9, *24*
  random sequence, 50–3
    folding, 50–3
    molecular design, 53–6
  ribosomal, 241
  TNA pairing conformation, *29*
  XNA transcription duplex, **208**–10
RNA cell, 260–2
RNA Foster assay, 50, *51*
RNA polymerases, 220, 257–8
  T7, 212
RNA tectonics, 57–9
RNA world, 30, 233, 294–5
ROKKY protein structure suite, 166

S1 nuclease, 50–3
safety, 100, 202–3, 327–9, **328**, 336
sample receiving facilities, 329

screening, peptide libraries, 161–2
Search for Extra-Terrestrial Intelligence (SETI), 322–4
selection bias, 75
selenocysteine, 180
SELEX, 55
self-replicators, 86–7, 87
　artificial replicators, 302–6
　reaction rates, 297–9
　reactions, 293–4
　replicator networks, 306–11
Sievers network, 306–8
sigmoidal growth, 297–8
SimFit, 310, 314–15
size-exclusion chromatography, 125
smallpox virus, 100–1
Smith, Hamilton, 76
societal issues, 326–35
solid-phase synthesis, combinatorial libraries, 159
split and mix synthesis, 160–1
stromatolites, 234
structure theory, 71
synthetases, 234–6, **236**
synthetic biology (subject), 156–7
　aims, 48–9
　challenges, 71–3, 75–7
　definitions, 70–1, 74–5, 345–6, 346–9
　distinguished from artificial life, 345
　ethics, 331–5
　risks, 99–102
synthetic cells, 247–8, 248–50
　size, 262–3
　see also liposomes; minimal cell
synthetic genetics, 89–91, 202
　applications, 91–5
　Darwinian evolution and, 95–6
synthetic life, 77–9, 97–9, 192–3
　see also synthetic cells
Systematic Evolution of Ligands by Exponential Enrichment (SELEX), 55

systems chemistry, 290–3
　analysis tools, 314–15
　characterization methods, 311–12, 313–14
Szybalski, Waclaw, 71

tectoRNA, 58–9
teleology, 347
teleonomy, 347
Tetrahymena ribozyme, 57, *58*
Thermus aquaticus, **190**
Thermus thermophilus, 241
thioester peptide nucleic acid (tPNA), 115
thiophosphate, 216
threose nucleic acid (TNA), 28–33, *29*, 211, 213, 326
　self- and cross- pairing energies, *30*
　structural analogs, 38–9
thrombin, 167–8, 170–1
thymidylate synthase, 213
thymine, *89*
Titan, 78
TNA *see* threofuranosyl-oligonucleotides
tobacco mosaic virus, 353
top-down cell construction, 265
transfer RNA (tRNA), 181, 230–1
　aaRCS aminoacylation, **190**
　evolution, 234–6, *234–5*
transition state analog (TSA), 143–4
travelling salesman problem, 72
treble clef finger, 149–50
Treponema pallidum, **252**
tryptophan, 190
tyrosine kinases, 115

Ulmer, Kevin, 73
unnatural biochemical systems, 323–4
unSubtilis, 193
Ureaplasma urealyticum, **252**

VADEG proteins, 126–7, *128*
　characterization, 127–31, *128*
　see also random-sequence proteins

Venter, Craig, 76, 335
Venter-Smith artificial cell, 102
vesicles, 265–70
   polymerase chain reaction in, 268–9
   protein expression, 269–76, **272–3**
   self-reproduction, 275–6

Wheelis, Mark, 331

xeno-nucleic acids (XNA) *see* nucleic acid analogs
xenozyme, 207
XNA *see* nucleic acid analogs
XNA ligases, 221
XNA polymerases, 219–21

Z nucleobase, 96, 97